Updates in Surgery

Fulvio Calise • Luciano Casciola
Editors

Minimally Invasive Surgery of the Liver

In collaboration with Graziano Ceccarelli,
Antonio Giuliani, Alberto Patriti and Vincenzo Scuderi

Foreword by
Gianluigi Melotti

Springer

Editors
Fulvio Calise
Hepatobiliary Surgery and Liver Transplant Center
A. Cardarelli Hospital
Naples, Italy

Luciano Casciola
Department of Surgery
Division of General, Minimally Invasive and Robotic Surgery, ASL3 Umbria
San Matteo degli Infermi Hospital
Spoleto, Italy

In collaboration with:
Graziano Ceccarelli, Antonio Giuliani, Alberto Patriti and Vincenzo Scuderi

The publication and the distribution of this volume have been supported by the Italian Society of Surgery

The volume Editors and the publisher gratefully acknowledge the educational contribution kindly offered by Johnson & Johnson Medical S.p.A. *and* ab medica S.p.A., Baxter S.p.A., Covidien Italia S.p.A., Euro Hospitek S.r.l., M.D.H. Forniture Ospedaliere S.r.l., Nycomed: a Takeda Company, Tecsud S.r.l., Tekmed Instruments S.p.A.

Additional material to this book can be downloaded from http://extras.springer.com.

ISBN 978-88-470-5835-4 ISBN 978-88-470-2664-3 (eBook)
DOI 10.1007/978-88-470-2664-3

Springer Milan Dordrecht Heidelberg London New York

Softcover reprint of the hardcover 1st edition 2013

9 8 7 6 5 4 3 2 1 2013 2014 2015 2016

Cover design: Ikona S.r.l., Milan, Italy
Typesetting: Graphostudio, Milan, Italy

Springer-Verlag Italia S.r.l. – Via Decembrio 28 – I-20137 Milan
Springer is a part of Springer Science+Business Media (www.springer.com)

Foreword

New technologies and a generation of surgeons mastering advanced minimally invasive surgery have greatly contributed to the development and growth of laparoscopic hepatic surgery.

Skepticism related to technical and oncological concerns initially limited the widespread acceptance and growth of this approach. However, the positive results obtained by skilled surgeons using the laparoscopic technique clearly demonstrate its advantages, as does the rate of oncologic clearance and disease-free interval for primary and secondary tumors, which is comparable with that of open procedures. Now, both minor and major hepatectomies are performed in specialized centers worldwide, and the slow diffusion rate of the technique itself guarantees that safe procedures are implemented.

Robot-assisted surgery is also developing as an important contribution to hepatic surgery: the more accurate and precise movements provided by robotic instruments, coupled with a much greater dexterity, facilitate complex posterior and superior resections and procedures involving biliary reconstructions.

This book meticulously describes the lights and shadows, indications and contraindications, pros and cons, advantages and limitations of minimally invasive surgery of the liver, analyzing its progression from technical evolution to highly delicate procedures and projecting into the future.

Fulvio Calise and Luciano Casciola, first-class leaders of the Italian surgical school and community, beyond their significant personal contributions, engaged the most distinguished domestic and international experts in minimally invasive hepatic surgery in the creation of this highly practical and enjoyable manual. The significant experience and scientific excellence of the contributing authors have culminated in an accurate description of laparoscopic and robot-assisted procedures related to all types of hepatic resections, defining in detail the necessary learning curve and clarifying the technological aspects of the procedures.

Thank you, Fulvio and Luciano, for this remarkable effort, and I hereby express my personal gratitude and that on behalf of the Italian Surgical Society.

Rome, September 2012

Gianluigi Melotti
President, Italian Society of Surgery

Preface

This book is meant as a challenge to the usual.

Scientific books usually certify and codify what has already been acknowledged and shared in clinical practice as they serve the purpose of improving what the scientific community has already established in the treatment of patients. In contrast, this book tries to indicate what could be safely done in the near future in a wider surgical context due to the extensive and increasing application of the laparoscopic approach to abdominal diseases.

We strongly believe that, thanks to the new technical modalities, the very near future will witness rapid change in the way of approaching liver diseases and, consequently, of learning how to carry out minimally invasive liver surgery. Although the latter is a highly questionable point, we think that learning will inevitably be modified by the swift advance of technology and that there will be a compelling need to change the rules of surgical training. Robotics will play a key role in forming young surgeons as soon as costs fall following the expiry of the original patents. Robotic surgery will then be able to achieve much wider and well-deserved diffusion.

We have therefore asked our colleagues to present what they consider to be best practices in their current activity and also what today, in some cases, is still regarded as somehow controversial in hepatobiliary and advanced general laparoscopic surgery. Thirty-one videos presenting all the techniques and technicalities related to a mini-invasive approach complete this book, together with a final survey of the last seven years of activity in this field in Italy. We think that the survey shows interesting practices both in dedicated hepatobiliary units and in general surgery units.

We sincerely thank all our co-authors for their work and for believing that this effort might contribute to a better insight into the way ahead for new generations of surgeons.

Naples and Spoleto, September 2012

Fulvio Calise
Luciano Casciola

Contents

Contributors

Giulio Angelini General and Hepatobiliary Surgery, Galliera Hospitals, Genoa, Italy

Luca Aldrighetti Department of Surgery, Hepatobiliary Surgery Unit, San Raffaele Hospital, Vita-Salute San Raffaele University, Milan, Italy

Mario Annecchiarico Department of Surgery, General Surgery Unit, Misericordia Hospital Grosseto, Grosseto Italy

Giuseppe Aragiusto Hepatobiliary Surgery and Liver Transplant Center, A. Cardarelli Hospital, Naples, Italy

Francesco Ardito Hepatobiliary Surgery Unit, Università Cattolica del Sacro Cuore, Rome, Italy

Piero Atelli Hepatobiliary Surgery and Liver Transplant Center, A. Cardarelli Hospital, Naples, Italy

Claudio Bagarolo General and Hepatobiliary Surgery, Galliera Hospitals, Genoa, Italy

Andrea Barberis General and Hepatobiliary Surgery, Galliera Hospitals, Genoa, Italy

Alberto Bartoli Department of Surgery, Division of General, Minimally Invasive and Robotic Surgery, ASL3 Umbria, San Matteo degli Infermi Hospital, Spoleto, Italy

Andrea Belli General and HPB Surgery, Loreto Nuovo Hospital, Naples, Italy

Fiorenza Belli General and Hepatobiliary Surgery, Galliera Hospitals, Genoa, Italy

Giulio Belli General and HPB Surgery, Loreto Nuovo Hospital, Naples, Italy

Raffaele Bellochi Department of Surgery, Division of General, Minimally Invasive and Robotic Surgery, ASL3 Umbria, San Matteo degli Infermi Hospital, Spoleto, Italy

Mario Belluomini Department of Oncology, Transplants, and New Technologies, AOU Pisana, University of Pisa, Pisa, Italy

Riccardo Boetto Hepatobiliary Surgery Unit and Liver Transplant Center, University of Padua, Padua, Italy

Ugo Boggi Department of Oncology, Transplants, and New Technologies, University of Pisa, Pisa, Italy

Joseph F. Buell Tulane Transplant Institute, Department of Surgery, Tulane University, New Orleans, Louisiana, USA

David Calatayud Division of General, Minimally Invasive and Robotic Surgery, Department of Surgery, University of Illinois at Chicago, Chicago, USA

Fulvio Calise Hepatobiliary Surgery and Liver Transplant Center, A. Cardarelli Hospital, Naples, Italy

Fabio Caniglia Department of Oncology, Transplants, and New Technologies, AOU Pisana, University of Pisa, Pisa, Italy

Robert Cannon Department of Surgery, University of Louisville, Louisville, Kentucky, USA

Carla Cappelli Department of Radiology, AOU Pisana, Pisa, Italy

Lorenzo Capussotti Department of HPB and Digestive Surgery, Ospedale Mauriziano Umberto I, Turin, Italy

Luciano Casciola Department of Surgery, Division of General, Minimally Invasive and Robotic Surgery, ASL3 Umbria, San Matteo degli Infermi Hospital, Spoleto, Italy

Graziano Ceccarelli Department of Surgery, Division of General, Minimally Invasive and Robotic Surgery, ASL3 Umbria, San Matteo degli Infermi Hospital, Spoleto, Italy

Antonio Ceriello Hepatobiliary Surgery and Liver Transplant Center, A. Cardarelli Hospital, Naples, Italy

Daniel Cherqui Section of Hepatobiliary Surgery and Liver Transplantation, New York Presbyterian Hospital – Weill Cornell Medical Center, New York, USA

Umberto Cillo Hepatobiliary Surgery Unit and Liver Transplant Center, University of Padua, Padua, Italy

Luigi Cioffi General and HPB Surgery, Loreto Nuovo Hospital, Naples, Italy

Federica Cipriani Department of Surgery, Hepatobiliary Surgery Unit, San Raffaele Hospital, Vita-Salute San Raffaele University, Milan, Italy

Marco Colasanti General Surgery and Transplantation, San Camillo Hospital, Rome, Italy

Andrea Coratti Department of Surgery, General Surgery Unit, Misericordia Hospital Grosseto, Grosseto Italy

Santolo Cozzolino Center of Biotechnologies, A. Cardarelli Hospital, Naples, Italy

Alberto D'Agostino General and HPB Surgery, Loreto Nuovo Hospital, Naples, Italy

Francesco D'Amico Hepatobiliary Surgery Unit and Liver Transplant Center, University of Padua, Padua, Italy

Ada De Felice Anesthesia and Intensive Care Unit, Transplants Department, A. Cardarelli Hospital, Naples, Italy

Maurizio Defez Hepatobiliary Surgery and Liver Transplant Center, A. Cardarelli Hospital, Naples, Italy

Giuseppe De Simone Anesthesia and Intensive Care Unit, Transplants Department, A. Cardarelli Hospital, Naples, Italy

Bjørn Edwin The Interventional Center and Surgical Department, Oslo University Hospital, Rikshospitalet, Oslo, Norway

Giorgio Ercolani Liver and Multiorgan Transplant Unit, S. Orsola-Malpighi Hospital, University of Bologna, Bologna, Italy

Ciro Esposito Anesthesia and Intensive Care Unit, Transplants Department, A. Cardarelli Hospital, Naples, Italy

Giuseppe Maria Ettorre General Surgery and Transplantation, San Camillo Hospital, Rome, Italy

Federico Famiglietti Digestive and Colorectal Surgery, University of Turin, Turin, Italy

Corrado Fantini General and HPB Surgery, Loreto Nuovo Hospital, Naples, Italy

Gianfranco Ferla Department of Surgery, Hepatobiliary Surgery Unit, San Raffaele Hospital, Vita-Salute San Raffaele University, Milan, Italy

Alessandro Ferrero Department of HPB and Digestive Surgery, Ospedale Mauriziano Umberto I, Turin, Italy

Marco Filauro General and Hepatobiliary Surgery, Galliera Hospitals, Genoa, Italy

Brice Gayet Department of Digestive Diseases, Institut Mutualiste Montsouris, University Paris Descartes, Paris, France

Antonio Giuliani Hepatobiliary Surgery and Liver Transplant Center, A. Cardarelli Hospital, Naples, Italy

Pier Cristoforo Giulianotti Division of General, Minimally Invasive and Robotic Surgery, Department of Surgery, University of Illinois at Chicago, Chicago, USA

Felice Giuliante Hepatobiliary Surgery Unit, Università Cattolica del Sacro Cuore, Rome, Italy

Bruno Gridelli Department of Abdominal and Transplantation Surgery, Istituto Mediterraneo per i Trapianti e Terapie ad Alta Specializzazione (IsMeTT) - University of Pittsburgh Medical Center (UPMC) Italy, Palermo, Italy

Enrico Gringeri Hepatobiliary Surgery Unit and Liver Transplant Center, University of Padua, Padua, Italy

Salvatore Gruttadauria Department of Abdominal and Transplantation Surgery, Istituto Mediterraneo per i Trapianti e Terapie ad Alta Specializzazione (IsMeTT) - University of Pittsburgh Medical Center (UPMC) Italy, Palermo, Italy

Michael D. Kluger Assistant Professor of Surgery, Section of Hepatobiliary Surgery and Liver Transplantation, New York Presbyterian Hospital – Weill Cornell Medical Center, New York, USA

Alan Koffron Department of Surgery, Oakland University William Beaumont School of Medicine, Detroit, Michigan, USA

Pasquale Lepiane General Surgery and Transplantation, San Camillo Hospital, Rome, Italy

Paolo Limongelli General and HPB Surgery, Loreto Nuovo Hospital, Naples, Italy

Roberto Lo Tesoriere Department of HPB and Digestive Surgery, Ospedale Mauriziano Umberto I, Turin, Italy

Andrea Laurenzi General Surgery and Transplantation, San Camillo Hospital, Rome, Italy

Marcella Marracino Anesthesia and Intensive Care Unit, National Cancer Institute "Pascale", Naples, Italy

Gianpaolo Marte Hepatobiliary Surgery and Liver Transplant Center, A. Cardarelli Hospital, Naples, Italy

Carla Migliaccio Hepatobiliary Surgery and Liver Transplant Center, A. Cardarelli Hospital, Naples, Italy

Giangiacomo Nunzio Monti Hepatobiliary Surgery and Liver Transplant Center, A. Cardarelli Hospital, Naples, Italy

Mario Morino Digestive and Colorectal Surgery, University of Turin, Turin, Italy

Daniele Neri Hepatobiliary Surgery Unit and Liver Transplant Center, University of Padua, Padua, Italy

Raffaella Niola Departement of Diagnostic Imaging, Section of Emergency Radiology, A. Cardarelli Hospital, Naples, Italy

Gennaro Nuzzo Hepatobiliary Surgery Unit, Università Cattolica del Sacro Cuore, Rome, Italy

Enrico Opocher Hepatobiliary, Pancreatic and Digestive Surgery, San Paolo Hospital, University of Milan, Milan, Italy

Michele Paganelli Department of Surgery, Hepatobiliary Surgery Unit, San Raffaele Hospital, Vita-Salute San Raffaele University, Milan, Italy

Duilio Pagano Department of Abdominal and Transplantation Surgery, Istituto Mediterraneo per i Trapianti e Terapie ad Alta Specializzazione (IsMeTT) - University of Pittsburgh Medical Center (UPMC) Italy, Palermo, Italy

Alberto Patriti Department of Surgery, Division of General, Minimally Invasive and Robotic Surgery, ASL3 Umbria, San Matteo degli Infermi Hospital, Spoleto, Italy

Antonio Daniele Pinna Liver and Multiorgan Transplant Unit, S. Orsola-Malpighi Hospital, University of Bologna, Bologna, Italy

Marina Polacco Hepatobiliary Surgery Unit and Liver Transplant Center, University of Padua, Padua, Italy

Francesca Ratti Department of Surgery, Hepatobiliary Surgery Unit, San Raffaele Hospital, Vita-Salute San Raffaele University, Milan, Italy

Calogero Ricotta Department of Abdominal and Transplantation Surgery, Istituto Mediterraneo per i Trapianti e Terapie ad Alta Specializzazione (IsMeTT) - University of Pittsburgh Medical Center (UPMC) Italy, Palermo, Italy

Aldo Rocca Hepatobiliary Surgery and Liver Transplant Center, A. Cardarelli Hospital, Naples, Italy

Marina Romano Hepatobiliary Surgery and Liver Transplant Center, A. Cardarelli Hospital, Naples, Italy

Gianluca Russo General and HPB Surgery, Loreto Nuovo Hospital, Naples, Italy

Nadia Russolillo Department of HPB and Digestive Surgery, Ospedale Mauriziano Umberto I, Turin, Italy

Bob Saggi Tulane Transplant Institute, Department of Surgery, Tulane University, New Orleans, Louisiana, USA

Roberto Santambrogio Hepatobiliary, Pancreatic and Digestive Surgery, San Paolo Hospital, University of Milan, Milan, Italy

Walter Santaniello Hepatobiliary Surgery and Liver Transplant Center, A. Cardarelli Hospital, Naples, Italy

Roberto Santoro General Surgery and Transplantation, San Camillo Hospital, Rome, Italy

Vincenzo Scuderi Hepatobiliary Surgery and Liver Transplant Unit, A. Cardarelli Hospital, Naples, Italy

Francesco Sicoli Hepatobiliary Surgery and Liver Transplant Center, A. Cardarelli Hospital, Naples, Italy

Stefano Signori Department of Oncology, Transplants, and New Technologies, AOU Pisana, University of Pisa, Pisa, Italy

Marco Spada Department of Abdominal and Transplantation Surgery, Istituto Mediterraneo per i Trapianti e Terapie ad Alta Specializzazione (IsMeTT) - University of Pittsburgh Medical Center (UPMC) Italy, Palermo, Italy

Alessandro Spaziani Department of Surgery, Division of General, Minimally Invasive and Robotic Surgery, ASL3 Umbria, San Matteo degli Infermi Hospital, Spoleto, Italy

Giuseppe Surfaro Hepatobiliary Surgery and Liver Transplant Center, A. Cardarelli Hospital, Naples, Italy

Maria Vellone Hepatobiliary Surgery Unit, Università Cattolica del Sacro Cuore, Rome, Italy

Giovanni Vennarecci General Surgery and Transplantation, San Camillo Hospital, Rome, Italy

Luca Viganò Department of HPB and Digestive Surgery, Ospedale Mauriziano Umberto I, Turin, Italy

Alessandro Vitale Hepatobiliary Surgery Unit and Liver Transplant Center, University of Padua, Padua, Italy

Minimally Invasive Surgery: An Update 1

Luciano Casciola

1.1 Lights and Shadows of Laparoscopic Surgery

More than quarter of a century of minimally invasive surgery has most certainly not passed in vain [1]. It has enabled us to reflect and make considerations and assessments that are certainly more objective than those of the past. An anecdote often circulated in those early days of laparoscopic surgery, even in influential editorials, about the gynecologist carrying out a pelvic laparoscopy who turned the optic upward to perform the first laparoscopic cholecystectomy and how later he would curse that moment [2].

Actually, from the 1960s onward, the expert Philippe Mouret laid claim to the technique of exploring upward and accused general surgeons of having snubbed and underestimated the exploratory technique of laparoscopy simply because it was designed for and practiced by gynecologists. His innovative and philosophical thought highlights the idea that "if specialization presents an advantage, it is that of adding specialist or ultra-specialist knowledge to sound general knowledge and not deliberately ignoring everything that does not pertain to one's own privileged practice. The practice of laparoscopy strikingly demonstrates the artificial nature of our classifications and the difficulties pathologies will certainly have in adapting to them" [3].

It is difficult not to agree with Mouret about this. Laparoscopy, a surgery performed in the abdomen facilitated by the use of a video camera, is a practice that has been gradually perfected by committed surgeons. Minimally invasive colorectal surgery, unduly contested in the beginning, is now considered to be a kind of litmus test, an assessment of laparoscopy's inexplicably

L. Casciola (✉)
Department of Surgery
Division of General, Minimally Invasive and Robotic Surgery,
ASL3 Umbria, San Matteo degli Infermi Hospital, Spoleto, Italy
e-mail: casciolas@libero.it

F. Calise and G. Casciola (eds.), *Minimally Invasive Surgery of the Liver,*
Updates in Surgery
DOI: 10.1007/978-88-470-2664-3_1, © Springer-Verlag Italia 2013

limited growth, particularly considering its recognized feasibility, which is evident from oncological results. If the obviously justified initial scruples concerning oncological limits in some sense curbed the growth of colorectal laparoscopic surgery, the same should not have applied for a benign pathology. However, other preclusions came into force – the same preclusions that emphasized the limits more than the advantages; the contraindications more than the indications. The experts contested cholecystectomy performed through videolaparoscopy, which occurred as late as 1987, and the technique's mainly young supporters with their idea of calling into question something like surgery of the biliary tract – an issue about which everything was considered to have been said. Nothing could have been further from the truth. For various reasons, the delayed growth of this new surgical technique – essentially, colorectal surgery – is interpreted today as failed growth.

The story repeats itself: Minimally invasive surgery of the colon has not developed at the same rate as it has in other areas of pathology – appendix, inguinal hernia, spleen, stomach, to mention but a few [4–6]. Attention is now focused, above all, on colorectal surgery because, all of a sudden, it has changed from being the most contested to the most legitimized form of minimally invasive surgery [7]. However, there are probably many surgeons operating with insufficient experience in general laparoscopy, which Mouret held so dear. The systematic adoption of laparoscopy as an ultimate diagnostic tool should, in time, have led to the decrease in useless laparotomies. This did not happen to a significant degree. This can partly be justified by growth in the preoperative workup of diagnostic technologies, but routine practice remains irreplaceable for the development of the minimally invasive discipline. Routine practice has probably failed to be widespread.

However, it would be unjust not to recognize the significant role of minimally invasive surgery in modern practice and its exponential growth with regard to certain pathologies, for which it is now considered to be a gold standard, particularly regarding the esophagogastric junction, the left and right adrenals, the distal pancreas, and the spleen [5, 8, 9]. Obviously, reservations exist about the yet to be fully defined meaning of the gold standard.

1.2 Historical Limitations of Laparoscopic Surgery

Two almost absolute contraindications have always been considered with regard to minimally invasive surgery: previous abdominal surgeries, and parenchyma surgery. Surgery on the operated abdomen also means surgery on the abdominal wall and its defects. Minimally invasive laparoscopic hernia surgery, essentially initiated in Spain, has achieved a primary role in less than a decade (thanks also to the growth in prosthetic materials), overcoming all obstacles in a short time: myth or reality; cost–benefit; made-to-measure treatment; extreme cases [10, 11]. Results of laparoscopic hernia surgery are good and indisputable. Certainly, the minimally invasive approach has all the prerequisites necessary to overturn the natural history of this illness. Damage to the abdominal wall is primarily the result of a long-standing, incisional hernia that has not been treated by traditional surgery, which is historically associated with a poor outcome [12]. Minimally invasive surgery, in itself less encumbered by wall complications and perfectly in keeping with prosthetic treatment of the hernia, will allow treatment of initial lesions earlier and with better results. However, hernia signifies an operated abdomen, and an operated abdomen implies adhesions. There cannot be surgery on the abdominal-wall defect without the viaticum of adhesiolysis. Treating adhesions, whether symptomatic or not, represents a kind of "credit card" for the following wall-defect treatment, which thus becomes a not always smooth, combined operation. Learning the technique is what makes the difference.

Acceptance of laparotomy once again renders justice to Mouret; others debated whether or not videolaparoscopic cholecystectomy presented opportunities, discussed its limitations and contraindications, researched its role in adhesiolysis, catalogued it, and rendered it workable in an attempt to overcome the initial limitation: the operated abdomen. It is through this process that the incisional hernia is now approached laparoscopically. It is difficult to leap beyond a generation; however, belief and application make things possible. Even if, in the beginning, minimally invasive surgery did not represent a real, progressive model in its technological innovation, and even if it got off the ground in minor, nonacademic centers, isolated learning curves have developed around it that are respectful of the principles of traditional surgery. It is true that the use of the Veress needle is still a topic of debate, but at the same time, we are discussing minimally invasive surgery of the liver, which once certainly was impossible. And we are now reporting large case series, as this monograph demonstrates.

Perhaps we are still far from a unified perspective that defines laparoscopy either as a choice or an obligation for the surgeon, but in the meantime, use of this technique in every area of pathology has been investigated, and innovation has closely followed. We are at the point of no return.

With its proactive potential, laparoscopy also entered the practice of major vascular surgery, with the total respect for anatomical and conceptual rules that we ourselves guarantee. However, its use in that pathology was abandoned when confronted by undisputed difficulties, because the ideal case of aortic aneurism was also the ideal case for the emerging endovascular treatment. The Marseille meeting in 1999, dedicated to improving use in this specific arena, is no more than a distant memory: laparoscopy had to give way to other forces.

Had it not already existed, and although treated with contempt for a long period, laparoscopic surgery of the hernia would have had to have been invented in certain clinical situations. Reference is made here to the recurrences or plurirecurrences of open prosthesis surgery, which highlight the physiological limitations of totipotency. In this case, the smoothest path is the posterior approach. If the hernia is associated with bilaterality and concomitant pathologies, indications for a laparoscopic approach exceed 30%. For those of us who have been advocates for so long, substantiating the technique with important figures and experience dating back 20 years is like being recognized by evidence-based medicine (EBM). Formal medicine always urges us to work on the basis of evidence, but events are such that before useful case series are available, new procedures quickly develop, other techniques take over from previous ones, and one begins again with the same wariness of doing something without evidence: the dog chasing its own tail. On one hand, certainly the opinion of the majority, we must modulate what we do on the basis of evidence, On the other hand, we are witnesses to the development of new procedures, such as Natural Orifice Transluminal Endoscopic Surgery (NOTES) or the single-port technique – the former being somewhat in line with the concept of minimally invasive surgery; the latter risking making difficult what seems by now to have become easy – both techniques without the support of evidence. In the meantime, laparoscopic surgery of the hernia has not developed as it should, even though indications for its use are, absurdly, agreed upon.

1.3 The Case of Robotic Surgery

Robotic surgery, with its claim through minimally invasive surgery to accommodate the needs of both patient and surgeon – the latter wishing to lay hands on the patient's abdomen without opening it up – from a purely theoretical point of view should have facilitated the growth of a type of surgery that was at that time relatively new. The amazing technology involved – from the high-definition, three-dimensional vision and articulation of its instruments, to the great stability of those same instruments guaranteeing the absence of tremors – seemed to have the right credentials for a further leap forward in surgical quality or even a definitive takeoff [13]. Even earlier, remote surgery could have appeared as a change in gear for minimally invasive surgery, making even smoother what was already consolidated and opening up new frontiers. So far, this has not happened. Although devised for cardiothoracic surgery, the technique has benefited the decidedly specialized branch of urology, particularly with regard to prostate cancer. In fact, robotic technology does not permit those with access to it to use it to make up for important deficiencies in general minimally invasive surgery. Whereas one is the natural evolution of the other, they share the same problems of the closed abdomen, of the pneumoperitoneum, of the port sites, of the orthogonality, of the instruments, all of which are and remain common problems. Robotic surgery does not allow leaps in procedure but increases contradictory elements. How else can its successes in urology, achieved rapidly and without an extensive background experience, be explained? Should it not have been the same scenario for endocrine surgeons, who saw the adrenal in the hands of general laparoscopy surgeons?

On one hand, encouraged by defensive medicine, there is an inescapable movement toward surgery that is increasingly more specialized. On the other hand, new demands emerge in specific areas of pathology. Our times have seen the emerging possibility of surgical treatment for colorectal liver metastases, thanks in part to the growing possibilities of chemotherapy together with the emerging potential of operative endoscopy. What possibilities can a patient have of being treated effectively with up-to-date approaches, regardless of a hospital's geographic location, while waiting for the strategic reorganization of hospitals and clinics and specializations, which has always been talked about but never defined? Without doubt, this is a period of change, but every era undergoes its changes. It is deceptive to think that everything can be tied to EBM, just as everyone should not shape their evidence and modulate their gestures – because surgery is nothing more than a gesture. Literature, although often late, is of great help, if not indispensable, in facilitating change.

This change is difficult to manage. However, the pioneering of the early days is now contrasted by didactics, which is certainly in need of improvement but extremely effective, like that of the historical, 20-year-old Associazione Chirurghi Ospedalieri Italiani (*National Congress of Italian Hospital Surgeons*; ACOI) school and that of the more recent institution of the Società Italiana di Chirurgia (International Society of Surgery; SIC) school. The schools and specialist courses cannot only be the showcases for those who propose them; they must emphasize that everyone can and must decline, when necessary, for the good of the patient.

There is no greater gap to fill than that of minimally invasive surgery, which has failed to fully develop despite its now universally legitimized status. We welcome the productive collaborations with their sharing of experiences, from which we have all benefited. Thanks to his willingness and that of his team, we have established a reciprocal project of collaboration with the coeditor of this monograph, Fulvio Calise, director of Hepatobiliary and Transplantation Surgery at the Cardarelli Hospital of Naples, Italy. We have shared our experience in colorectal laparoscopic surgery and, thanks to them, have been able to supplement it with knowledge regarding hepatic surgery, making efforts to improve it further. The results are gratifying. Whenever possible, we deal with synchronous colorectal liver metastases, approaching whenever possible both conditions at the same time. Similarly, the Naples group has developed autonomy in their use of laparoscopic surgery in the large intestine. Credit must be given to Calise, a renowned expert, for having recognized the importance of integration with general surgery, an approach he has always maintained achieves the best result for the patient.

Robotic surgery, which we have been practicing for a decade, has enabled us to finely tune minimally invasive surgery of the liver, even for the posterolateral segments [14]. With regard to parenchyma surgery, always considered taboo for minimally invasive surgery, robotic surgery has enabled us to perform a greater number of major pancreatic surgery, with results to be confirmed, but certainly, in our experience, no less significant than open surgery [15]. It cannot be hypothesized that minimally invasive surgery can change the natural course of pancreatic cancer, but the minimal approach could, in a short time, integrate with emerging alternative chemotherapy or locoregional therapies. It is with regard to the latter that the current scenario could undergo changes for certain diseases (rectum and cardia carcinoma) for which positive results are usually difficult to achieve. Minimally invasive surgery, robotic and nonrobotic, will effectively perform its role, as it is ready and waiting [14, 16, 17–20].

These major surgeries with a minimally invasive approach are limited thus far to isolated groups, but they are developing into significant numbers. These groups are certainly important for the overall growth of the techniques in that they are proactive in the destiny of greater pathologies and offer stimulus for those who still hesitate or who have yet to start – perhaps obstructed by the problems of costs. This is another age-old problem, certainly real today, against which the evolution of what is new has always had to struggle. However, I am pleased to observe that in the long run, costs have never stopped great innovation, although at times they may have slowed it down, and for those who are less enlightened, costs can even constitute a sort of alibi in the beginning.

It is not possible to deny the great results achieved by minimally invasive surgery, despite its commonly considered insufficient growth relative to the sum total of surgery that is carried out. Compared with procedures that have been absolutely unsurpassable until now (the transplant, for example), an infinite skepticism has held back progress in this field. The generational change in surgery and the driving force of new technology will make the difference in the evolution of minimally invasive surgery, which represents, above all, a cultural leap forward in its rekindling of debate and its promise of new solutions to illnesses in relation to the old and established surgical techniques.

References

1. Mouret P (1996) How I developed laparoscopic cholecystectomy. Ann Acad Med Singapore 25(5):744–747
2. Terblanche J (1991) Laparoscopic cholecystectomy: a new milestone or a dangerous innovation? HPB Surg 3(3):177–180
3. Meinero M, Melotti G, Mouret P (1993) Chirurgia laparoscopica, Masson, Milano
4. Vettoretto N, Gobbi S, Corradi A et al (2011) Consensus conference on laparoscopic appendectomy: development of guidelines. Colorectal Dis 13(7):748–754. doi:10.1111/j.1463-1318.2011.02557.x
5. Casaccia M, Torelli P, Squarcia S et al (2006) The Italian Registry of Laparoscopic Surgery of the Spleen (IRLSS). A retrospective review of 379 patients undergoing laparoscopic splenectomy. Chir Ital 58(6):697–707
6. Bittner R, Arregui ME, Bisgaard T et al (2011) Guidelines for laparoscopic (TAPP) and endoscopic (TEP) treatment of inguinal hernia [International Endohernia Society (IEHS)]. Surg Endosc 25(9):2773–2843. doi:10.1007/s00464-011-1799-6
7. Hazebroek EJ (2002) COLOR: a randomized clinical trial comparing laparoscopic and open resection for colon cancer. Surg Endosc 16(6):949–953. doi:10.1007/s00464-001-8165-z
8. Cavallini A, Butturini G, Daskalaki D et al (2011) Laparoscopic pancreatectomy for solid pseudo-papillary tumors of the pancreas is a suitable technique; our experience with long-term fol-

low-up and review of the literature. Ann Surg Oncol 18(2):352–357. doi:10.1245/s10434-010-1332-5
9. Salminen P, Hurme S, Ovaska J (2012) Fifteen-year outcome of laparoscopic and open Nissen fundoplication: a randomized clinical trial. Ann Thorac Surg 93(1):228–233. doi:10.1016/j.athoracsur.2011.08.066
10. Carbajo MA, Martp del Olmo JC, Blanco JI et al (2003) Laparoscopic approach to incisional hernia. Surg Endosc 17(1):118–122. doi:10.1007/s00464-002-9079-0
11. Carbajo MA, Martin del Olmo JC, Blanco JI et al (1999) Laparoscopic treatment vs open surgery in the solution of major incisional and abdominal wall hernias with mesh. Surg Endosc 13(3):250–252
12. Cassar K, Munro A (2002) Surgical treatment of incisional hernia. Br J Surg 89(5):534–545. doi:10.1046/j.1365-2168.2002.02083.x
13. Giulianotti PC, Coratti A, Angelini M et al (2003) Robotics in general surgery: personal experience in a large community hospital. Arch Surg 138(7):777–784. doi:10.1001/archsurg.138.7.777
14. Casciola L, Patriti A, Ceccarelli G et al (2011) Robot-assisted parenchymal-sparing liver surgery including lesions located in the posterosuperior segments. Surg Endosc 25(12):3815–3824. doi:10.1007/s00464-011-1796-9
15. Giulianotti PC, Addeo P, Buchs NC, Ayloo SM, Bianco FM (2011) Robotic extended pancreatectomy with vascular resection for locally advanced pancreatic tumors. Pancreas 40(8):1264–1270. doi:10.1097/MPA.0b013e318220e3a4
16. Caruso S, Patriti A, Marrelli D et al (2011) Open vs robot-assisted laparoscopic gastric resection with D2 lymph node dissection for adenocarcinoma: a case-control study. Int J Med Robot 7(4):452–458. doi:10.1002/rcs.416
17. Patriti A, Ceccarelli G, Bartoli A et al (2009) Short- and medium-term outcome of robot-assisted and traditional laparoscopic rectal resection. JSLS 13(2):176–183
18. Patriti A, Ceccarelli G, Bartoli A et al (2009) Laparoscopic and robot-assisted one-stage resection of colorectal cancer with synchronous liver metastases: a pilot study. J Hepatobiliary Pancreat Surg 16(4):450–457. doi:10.1007/s00534-009-0073-y
19. Patriti A, Ceccarelli G, Bellochi R et al (2008) Robot-assisted laparoscopic total and partial gastric resection with D2 lymph node dissection for adenocarcinoma. Surg Endosc 22(12):2753–2760. doi:10.1007/s00464-008-0129-0
20. Patriti A, Ceccarelli G, Ceribelli C et al (2011) Robot-assisted laparoscopic management of cardia carcinoma according to Siewert recommendations. Int J Med Robot 7(2):170–177. doi:10.1002/rcs.385

Minimally Invasive Surgery of the Liver: An Update

2

Fulvio Calise and Carla Migliaccio

Your old road is rapidly agin'.
Please get out of the new one
If you can't lend your hand
For the times they are a-changin'.
Bob Dylan, 1964

2.1 Introduction

These prophetic words, written and sung by Bob Dylan almost 50 years ago, may well apply to the (relatively) new field of minimally invasive surgery of the liver (MISL). Nowadays the open approach to liver surgery (OLS) quickly falls behind in comparison with MISL. Older surgeons still have ringing in their ears the repeated warnings concerning risks related to laparoscopic cholecystectomy (LC). Prominent remarks included a possible increased risk of bile duct injuries [1], a prolonged learning curve [2], and the need for young surgeons to have well-established training in OLS before starting to perform LC. None of these alarms has overcome the tide of LC diffusion, and none of them proved to be effectively true. Thanks to many training modalities, such as dry lab, wet lab, simulators, virtual realities, and practice in large animals (pig), young surgeons directly enter into liver surgery using the laparoscopic approach, and open cholecystectomy is almost always and everywhere an embarrassing memory.

So what is the future for MISL? The answer is bright if we take into account the following considerations:

- In 1991, the first report of a hepatic resection appeared by Reich, followed by Gagner [3], which was followed by the first multicenter report, by Azagra, concerning resection of benign lesions [4]. From then on, in PubMed, the number of recorded publications rose to more than 150 in 2011 as an increasing number of centers began practising MISL (Fig. 2.1).

F. Calise (✉)
Hepatobiliary Surgery and Liver Transplant Center, A. Cardarelli Hospital
Naples, Italy
e-mail: fulvio.calise@libero.it

F. Calise and G. Casciola (eds.), *Minimally Invasive Surgery of the Liver,*
Updates in Surgery
DOI: 10.1007/978-88-470-2664-3_2, © Springer-Verlag Italia 2013

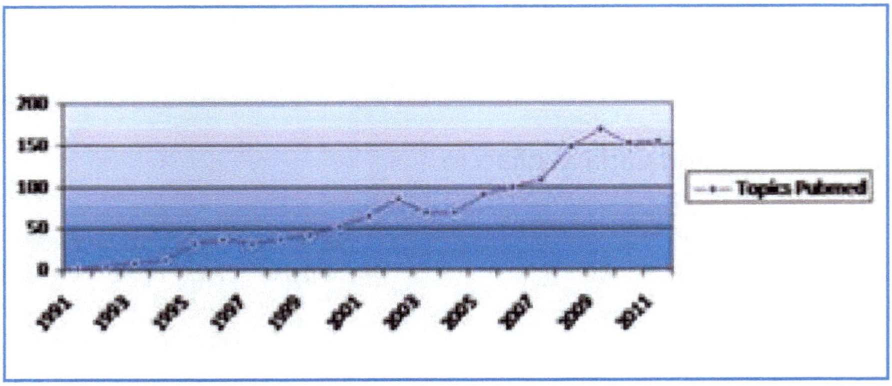

Fig. 2.1 Minimally invasive liver surgery studies listed in PubMed

- In the report by Aldrighetti et al. in this volume (Chap. 35), more than 1600 minimally invasive liver resections (LR) have been performed in Italy in 39 centers in the last 7 years, with only one third of them in surgical units thoroughly dedicated to hepatobiliary surgery.
- A national school of hepatic surgery in Italy was established by Lorenzo Capussotti, and in some affiliated centers, such as ours, MISL is taught and practiced by attending students.
- In a few centers scattered through the country, including ours, practical teaching using pigs allows basic and advanced procedures to be performed.
- Industry strongly supports the development of new transection devices and staplers that have, de facto, modified the technical approach to LR, facilitating accomplishment of the procedure

Since MISL was first introduced, laparoscopic liver surgery (LLS) has been considered a promising technique due to fact that no reconstruction is demanded for resections – with the exception of Klatskin tumors that are, thus far, not considered among indications for LLS. Also, problems related to hypothetical air embolism have been overcome by anesthesiological management using low caval pressure and attention to appropriate indications. Hemorrhage in the transection plane is consistently diminished due to the intra-abdominal pressure induced by the gas, although we should consider that care for bleeding sources must be taken for both resection surfaces.

- The Louisville Consensus Conference [5], held in 2008, clearly stated indications and safety limits for MISL codifying (Table 2.1):
- Liver segments: all segments but 7 and 8 can be approached because of the inability to date of instruments to overcome axial projection on the surface plane. In this regard, robot-assisted surgery may contribute to including surgical indications to these two posterior segments also
- Lesion size
- Appropriate benign lesions

Table 2.1 The international position on laparoscopic liver surgery

The Louisville Statement, 2008
Currently acceptable indications for laparoscopic liver resection (LLR): solitary lesions 5 cm or less located in liver segments 2–6
LLR approach to left lateral sectionectomy should be considered standard practice
Although all types of liver resection can be performed laparoscopically, major liver resections (e.g., right or left hepatectomies) should be reserved for experienced surgeons
Conversion should be considered prudent surgical practice rather than failure
Indications for surgery for benign hepatic lesions should not be widened simply because the surgery can be done laparoscopically
Although data presented on colorectal metastases did not reveal an adverse effect of the laparoscopic approach on oncological outcomes in terms of margins or survival, adequacy of margins and ability to detect occult lesions are concerns

- Conversion modalities
- Certain surgical cases to be appointed to expert surgeons/dedicated centers, only for patient safety
- Left-lateral sectionectomy as a standard of care procedure. A definition of gold standard for left sectionectomy has been proposed by some authors [6, 7], and clinical results seem to confirm this definition.

A growing number of reports demonstrate that no difference exists between OLS and MISL in terms of oncological results, resection margins, blood transfusions, and complications, whereas MISL encompasses less postoperative pain, shorter hospital stay [8, 14], and greater comfort and acceptance by the patient in the first year after surgery (unpublished personal results). Also, the combined one-stage laparoscopic surgery of colorectal tumors and liver metastasis may avoid skin incisions, making resection easier in the case of metastasis recurrence. As the ultimate step in the learning curve, laparoscopic left lobectomy, without pedicle clamping, could be routinely proposed for implementation from a living donor for liver transplantation [15].

LLR is feasible and safe in selected patients with hepatocellular carcinoma (HCC) in cirrhotic liver disease and attains long-term outcomes similar to those reported for the traditional open approach [16]. LLR will probably very soon prove to be as effective as, or more effective than, OLS in HCC resection as a bridge to transplantation.

Thus far, the real remaining limitation in MISL is the quality of laparoscopic ultrasound (US), as devices do not yet provide easy and reliable use of open probes. Progress continues, however, as discussed by Ferrero et al. in Chap. 13 of this volume. Remaining questions concerning the learning curve for either general laparoscopic surgeons or surgeons already expert in OLS are treated specifically in a dedicated chapter.

2.2 Robot-Assisted Liver Surgery

The introduction of robot-assisted liver surgery (RALS) into the field of MISL began in 2001 with the first cholecystectomy [17] and continued in 2003 with the first LR [18]. This new approach once more modifies the scenario. The 360° rotational ability of robotic forceps allows trained surgeons to resect posterior segments 7 and 8, with promising results regarding patient safety and security [18]. Moreover, as in open liver surgery [19], RALS allows the surgeon to perform a multiple-wedge resection, especially for bilobar multiple colorectal metastases, which is still debated for the pure laparoscopic approach.

Furthermore, the da Vinci Si HD© system may now provide a new technology through a second console to facilitate the surgeon's collaboration with the trainee during surgery. The mentoring console provides two options: the "swap" mode, allowing the mentor and the trainee to operate simultaneously; and the "nudge" mode, allowing both to control two robotic arms at the same time (see report by Coratti and Annecchiarico in Chap. 4 in this volume).

2.3 Two Questions Remain to Be Debated

2.3.1 Costs

In many reports, the higher expenses associated with LLR due to use of more transection devices and staplers is largely overcome by a shorter postoperative stay and fewer complications, particularly in case of repeated surgery [18, 20]

Questions still are raised, however, concerning RALS. Answers to this item necessarily must take into account a suggested interdisciplinary use of the robot system (by urologist, gynecologist, vascular surgeon, etc.) that may contribute to a sharp decrease in the effective general cost of the instrument.

2.3.2 Where Should LLC Be Performed?

There is no doubt that in high-volume hepatobiliary units LLS is now a requirement. Major hepatectomies are carried out in many centers around the world with comparable results to those of OLS [21–25]. LS is increasingly performed everywhere in the world, especially for colorectal and gastric diseases. An interdisciplinary approach is strictly required to treat, in particular, synchronous metastasis. Therefore, in a general hospital unit, the need for MISL is increasingly required. With proper training, surgeons with a high expertise in LS may safely approach simple procedures such as resection of anterior segment metastasis in left and right lobes and, eventually, ligature of a portal branch for preparing a two-stage resection, which is to be eventually referred to tertiary centers for open procedures.

2.4 Conclusions

We believe that MISL is entering a new era in the surgical management of the liver diseases. We even dare to say a new, democratic era, because at least minor hepatic surgery is in the hands of a well-trained laparoscopic surgeon working in a general surgery unit. Considering that robot-assisted surgery costs in Europe are twice as much as in the United States and that costs will probably decrease after patent expiration, there really will be no limits to the development of MISL. What will happen is strictly related to the training ability of leading centers.

The secret to advancing these changes in the era of devolution, which the Western world is facing at the present time, probably lies in the willingness of centers to exchange experience and practice. Proctoring, second opinion, may represent the key to saving money and increasing surgeon's knowledge and ability. More data are surely needed to confirm the safety and oncological value of MISL. However, the trail is already broken, and an increasing number of surgeons will follow it, happily and safely, together with their patients.

References

1. Talamini MA, Gadacz TR (1992) Laparoscopic approach to cholecystectomy. Adv Surg 25:1–20
2. Marescaux J, Evrard S, Keller P, Miranda E, Mutter D, Van Haaften K (1992) Is cholecystectomy by celiovideoscopy dangerous in the initial period? Prospective study of 100 initial cases. Gastroenterol Clin Biol 16:875–878
3. Gagner M, Rheault M, Dubuc J (1992) Laparoscopic partial hepatectomy for liver tumor. Surg Endosc 6:99
4. Azagra JS, Goergen M, Gilbart E, Jacobs D (1996) Laparoscopic anatomical (hepatic) left lateral segmentectomy-technical aspects. Surg Endosc 10:758-61
5. Buell JF, Cherqui D, Geller DA, O'Rourke N et al (2009) The international position on laparoscopic liver surgery: The Louisville Statement, 2008. Ann Surg 250:825–830
6. Carswell KA, Sagias FG, Murgatroyd B, Rela M, Heaton N, Patel AG (2009) Laparoscopic versus open left lateral segmentectomy. BMC Surg 7;9:14
7. Chang S, Laurent A, Tayar C, Karoui M, Cherqui D (2007) Laparoscopy as a routine approach for left lateral sectionectomy, Br J Surg 94:58–63
8. Cherqui D, Husson E, Hammoud R, Malassagne B, Stephan F, Bensaid S et al (2000) Laparoscopic liver resections: a feasibility study in 30 patients. Ann Surg 232:753–762
9. Gagner M, Rogula T, Selzer D (2004) Laparoscopic liver resection: benefits and controversies. Surg Clin North Am 84:451–462
10. Cherqui D (2003) Laparoscopic liver resection. Br J Surg 90:644–646
11. Van Gulik T (2009) Open versus laparoscopic resection for liver. HBP (Oxford) 11:465–468
12. Edwin B, Nordin A, Kazaryan AM (2011) Laparoscopic liver surgery: new frontiers. Scand J Surg 100(1):54–65
13. Nguyen KT, Geller DA (2010) Laparoscopic liver resection-current update. Surg Clin North Am 90(4):749–760
14. Hu BS, Chen K, Tan HM, Ding XM, Tan JW (2011) Comparison of laparoscopic vs open liver lobectomy (segmentectomy) for hepatocellular carcinoma. World J Gastroenterol 17(42):4725–4728

15. Troisi RI, Van Huysse J, Berrevoet F et al (2011) Evolution of laparoscopic left lateral sectionectomy without the Pringle maneuver: through resection of benign and malignant tumors to living liver donation. Surg Endosc 25(1):79–87
16. Belli G, Fantini C, Belli A, Limongelli P (2011) Laparoscopic liver resection for hepatocellular carcinoma in cirrhosis: long-term outcomes. Dig Surg 28(2):134–40. doi: 10.1159/000323824
17. Lomanto D, Cheah WK, So JB, Goh PM (2001) Robotically assisted laparoscopic cholecystectomy: a pilot study. Arch Surg 136:1106–1108
18. Casciola L, Patriti A, Ceccarelli G, Bartoli A, Ceribelli C, Spaziani A (2011) Robot-assisted parenchymal-sparing liver surgery including lesions located in the posterosuperior segments. Surg Endosc 25:3815–3824
19. Torzilli G, Procopio F, Botea F et al (2009) One-stage ultrasonographically guided hepatectomy for multiple bilobar colorectal metastases: a feasible and effective alternative to the 2-stage approach. Surgery 146:60-71
20. Vibert E, Denet C, Gayet B (2003) Major digestive surgery using a remote-controlled robot: the next revolution. Arch Surg 138:1002–1006
21. Bhojani FD, Fox A et al (2012) Clinical and economic comparison of laparoscopic to open liver resections using a 2-to-1 matched pair analysis: an institutional experience. J Am Coll Surg 214:184–195
22. Vanounou T, Steel JL, Nguyen KT et al (2010) Comparing the clinical and economic impact of laparoscopic versus open liver resection. Ann Surg Oncol 17:998–1009
23. Topal H, Tiek J, Aerts R, Topal B (2012) Outcome of laparoscopic major liver resection for colorectal metastases. Surg Endosc [Epub ahead of print, February 23, PMID: 22358126]
24. Pearce NW, Di Fabio F et al (2011) Laparoscopic right hepatectomy: a challenging, but feasible, safe and efficient procedure. Am J Surg 202:52–58
25. Dagher I, O'Rourke N, Geller DA et al (2009) Laparoscopic major hepatectomy: an evolution in standard of care. Ann Surg 250:856–860

Part I

Requirements

The Learning Curve in Laparoscopic Liver Surgery

3

Fulvio Calise, Marina Romano, Antonio Ceriello,
Antonio Giuliani and Santolo Cozzolino

3.1 Introduction

The Shakespearian-like dilemma facing the general surgeon – to open or not to open [the abdomen] – before being ready to perform a laparoscopic cholecystectomy (LC) was hotly debated at round tables and symposia in the early days of the laparoscopic era. Reality, as often happens in life, quickly resolved the many discussions and conflicting reports around this subject [1].

Although the aphorism "see one, do one, teach one" is a long-standing phrase heard throughout a general surgeon's education, never before has it been taken so literally. In the early 1990s, general surgeons worldwide signed up for 1- or 2-day courses to learn how to perform LC [2]. In many instances, the closest the surgeon ever came to "seeing one" was in the porcine laboratory model. He or she subsequently walked into the operating room to "do one" on a living human being (albeit while being proctored by the surgeon whose job it was to "teach one.") [3]. By 1999, these far-sighted words had become reality – more or less.

Could we have ever foreseen this scenario for laparoscopic liver surgery (LLS)? Evidently not, but the worldwide diffusion of laparoscopic surgery (LS), the extraordinary technological progress in surgical devices, and the relatively marginal role played by training programs raise questions that are still being discussed. The main criticism to laparoscopic liver resection (LLR) is that it is a long and difficult procedure. It requires considerable expertise in both hepatic and laparoscopic surgery, dedication, specific training, and the availability of appropriate technology. Moreover, it seems unreasonable that even an experienced laparoscopic surgeon should perform LLR outside a reg-

F. Calise (✉)
Hepatobiliary Surgery and Liver Transplant Center, A. Cardarelli Hospital
Naples, Italy
e-mail: fulvio.calise@libero.it

F. Calise and G. Casciola (eds.), *Minimally Invasive Surgery of the Liver*,
Updates in Surgery
DOI: 10.1007/978-88-470-2664-3_3, © Springer-Verlag Italia 2013

ular practice of open liver surgery [2]. Cherqui argues that the collaboration of two surgeons, one expert in each field, seems desirable when initiating an LLR program [4].

3.2 Surgeon Training

3.2.1 The Surgeon in a General Surgery Unit

LS is more than 25 years old. Most of the surgeons active nowadays were educated in the aftermath of this revolutionary approach, and surgical expertise in LS is now virtually mandatory. Training in LS is therefore a prerequisite for any surgeon wishing to perform LLS. The LS training program is based on traditional and well-established simulation and practice in the operating theater. So the first conclusion we may draw is that no surgeon should recommend LLS if he or she has not gained sufficient experience in advanced LS (colorectal surgery, gastric surgery).

In fact, the primary need for LLS in a general surgical unit is for treating synchronous metastases and/or ligating a portal branch to prepare the patient for a two-stage hepatectomy, which will be performed in a liver unit. Resections of lesions <2 cm located in the anterior liver segments require little additional skill to that needed for a complex LC [4–6].

Chapter 35 by Aldrighetti et al. discusses the first Italian LLS survey, reporting on more than 1,600 cases in the last 7 years and only one third of these centers is thoroughly dedicated to hepatobiliary surgery. These numbers strongly suggest that proper training in LS may allow the surgeon to perform minor LLS – and in some cases major resections – without a significant background in open liver surgery (OLS).

3.2.2 The Surgeon in a Hepatobiliary Unit

Twenty years after the first report, major LLRs are still limited to a few expert centers, and only a small percentage of patients are considered by the majority of authors to be suitable for the laparoscopic approach. Some encouraging data come from recently published large series, but reproducibility and routine feasibility of this technique remain questionable [7].

Adequate training during the learning curve in LLS results in improved patient outcomes in terms of operative time, rate of pedicle clamping and conversion, blood loss, morbidity, and hospital stay [4–8]. With the learning curve completed, blood loss and operative time progressively decrease and pedicle clamping is less commonly required, and if so, is required for a shorter duration. Simultaneously, reduced blood loss and need for pedicle clamping confirms the safety of LLR despite the increased complexity of the resection.

A further improvement attributable to LLR is in postoperative complications. Many articles report lower morbidity rates than with the open approach [8]. There are two reasons for these results: strict patient selection is commonly adopted, LLR is a more accurate technique, and magnification of by the laparoscope may allow more precise hemostasis. Conversion rate is considered a criterion of quality in LS. The literature reports that there are essentially two factors influencing the conversion rate: bleeding, and technical problems such as difficult exposure, insufficient or poor-quality view, a fragile tumor with risk of rupture, and uncertainty regarding the distance between the tumor and the transection plane [9]. With an appropriate learning curve and patient selection, conversions due to technical problems are reduced. However, conversion should be considered prudent surgical practice rather than LS failure [5].

The duration of the learning curve depends on the trainee surgeon's previous training and types of resections performed. For major resections (right or left hepatectomy), the bar is raised significantly, and these resections should only be attempted by surgeons regarded as laparoscopic hepatobiliary and pancreatic (HPB) specialists who regularly perform complex laparoscopic procedures. The main limiting factor for these resections is technical difficulty and surgical access. Some reports suggest that increased tumor size is not a limiting factor, but this is not what is recommended in the Louisville Statement [5]. Huge tumors have proven to be more difficult to manipulate laparoscopically and require much larger incisions to remove, consequently eliminating the benefits of a minimally invasive procedure [6].

Major LLR is anyway technically demanding and obviously requires expertise in conventional OLS as well as in advanced laparoscopic techniques [4–5]. A slow, but constant evolution of LLR is occurring: indications for and the magnitude of such procedures have increased, and technical outcomes have improved. The learning curve demonstrated in this study suggests that LLR is reproducible in liver units, but specific training in advanced laparoscopy is required [2, 7].

3.3 Training Modalities

Several training modalities can be used when constructing an LS training program. Direct technical training is available either from virtual or animal models [11]. In general, a program starts with knowledge development, followed by skills training using a combination of simulation modalities, which, in turn, is followed by real-life case observation in the operating room. When starting an LS training program, the trainee/surgeon needs to attain knowledge of and understand laparoscopic technology, device functions, basic troubleshooting, device parameters, and limitations of the system. The next step is to develop knowledge for specific surgical procedures, which include patient selection and indication, preoperative preparation, patient and system positioning, port placement, procedural steps, and complications and their management [12].

3.4 Dry Lab

Training for LS can be scheduled in a skills laboratory. A dry laboratory (Dry Lab) is specific to work with dry stored materials, electronic and/or large instruments. Unlike wet laboratories, biological tissues (living or dead) are not utilized. These dry labs are equipped with workstations for practicing endoscopic techniques, in a realistic setting, on phantoms and organ models [13].

3.4.1 Virtual Reality

Simulators offer standardized and reproducible laparoscopic tasks, ranging from simple basic skills training to complex procedures. The development of virtual reality (VR) laparoscopic simulators actually began in the early 1990s, the first such simulator being introduced in the mid-1990s. Progress has been rapid, and simulators and simulation technology have recently been the subject of reviews and editorials in major surgical and medical journals. The notion is to create "a realistic" operating environment that offers unlimited laparoscopic practice without detrimental consequences when trainees make mistakes. Two main contributions arise from this global aim: (1) specification of didactic exercises that has developed into the creation of a new training environment; (2) appropriate use of simulation technologies that significantly improve overall simulation [15].

The simulator comprises these didactic units: hand–eye coordination, camera manipulation, grasping, pulling, cutting, dissecting, and suturing, with one or more tasks required in each unit. With the development of optic virtual simulators based on the science of "the study of sensing through touch," these new simulators provide an immersive environment for surgeons to touch, feel, and manipulate computer-generated 3D images of tissue and organs with tools used in actual operating theaters. The LS simulator is developed for surgical education with an expandable surgical platform that includes life-like surgical anatomy and the ability to record the learner's performance during each session. It also offers an immediate review and critique of the performed laparoscopic task. This allows trainees to develop and test their technical, cognitive, and medical decision-making skills in a safe virtual environment [16].

3.5 Wet Lab

3.5.1 Animal Models

Although not always replicable, models of actual surgical procedures are increasingly important in the surgical environment. Such models have been used for many purposes, ranging from a source of experience for students, residents, and surgeons to the development of new surgical techniques [17].

These models reduce the learning curves for various procedures and make it easier to reproduce the learned techniques and continue to develop them in the operating room. Laparoscopic procedures are generally performed in the porcine model in a manner comparable with actual surgery in humans [18]. The main limitations of using living animals are ethical considerations and the relatively high cost to training centers of using large animals.

3.5.2 Cadaver Models

The human cadaver remains the gold standard for anatomic training and is highly useful during minimally invasive surgery training programs. As a matter of fact, it may be used in a multidisciplinary cooperation. Many skills can be learned through tissue manipulation and hemostasis exercises in animal training courses, but the human anatomical material still provides the best instruction for surgeons participating in robotic and laparoscopic training courses [19].

However, a widespread use of human anatomical material for surgical skills training is limited by a restricted supply of cadavers, by the fact that there is a difference in tissue quality between cadaveric and blood-perfused tissue, and the fact that cadavers may be only used once.

At the moment, in Italy the law allows to use only frozen cadavers. The Center of Biotechnologies of Naples has hosted the Donor's Surgeon course since 2009. In this course the frozen cadavers are used to train young surgeons for multi-organ procurement, liver resection and split liver, and transplant techniques for living donor were taught.

3.6 Training Centers

Following the experience initiated in Strasbourg many training centers throughout Europe have been created. These centers were advanced mainly by private companies interested in the diffusion of laparoscopic techniques and instruments. Although the government of such centers is the hands of the leading surgeons who promoted them, the company's financial support is essential to their survival. In some cases, national societies of surgeons, such as the Associazione Chirurghi Ospedalieri Italiani (ACOI; National Congress of Italian Hospital Surgeons and the Società Italiana di Chirurgia(Italian Society of Surgery) and the Società Italiana di Chirurgia della Mano (Italian Society of Surgery of the Hand) (SIC-SICM), encouraged the foundation of national schools of surgery primarily devoted to the advancement of open and laparoscopic training.

In the Cardarelli Hospital in Naples, the Center of Biotechnologies has been active since 1988. The center serves many purposes: experimental preclinical research in small (mouse, rat, rabbit) and large (pig) animals, and engages in international cooperation and educational training in surgery, microsurgery, laparoscopy, and other specialties using either rats and pigs or frozen cadavers. The center has run more than 207 courses in all fields of specialized surgery, for a total of 583 educational days and 2,881 students. Since 2007, an average of 50 courses/year have been presented, with nine of them providing the opportunity for students to first practice on animals and subsequently on human frozen specimens from the USA (trunk, head, spine). Moreover, in 2012, a larger, three-story building will become the new location for all surgical, microsurgical, and laparoscopic activities. The Center of Biotechnologies also hosts the national ACOI school on microsurgery and experimental surgery and an advanced course on microsurgery of the SIC-SICM.

Other solutions aiming at providing an appropriate learning curve have been spontaneously devised to overcome budget restrictions for continuing education:

- Proctoring by expert surgeons in centers willing to start or strengthen liver surgery programs
- Exchange skills and know-how in LS, as done by the editors of this volume: colorectal (Casciola) versus hepatic (Calise). This free exchange of surgical competence allows training centers to ultimately become autonomous in both surgical fields.

3.7 Conclusions

The increasing use of LS and consequently of LLS is changing the characteristics of the traditional learning curve. The axiom that surgery is a wise mix of seeing much and practising much may not be so true in the near future. Children learn to use a computer and explore the Internet at the age of 3 years, or even younger. The new generation of surgeons, thanks to the astonishing progresses in technology, will probably accelerate their learning curve and jump many steps that older surgeons would never have imagined as possible.

Giacomo Leopardi wrote in his poem "L'infinito" 200 years ago: "*Naufragar m'è dolce in questo mar*" – "And to wreck in this sea is sweet to me."

The sea of the Internet, with all its delights and devilries, now surrounds us.

References

1. Miller TA (1990) Laparoscopic cholecystectomy: passing fancy or legitimate treatment option? Gastroenterology 99:1527-9
2. Cherqui D (2003) Laparoscopic liver resection. Br J Surg 90:644–646
3. Shayani V, Jacobs L, Jonathan M (2005) Laparoscopic Cholecystectomy II. In: Wetter PA, Kavic MS et al (eds) Prevention and Management of Laparoendoscopic Surgical Complications, 2nd Edition. Society of Laparoendoscopic Surgeons
4. Vigano L, Laurent A, Tayar C et al (2009) The learning curve in laparoscopic liver resection: improved feasibility and reproducibility. Ann Surg 250:772-82

5. Buell JF, Cherqui D, Geller DA et al (2009) The international position on laparoscopic liver surgery: The Louisville Statement Ann Surg 250:825-30
6. Van Gulik T (2009) Open versus laparoscopic resection for liver tumours. HPB 11:465-468
7. Chang A, Laurent C et al (2007) Laparoscopy as a routine approach for left lateral sectionectomy. Br J Surg 94:58–63
8. Robinson SM, Hui KY, Amer A, Manas DM et al (2012) Laparoscopic liver resection: is there a learning curve? Dig Surg 29:62-9
9. Giulianotti PC, Sbrana F, Coratti A et al (2011) Totally robotic right hepatectomy: surgical technique and outcomes. Arch Surg 146:844-50
10. Schreuder HW, Wolswijk R, Zweemer RP et al (2012) Training and learning robotic surgery, time for a more structured approach: a systematic review. BJOG 119:137-49
11. Rodríguez-Sanjuán JC, Manuel-Palazuelos C, Fernández-Díez MJ et al (2010) Spanish Assessment of resident training in laparoscopic surgery based on a digestive system anastomosis model in the laboratory. Cir Esp 87:20-25
12. Brinkman WM, Havermans SY, Buzink SN et al (2012) Single versus multimodality training basic laparoscopic skills. Surg Endosc PMID: 22350237
13. Stelzer MK, Abdel MP, Sloan MP, Gould JC (2009) Dry lab practice leads to improved laparoscopic performance in the operating room. J Surg Res 154:163-6
14. Diesen DL, Erhunmwunsee L, Bennett KM et al (2011) Effectiveness of laparoscopic computer simulator versus usage of box trainer for endoscopic surgery training of novices. J Surg Educ 68:282-9
15. Oestergaard J, Bjerrum F, Maagaard M et al (2012) Instructor feedback versus no instructor feedback on performance in a laparoscopic virtual reality simulator: a randomized educational trial. BMC Med Educ 28:7
16. Stefanidis D, Scerbo MW, Montero PN et al (2012) Simulator training to automaticity leads to improved skill transfer compared with traditional proficiency-based training: a randomized controlled trial. Ann Surg 255:30-7
17. Cruz JA, Passerotti CC, Frati RM et al (2012) Surgical performance during laparoscopic radical nephrectomy is improved with training in a porcine model. J Endourol 26:278-82
18. Valero RJ, Moanack J, Cruz G et al (2012) Animal model for training in laparoscopic pyeloplasty. Acta Urol Esp 36:54-9
19. Sharma M, Horgan A (2012) Comparison of Fresh-Frozen Cadaver and High-Fidelity Virtual Reality Simulator as Methods of Laparoscopic Training. World J Surg 36:1732-7

The Learning Curve: Teaching in Robotic Surgery

4

Andrea Coratti and Mario Annecchiarico

4.1 Introduction

The introduction of robot-assisted surgery has radically changed the approach to minimally invasive surgery, and this new technique is growing rapidly in various fields of surgery [1–3].

The rapid introduction of robotic procedures necessitates new training methods. Next to the more traditional forms of surgical teaching, the robotic system seems ideal for integrating various forms of simulation [4]. While using simulation, surgeons can develop their skills and pass their basic learning curve on a simulator, hence avoiding the medicolegal aspects of surgical training. Implementing simulation has the potential to create high-quality and competence-based robotic training programs. This could shorten the learning curve and thereby ensure safety and surgical outcomes for the patients [5]. Next, simulation allows experienced surgeons to develop or familiarize themselves with new instruments in a virtual environment [6]. Many hospitals had insufficient criteria for the surgeon's competence before starting with robotic surgery. This is indicative for the growing need for competence-based training and assessment criteria. In 2007, an international multidisciplinary consensus group published a consensus statement on robotic surgery. Training and credentialing was one of the four main items addressed in this statement [7].

In our opinion, the learning process of robotic surgery starts with training, and it continues during the clinical practice as a learning curve that can be variable depending on the type of procedure, previous surgeon exposure to minimally invasive surgery, protocol availability, and caseload.

A. Coratti (✉)
Department of Surgery, General Surgery Unit, Misericordia Hospital Grosseto,
Grosseto, Italy
e-mail: a.coratti@usl9.toscana.it

F. Calise and G. Casciola (eds.), *Minimally Invasive Surgery of the Liver,*
Updates in Surgery
DOI: 10.1007/978-88-470-2664-3_4, © Springer-Verlag Italia 2013

Several training modalities can be used to construct a training program for robotic surgery. In general, a program starts with knowledge development, followed by skills training using a combination of simulation modalities such as dry lab, wet lab and virtual simulators, which are finally followed by real-life case observation in the operating room. When starting with robotic surgery, there is a role for bedside assisting, proctoring, and the mentoring console. In this training program the role played by the training center is very important, as with our International School of Robotic Surgery in Grosseto, Italy, which provides the opportunity for surgeons to start their robotic program training with specific courses.

4.2 Training

4.2.1 Surgeon Training

The robotic training program for the surgeon can be divided into steps: basic training, advanced training, and learning curve.

4.2.1.1 Basic Training

When starting with robotic surgery, the trainee surgeon needs to gain knowledge and understand the robotic technology, device functions, basic troubleshooting, device parameters, and system limitations. The next is the development of knowledge for specific surgical procedures. This includes patient selection and indications, preoperative preparation, patient and system positioning, port placement, procedural steps, and complications and their management. To make sure every surgeon starting with robotic surgery has a basic level of theoretical knowledge, a theoretical examination on these items could be helpful. Just as in laparoscopy, training for robotic surgery can be scheduled in a skills laboratory. In such facilities, exercises on pelvic trainers and other exercises can be performed. A skills laboratory usually has the advantage of high accessibility, but a disadvantage is the need for an expensive robot for dedicated use in the training facility. With this in mind, most hospitals could probably not afford a separate robot for use in a skills laboratory only. In these cases, the available robot at the operating room could be used for training after working hours or at scheduled times when no surgery is planned. Conventional exercises for laparoscopy can be used and can actually be performed faster and more accurately with robotic surgery. The exercises have a shorter learning curve and are performed more accurately with robot assistance. At our institution, residents without any laparoscopic experience demonstrated the capacity to rapidly learn basic surgical maneuvers using the robot.

It is important that training exercises are validated and have a proper goal. Several levels of validation can be distinguished. Exercises should at least have face (the simulation resembles the real task) and construct (the ability to differentiate between groups with different levels of competence) validity and translate well to the clinical setting before they are used in a robotic training program. However, there are only a few reports of validated exercises. During this step of training, in a center with a robot dedicated to the laboratory, animal or cadaver training can be highly useful for performing the same basic procedures, such as dissecting the porcine ureter, followed by section and reconstruction, dissection of the renal hilum and nephrectomy, cholecystectomy, Nissen fundoplication, and an entero-entheric anastomosis [8, 9]. In this phase of training, didactic lectures given by expert robotic surgeons are important.

Another important aspect of skills laboratory training is the transferability of basic skills acquisition to real surgical performance [10]. Surgeons tend to move slower, make more curved movements, and use more grip force during human surgery. During robotic training, it is possible to record objective measures of the robotic instruments. These parameters can be used to describe aspects of robotic surgical performance. In addition, using real-time augmented visual feedback during training can enhance the actual surgical performance.

Virtual reality (VR) training could play an important role in teaching and learning robotic surgery [11]. Since 2006, several, mainly small, studies with respect to VR systems for robotic surgery have been published. Depending on budget and training purposes, several simulators are commercially available, all yet to be validated. With the introduction of simulators for robotic surgery, the problem of system laboratory availability and issues related to costs can be partially overcome, even if the cost of simulators is also expensive. Two systems are in the advanced stage of development and validation: Mimic (Seattle, WA, USA), with the dV Trainer™, and Simulated Surgical System (Williamsville, NY, USA) with the RoSS™. Surgical-skills training in a virtual environment provides a significant learning effect, and the learned skills are consistent with and transferable to actual robot-assisted procedures. Surgical simulators facilitate familiarization with the console and the way it operates, including basic troubleshooting, skill development in regard to camera and instrument operation, and compensation for the loss of tactile feedback. However, further research is needed to develop this as an effective and reliable VR environment.

4.2.1.2 Advanced Training

After the conclusion of basic training, there are different options to continue the surgeon's training program, and operating room observation is an important component. There are several ways to implement operating room observation in a training program: watching live surgery and actually being present in the operating room; watching surgery in another room with the possibility of communicating with the surgeon gives a real-life experience; watching a video recording of an operation together with a teacher is another option. Video recording has the advantage that illustrative surgeries are selected in advance and the educational moments can be planned ahead.

Another important component of this step of training is the introduction of a specific proctorship program. Proctoring – that is, providing direct supervision of an expert – takes place in the initial phase of a learning curve, and the proctor is responsible for assessing the trainee's skills and knowledge. A review regarding proctoring emphasizes the importance of proctoring for robotic surgery and institutional credentialing and addresses the medicolegal aspects [13]. Although extended proctorship is an expensive way of training, it provides a relatively safe way to introduce a new technique and prevents surgeons from beginning to perform procedures before they have mastered the technique. If expert surgeons are already practicing in the trainee's institution, the surgeon in training can start a proctorship in that institution. If this is not the case, the recommendation is case observation at another hospital and proctorship organization in the trainee's home institution.

Usually, the proctor will visit the hospital of the trainee, and surgery is performed together, with the trainee receiving more responsibilities depending on his or her skills. Proctoring is a very time-consuming and expensive way of teaching, so it is interesting to look at alternatives. Modern communication technology, telementoring, and teleprocotoring will save time and travel. Alternatively, a trainee can make a video recording of the performed procedure and send it to a proctor; the evaluation can then be carried out by watching the video online together. Surprisingly, after a 5-day intensive robotic course, only 37.5% of attendees used the possibility of proctoring, even at no extra cost. This could be because most trainees attended as a team, and on returning to their hospital, they performed surgery together. However, there is an increased take-rate when a proctoring program follows the standard intuitive training: 100% of urologist who underwent an extended proctoring phase after training performed prostatectomies afterward. Another possibility is using the availability of the mentoring console. This is possible with new technology of the da Vinci Si HD system (Fig. 4.1). This is a second console that allows the surgeon to collaborate with the trainee during surgery. The mentoring console has two collaborative modes: the swap mode, which allows the mentor and the trainee to operate simultaneously and actively swap control of the robot arms; and the nudge mode, which allows them both to have control over two robot arms. The nudge mode seems to be particularly useful for guiding the trainee's hands during some steps of an operation. There is also the possibility for the trainee to sit at the mentoring console and passively follow the motions of the telemanipulators being used by the instructor (haptic learning).

Lack of mentors/proctors is one of the main causes of failure in establishing a new robotic program in a particular institution; the second is usually lack of volume [8, 13]. It is recommended that two surgeons from the same institution be trained at the same time, thus remaining in collaborating when returning to their home institution.

4.2.1.3 Learning Curve

The learning curve refers to the amount of surgical procedures performed before a surgeon reaches an accepted plateau in outcome parameters (operating time, blood loss, complication rate, surgery quality). More complex procedures have a relatively long learning curve. The length of a learning curve may also vary as a result of surgeon-related factors (surgical experience with a similar technology,

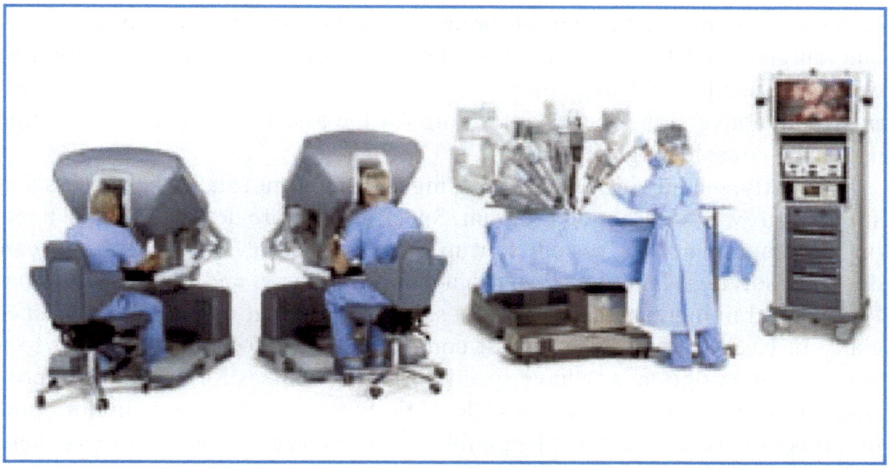

Fig. 4.1 da Vinci Surgical System Si HD (reproduced with permission from Intuitive Surgical, Inc.)

familiarity with the procedure) or hospital-related factors (availability of theater time, available case load). Many series of robot-assisted laparoscopic procedures have been reported, but only a minority addresses the aspect of the learning curve.

With respect to the outcome parameter operating time, there are the different phases of the operation. First, there is the aspect of time needed for the operating team to prepare and activate the robot system (setup time). Second, there is the time phase relating to positioning and installing the robot (docking time). Third, one can differentiate the actual time needed to complete the robotic surgery procedure (console time). Fourth, there is the entire time span in which the patient is in the theater (theater time). Setup and docking time can be reduced quickly when working in a high-volume setting with a dedicated team. Intraoperatively, outcome parameters are blood loss, complication rate, and conversion rate to open surgery. To maintain the quality of oncological surgery, parameters such as number of lymph nodes, tumour-free margins, and recurrence rate are known to be used. Instead of the learning curve, Sammon et al. suggest using the learning rate, defined as the percentage decrease in operative time (minutes) per doubling of cumulative procedure number [14].

It is difficult to estimate the specific learning curve for a specific procedure. For example, whereas robotic prostatectomy represents one of the most standardized robotic procedures, it still presents a high variability in terms of procedural learning.

4.2.2 Surgical Team Training

The experience of the team is proven to have a beneficial impact on overall surgery time, and this is confirmed by experience at our center [15]. With the da Vinci system, some important procedural steps are demanded of the assistant surgeon, such as trocar placement, robotic-cart docking, instrument exchange, suction and irrigation, retraction, staple use, solving troubleshooting, rapid conversion. The ideal training for the assistant surgeon is the same as for the console surgeon.

Similarly, the nursing team plays a highly important role in the success and speed of growth of a robotic program. Specific tasks are delegated to the nursing personnel, such as system startup, patient draping, camera setting, cart docking, solving vision tower or connection problems, and rapid conversion. Personnel training can be done at the same time as that of the surgeon, especially in basic training where most components are of common interest to a nurse. Another option is to have specific training courses for nurses, as is possible in our International School of Robotic Surgery in Grosseto. In our opinion, it is important that at the beginning of experience in robotic surgery there is an adequate number of sufficiently trained personnel to provide backup at all times, and once a core team is trained and proficient, new personnel can be added to shifts for training (Fig. 4.2).

The International School of Robotic Surgery in Grosseto was born in 2000 for clinical surgery and began experimental surgery and training with the purchase of a second dVSS in 2002. It is possible for staff to attend formal courses and participate in clinical activity. Different types of courses are available for surgeons. After completing formal courses, the school offers its cooperation as a proctor to other centers beginning a robotic surgery program.

Thanks to the collaboration with other robotic centers around the world, a new teaching method was recently introduced, with the birth of the Clinical Virtual University (CVU), which allows observing live surgery by streaming, with comment and interaction with the surgeon (Fig. 4.3).

Fig. 4.2 Robotic training process

Fig. 4.3 International School of Robotic Surgery, Grosseto, Italy

4.3 Discussion

With the increasing popularity of robotic surgery comes a growing need for sophisticated training programs for residents, fellows, and surgeons. Ideally, these training programs should be competence based. Courses are commonly used to share new information and/or learn new skills. Some are purely didactic, and other mainly consist of skills training, but many combine the two aspects. In contrast to open surgery, robotic skills can improve significantly in a relatively short time.

With the approval of the dVSS by the US FDA, the manufacturer was told to provide a comprehensive robotic training for all surgeons and their teams. The manufacturer has engaged 24 training centers around the world. Training comprises two parts: on-site training, which emphasizes key features of the system, preparation and management in individual hospitals, and off-site training consisting of a course to learn and practice procedural skills. From there on, surgical proctoring support is provided in the first cases. It is important to start quickly with regularly scheduled cases after completing a course; otherwise, newly learned skills can fade. At least one or two cases a week is recommended to overcome the first part of the learning curve. In addition to the registered training centers, some centers developed their own training program and thus function as training centers. These centers primarily focus on specific procedures.

Robotic surgery is still expensive, and several authors addressed or compared costs with laparoscopic or open surgical procedures. Most studies mainly focus on the costs of the robotic system, with an additional 10% per year of fixed service and instrument costs.

Rarely addressed are the costs of the learning curve of the surgeon and surgical team. These are substantial costs that are often underestimated. Steinberg et al. [17] constructed a theoretical model to describe costs associated with the learning curve of a single surgeon for robot-assisted laparoscopy radical prostatectomy (RALP). This study [17] illustrates the high costs involved with the learning curve for complex robotic procedures and emphasizes the need for sophisticated training programs together with a high case load to overcome the learning curve.

Implementing a new technology such as robot-assisted laparoscopic surgery in a safe and efficient way is demanding. There are many factors that influence successful implementation of a robotic training program. Issues such as training modalities, longer operative times, patient outcomes, cost, case volume, number of robotic procedures required for the surgeon to become proficient, and patient-quality parameters are all items that require attention. The exponential growth of robotic surgery, however, is not giving the surgical community much time to develop structured training programs for future robotic surgeons. In the near future, an increasing number of well-trained robotic surgeons will be needed.

Designing a competence-based training curriculum for robotic surgery remains a challenge, but with the exponential increase in robotic surgery, the need for such certified curricula is increasing rapidly. There is a lack of validated training tools for robotic-assisted laparoscopic surgery, and in the near future, further research in this field needs to be performed. With the increasing quality of virtual reality simulators for robotic surgery it is expected that this training modality will play an important role in training future robotic surgeons, even in terms of costs. Procedural training for robotic surgery needs to be carried out in a stepwise and systematic manner. In this way, introduction of this new technology can be performed in an efficient and safe way and without compromising results for our patients.

References

1. Ahmed K, Khan MS, Vats A et al. (2009) Current status of robotic assisted pelvic surgery and future developments. Int J Surg 7:431–440
2. Schreuder HW, Verheijen RH (2009) Robotic surgery. BJOG 116:198–213
3. Wilson EB (2009) The evolution of robotic general surgery. Scand J Surg 98:125–129
4. Kunkler K (2006) The role of medical simulation: an overview. Int J Med Robot 2:203–210
5. Hance J, Aggarwal R, Undre S, Darzi A (2005) Skills training in telerobotic surgery. Int J Med Robot 1:7–12
6. Gallagher AG, Traynor O (2008) Simulation in surgery: opportunity or threat? Ir J Med Sci 177:283–287
7. Herron DM, Marohn M (2008) A consensus document on robotic surgery. Surg Endosc 22:313–325
8. Intuitive Surgical Training manual (2009) Intuitive Surgical Inc
9. Mirheydar H, Jones M, Koeneman KS, Sweet RM (2009) Robotic surgical education: a collaborative approach to training postgraduate urologists and endourology fellows. JSLS 13:287–292

10. Judkins TN, Oleynikov D, Stergiou N (2008) Objective evaluation of expert performance during human robotic surgical procedures. J Robotic Surg 1:307–312
11. Albani JM, Lee DI (2007) Virtual reality-assisted robotic surgery simulation. J Endourol 21:285–287
12. Zorn KC, Gautam G, Shalhav AL et al. (2009) Training, credentialing, proctoring and medicolegal risks of robotic urological surgery: recommendations of the Society of Urologic Robotic Surgeons. J Urol 182:1126–1132
13. Rane A (2005) A training module for laparoscopic urology. JSLS 9:460–462
14. Sammon J, Perry A, Beaule L et al. (2010) Robot-assisted radical prostatectomy: learning rate analysis as an objective measure of the acquisition of surgical skill. BJU Int 106:855–860
15. Giulianotti PC, Sbrana F, Bianco FM et al. (2010) Robot-assisted laparoscopic pancreatic surgery: single-surgeon experience. Surg Endosc 24:1646–1657
16. Steinberg PL, Merguerian PA, Bihrle W III, Seigne JD (2008) The cost of learning robotic-assisted prostatectomy. Urology 72:1068–1072

The Laparoscopic Column

5

Alberto Bartoli and Alberto Patriti

5.1 Introduction

Operating rooms usually used for minimally invasive surgery (MIS) are not designed for this purpose. Hence, for this reason, all instrumental devices (screens, light sources, recording systems, electrocoagulators, insufflators) are located on dedicated trays. Because these devices take space, it is necessary to have a room appositely prepared for immediate and safe use. Operating room personnel need to be specifically prepared for this type of surgery; for example, it is necessary to know highly technical details of and sterilization methodology for all devices, which require more attention than conventional surgical instruments [1]. In the event of surgical complication or the impossibility of continuing with the laparoscopic approach, surgeons, assistants, and equipment need to be ready for conversion to an open procedure.

Modern operating rooms specifically designed for MIS are equipped with video screens installed on apposite mobile arms mounted to the ceiling; this setup greatly assists surgeons in their work. Moreover this design is easily manageable in case it is necessary for specific procedures or for particular patient positions. Recording systems are installed on either the cabinet or the perimeter wall, which allow staff to record without being directly involved. The primary tools necessary for general and liver MIS are briefly described.

A. Patriti (✉)
Department of Surgery, Division of General, Minimally Invasive and Robotic Surgery,
ASL3 Umbria, San Matteo degli Infermi Hospital,
Spoleto, Italy
e-mail: albertopatriti@gmail.com

F. Calise and G. Casciola (eds.), *Minimally Invasive Surgery of the Liver,*
Updates in Surgery
DOI: 10.1007/978-88-470-2664-3_5,

5.2 Laparoscopic Trolley

The mobile video cart is equipped with locking brakes and has four antistatic rollers. The trolley has a drawer and three shelves, the upper shelves have a tilt adjustment and are used for supporting the video monitor unit. Included on the trolley is an electrical terminal supply strip mounted on the rear of the second shelf. Ceiling-mounted trolleys have recently been introduced by many companies. They are ergonomically better and consume less space in the operating theater than the mobile cart.

5.3 Light Cable and Light Source

MIS depends on artificial light being made available in a closed body cavity. In 1867 Bruck, a dentist from Breslau, Prussia, made a platinum wire loop that he heated with electric current and used that as a light source for surgical purposes. Prior to 1954, endoscopy was performed using tungsten light bulbs to illuminate the body cavity. These bulbs emitted low-frequency red light. The bulb was so hot that there was always a risk of injury to bowel. In 1954, a major breakthrough in technology occurred with the development of fiber optic cables, the principle of which was based on the total internal reflection of light. Light is conducted along a curved glass rod by multiple totally internal reflections from the walls of the rod. Light enters at one end of the fiber and emerge at the other end, with virtually all its original strength, after numerous internal reflections [2]. Nowadays, there are two types of light cables available: (1) the fiber-optic cable, and (2) the liquid crystal gel cable.

Optic cables consist of a bundle of optical fiberglass thread swaged at both ends. The fiber size is usually between 10 and 25 mm in diameter. They have a very high quality of optical transmission but are fragile. In fact, as they are progressively used, some optical fibers break. The loss of optical fibers may be seen when one end of the cable is viewed in daylight: the broken fibers appear as black spots. Gel cables consist of a sheath filled with a clear optical gel (liquid crystal) [3]. Theoretically, they are capable of transmitting 30% more light than optic fibers. This type of cable poses three problems, however: (1) The quartz swaging at the ends is extremely fragile, especially when the cable is hot. (2) The slightest shock can cause the quartz end to crack and thus cause a loss in light transmission. (3) These cables transmit more heat than optical fiber cables and are made more rigid by a metal sheath, which makes them more difficult to maintain and to store.

Light cables are conventionally attached to a light source and to the camera through a right-angle connection. A typical light source consists of:

1. A lamp
2. A heat filter
3. A condensing lens
4. A manual or automatic intensity control circuit.

5.3.1 Lamp

The lamp, or bulb, is the most important part of the light source. The quality of light depends on the lamp used. Several modern types of light sources are currently available and mainly differ by the type of bulb used. Four types of lamp are usually used:

1. Quartz halogen
2. Incandescent
3. Xenon
4. Metal halide vapor arc.

5.3.1.1 Quartz Halogen

Halogen bulbs provide a highly efficient, almost crisp, white light source with excellent color rendering. Electrodes in halogen lamps are made of tungsten; this is the only metal with a sufficiently high melting temperature and sufficient vapor pressure at elevated temperatures. They use a halogen gas that allows bulbs to burn more intensely without sacrificing life. Halogen bulbs are low voltage and have an average life of 2,000 h. Color temperature of a halogen lamp is (5,000–5,600 K). These lamps are cheap and can be used for laparoscopic surgery if low-budget setup is required.

5.3.1.2 Xenon Lamps

Xenon lamps consist of a spherical or ellipsoidal envelope made of quartz glass, which can withstand high thermal loads and high internal pressure. For ultimate image quality, only the highest-grade clear fused silica quartz is used. It is typically doped, although not visible to the human eye, to absorb harmful UV radiation generated during operation. The color temperature of the Xenon lamp is 6,000–6,400 K. The operating pressures are tens of atmospheres at times, with surface temperatures exceeding 600 °C. The smaller, pointed electrode is called the cathode, which supplies the current to the lamp and facilitates the emission of electrons. To supply a sufficient amount of electrons, the cathode material is doped with thorium. The optimum operating temperature of the cathode tip is approximately 2,000 °C. To obtain this precise operating temperature, the cathode tip is pointed and in many cases has a groove on the pointed tip to act as a heat choke. This heat choke causes the tip to run at a higher temperature. This configuration allows for a very high concentration of light from the cathode tip and a highly stable arc. The anode, the larger electrode, receives electrons emitted by the cathode. Once the electrons penetrate the anode face, the resulting energy is converted to heat, most of which radiates away. The large, cylindrical shape of the anode helps to keep the temperature low by radiating the heat from the anode surface.

5.3.1.3 Metal Halide Vapor Arc Lamp

In metal halide lamps, the mix of compounds is carefully chosen to produce an output that approximates white light as perceived by the human eye. There are two types of metal halide lamps generally used: the iron iodide lamp, and the gallium iodide lamp.

The intensity of light delivered by any lamp also depends on the source power supply. However, increasing the power poses a real problem as concerns the heat created. At present, improvements made to cameras mean that it is possible to return to reasonable power levels, of the order of 250 W; however, 400-W units are preferable in order to guarantee sufficient illumination of the abdomen, even when bleeding causes strong light absorption.

The two most frequently used types of lamps are halogen and xenon. This may be due to the colors obtained. The xenon has a slightly bluish tint, and the light emitted is more natural than that of a halogen lamp. However, most cameras analyze and compensate for these variations by means of automatic equalization of whites (2,100–10,000 K), which allows the same quality image to be obtained with both light sources. Proper white balancing before beginning the operation is good practice for obtaining a natural color. The white light is composed of equal proportion of red, green, and blue (RGB), and at the time of white balancing, the camera sets its digital coding for these primary colors to equal proportion, assuming that the target is white. If at the time of white balancing the telescope is not seeing a perfectly white object, then the setup of the camera will be poor, resulting in poor color perception.

5.3.2 Heat Filter

For 100% of the energy consumed, a normal light source uses approximately 2% in light and 98% in heat. This heat is mainly due to the infrared spectrum of light and to obstruction in the light pathway. If infrared light travels through the light cable, then the cable will become intolerably hot. A heat filter is therefore introduced to filter this infrared light for travel through the fiber-optic cable. A cool light source lowers this ratio by creating more light, but it does not reduce heat produced by the energy to zero. This implies a significant dissipation of heat, which increases as the power rating increases. Sources are protected against transmitting too much heat; the heat is essentially dissipated in transport, along the cable, in the connection with the endoscope and along the endoscope.

5.3.3 Condensing Lens

The purpose of the condensing lens is to converge the light emitted by the lamp to the area of light-cable input. In most light sources, it is used for increasing light intensity per square centimeter of area.

5.4 Laparoscope and Camera Unit

The purpose of the laparoscope and camera unit is to transmit high-quality images from the surgical field to the signal processing unit, which relays the images to the video monitor. Traditional laparoscopes have two separate channels. One channel contains a series of glass rods (known as the Hopkins rod-lens system) that transmit images from the surgical field to the eyepiece of the laparoscope [4]. The second channel contains light fibers, which transmit light from an external source to the surgical field. Laparoscopes may be straight or angled at the tip. Straight (0°) laparoscopes offer a view of whatever lies directly ahead, whereas angled (e.g., 30° or 45°) scopes allow the surgeon to look around structures by rotating the scope. Typically, laparoscopes are either 5 or 10 mm in diameter and are usually 25–30 cm in length. Smaller diameter laparoscopes, such as 3-mm scopes, are also available. Longer laparoscopes (e.g., 40–45 cm) are also available and are frequently used in bariatric cases.

To transmit images from the laparoscope to the signal processing unit, a camera is attached to the eyepiece via a coupler [5]. First-generation cameras developed in the 1980s used a one-chip system to convert colors into electrical charges and relay those images. Most current-generation laparoscopic cameras use three-chip charge-coupled device systems, which sense primary colors (RGB) and convert these colors to electrical charges. This enhancement allows for a higher image resolution. Subsequent improvements in color detection and light sensitivity over the last 15 years have made three-chip systems the standard in many operating rooms. Modern high-definition three-chip systems can offer more than 1,000 lines of resolution.

5.5 Video Monitor

Most traditional video monitors used in the operating room will accept input from three types of analog signals: composite, brightness/color (Y/C,), and RGB. Older laparoscopic camera systems typically output either composite or super video signals, whereas many current-generation systems also generate RGB output, which often provides higher-resolution images. However, even with the use of RGB signals, the resolution provided by traditional cathode ray tube monitors is somewhat limited (typically 600 lines or less). For enhanced resolution, including high-definition formats, digital monitors that accept digital video interface (DVI) and serial digital interface (SDI) signals may be used with high-definition video cameras. These flat-panel digital monitors, which are also capable of accepting analog signals, are often ceiling-mounted displays that are capable of providing more than 1,000 lines of resolution (nearly twice the resolution of a typical television image) [6].

5.6 Insufflator

The creation of pneumoperitoneum is one of the essential steps of laparoscopic surgery because it distends the abdominal wall and significantly enhances visualization of intra-abdominal structures. The laparoscopic pioneers used room air primarily because it was readily available and cheap. Studies published as recently as 1980 suggest that the creation of pneumoperitoneum with room air was safe and cost effective [7]. However, its combustibility, poor solubility in blood, and possibility for creating venous air embolism have virtually eliminated the use of room air insufflation in today's operating rooms. Likewise, insufflation with nitrous oxide was popular throughout the 1970s [8]. However, reports of intra-abdominal explosions have subsequently limited its popularity [9].

Carbon dioxide is currently the most commonly used gas during insufflation. It is odorless, nonflammable, and highly soluble in blood, which reduces the likelihood of air embolisms. Although clinically significant hypercarbia and acidemia can occur, particularly in patients with cardiopulmonary comorbidities, most patients tolerate carbon dioxide pneumoperitoneum without any adverse effects.

To establish standard-pressure pneumoperitoneum (typically 12–14 mmHg), the insufflator must first be connected to the carbon dioxide cylinder. Tubing is then used to connect the insufflator to either a Veress needle or a laparoscopic port within the peritoneal cavity. Next, gas flow from the insufflator is initiated. Although the amount of gas needed to create adequate pneumoperitoneum varies according to patient size, abdominal wall compliance, and degree of gas leakage, several liters of carbon dioxide are usually adequate. To maintain pneumoperitoneum throughout a case, substantially more gas is needed. One report estimates that an average laparoscopic colectomy requires approximately 110–180 L of carbon dioxide [10].

Most conventional insufflators use an intermittent pressure monitoring system. This allows the insufflator to decrease the gas flow if the intra-abdominal pressure exceeds a preset value. Conversely, it can increase gas flow to account for a loss of gas either externally or through peritoneal absorption. Rather than cycling gas continuously, insufflators alternate between injecting gas every 2–3 s and monitoring intra-abdominal pressure.

The possibility of peritoneal contamination or disease transmission through the flow of intraperitoneal fluid or particulate matter from the patient to the insufflator (and then to the next patient) has been a concern since the early days of laparoscopy. To address this concern, insufflation filters composed of mesh with 0.1- to 0.3-μm have been developed. The literature surrounding the effectiveness of these filters is limited. One study published in 1989 reported that rust, dust, and metal filings could be detected on insufflator filters. Another by the same author found that the use of a 0.3-μm filter reduced microbial colonization of gas cylinders and insufflators [11]. However, a subsequent examination that used microscopy, mass spectrometry, and bacterial analysis showed no evidence of microbial or particulate matter trapping by the filters [12].

References

1. Girotti MJ, Nagy AG, Litwin DE, Mamazza J, Poulin EC (1992) Laparoscopic surgery–basic armamentarium. Can J Surg 35 (3):281–284
2. Bellina JH, Haas M (1984) Cold light sources. Are they really cold? J Reprod Med 29 (4):275–277
3. Nalwa HS (2001) Handbook of Advanced Electronic and Photonic Materials and Devices. Academic Press
4. Cockett WS, Cockett AT (1998) The Hopkins rod-lens system and the Storz cold light illumination system. Urology 51 (5A Suppl):1–2
5. Berber E, Siperstein AE (2001) Understanding and optimizing laparoscopic videosystems. Surg Endosc 15 (8):781–787. doi:10.1007/s004640000391
6. Soper N, Eubanks W, Swanstrom L (2008) Mastery of endoscopic and laparoscopic surgery. Lippincott, Williams and Wilkins
7. Diaz MO, Atwood RJ, Laufe LE (1980) Laparoscopic sterilization with room air insufflation: preliminary report. Int J Gynaecol Obstet 18 (2):119–122
8. Menes T, Spivak H (2000) Laparoscopy: searching for the proper insufflation gas. Surg Endosc 14 (11):1050–1056
9. El-Kady AA, Abd-El-Razek M (1976) Intraperitoneal explosion during female sterilization by laparoscopic electrocoagulation. A case report. Int J Gynaecol Obstet 14 (6):487–488
10. Sammour T, Kahokehr A, Hayes J, Hulme-Moir M, Hill AG (2010) Warming and humidification of insufflation carbon dioxide in laparoscopic colonic surgery: a double-blinded randomized controlled trial. Ann Surg 251 (6):1024–1033. doi:10.1097/SLA.0b013e3181d77a25
11. Ott DE (1997) Microbial colonization of laparoscopic gas delivery systems: a qualitative analysis. Jsls 1 (4):325–329
12. Frankel SM, Fitzgerald RL, Sackier JM (1998) Utility of insufflation filters in laparoscopic surgery. Surg Endosc 12 (9):1137–1140

Technology in the Operating Room: the Robot

6

Graziano Ceccarelli, Alberto Patriti, Alberto Bartoli,
Alessandro Spaziani and Luciano Casciola

6.1 Introduction

Robot-assisted and computer-assisted surgery are terms for technological devices using robotic systems to aid in surgical procedures. The surgeon, instead of directly using the surgical instruments, uses a computer-controlled telemanipulator that allows reproduction of finger and wrist movements during surgery, which is performed by articulated instruments at the tips of the robotic arms [1]. The term robot was used for the first time by the Czech playwright Karel Capek in 1921 in his play "Rossum's Universal Robots" (*robota* is a Czech word meaning forced labor) [2]. In 1985, the PUMA 560 robot was used to place a needle for a brain biopsy under computed tomography (CT) guidance [3]. In 1988, the PROBOT device, developed at the Imperial College of London, was used for prostate surgery. In 1993, the company Computer Motion developed the AESOP (a voice-controlled camera holder) and ZEUS systems [4]. The first robotic reconnection of fallopian tubes was performed in 1997 using this device. In 1999, the first robotically assisted heart bypass and the first beating-heart coronary bypass graft were performed using the same robotic system [5]. In September 2001, Marescaux et al. performed the first transatlantic surgical intervention, the Lindbergh Operation. Surgeons where in New York, USA, and the patient affected by gallbladder stones was in Strasburg, France [6].

The da Vinci Surgical System® was a real breakthrough. It was developed and marketed by Intuitive Surgical Inc. (Sunnyvale, CA, USA) in 1997 and was classified as a master–slave surgical system. The original telesurgery

G. Ceccarelli (✉)
Department of Surgery, Division of General, Minimally Invasive and Robotic Surgery,
ASL3 Umbria, San Matteo degli Infermi Hospital, Spoleto, Italy
e-mail: g.cecca2003@libero.it

F. Calise and G. Casciola (eds.), *Minimally Invasive Surgery of the Liver*,
Updates in Surgery
DOI: 10.1007/978-88-470-2664-3_6, © Springer-Verlag Italia 2013

robotic system at the basis of da Vinci was developed with grant support from Defence Advanced Research Projects Agency (DARPA) and the National Aeronautics and Space Administration (NASA). The telesurgical robot was originally intended to facilitate remote surgery in the battlefield. The first machine was set up in Europe in 1997, and the first surgical procedure was reported by Himpens et al. in 1997 [20]. In 1998 the first robotic heart bypass was performed in Leipzig, Germany. The da Vinci was cleared by the US Food and Drug Administration (FDA) in 2000 for general laparoscopic surgery. In 2003. the Intuitive Surgical Company bought the Computer Motion system, and ZEUS was no longer marketed. In January 2009, the first all-robot-assisted kidney transplant was performed in Livingston, NJ, USA. Robotics in hepatobiliary and pancreatic surgery continues to develop with interesting results, above all the possibility of overcoming laparoscopic limitations [7–9]. Recently, even a right lobe donor hepatectomy was performed robotically [10].

6.2 The da Vinci System

The da Vinci system has been gradually upgraded from the first three-arm system to the current light-weight da Vinci Si with four arms and a high-definition (HD) camera. The da Vinci system basically has three components: the surgeon console, the robotic cart (with four arms), and the endoscopic column or vision cart (Fig. 6.1a). The robotic instruments (hook, scissors, grasper, bipolar forceps, etc) are articulated with 7 degrees of freedom. In the more recent versions, an HD, 3D camera is available. In the typical robotic operating room, the surgeon sits at the console and his or her finger movements are translated electromechanically into precise and scaled micromovements of the instruments; the computer detects and filters out any tremors in the surgeon's hand movements [11].

6.2.1 Surgeon Console

The console consists of the binocular viewer, instrument controllers, two control panels (for system setup), and foot-control pedals (Fig. 6.2a). It contains the computer hardware and software. The surgeon operates while seated, viewing in an HD, 3D view. The surgeon's thumb and index finger of each hand are placed in loops (master controllers). The system translates the surgeon's wrist and finger movements into real-time movements to the surgical instruments. There are five foot-control pedals in the da Vinci Si model: the clutch pedal, which disengages the instruments from the controllers; the camera pedal, allowing adjustment of the view; the pedal to adjust focus; and the bipolar and monopolar coagulation pedals. Two panels are positioned on each side of the surgeon: on the left is the camera and endoscopic calibration and motion scaling; on the right is the system start control, emergency stop control, and standby buttons. If necessary, the system can be rapidly disengaged by placing it on standby mode.

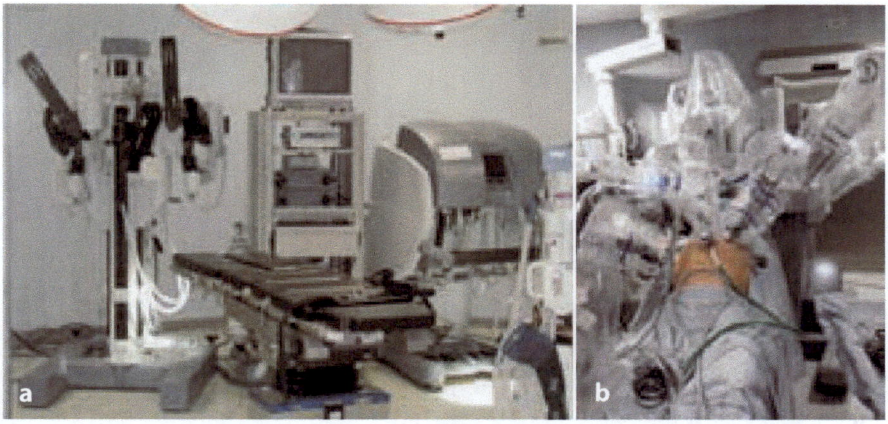

Fig. 6.1 a The three components of the da Vinci Surgical System; b setup of the robotic theater

6.2.2 Patient-side Cart

The patient-side cart (Fig. 6.1a) consist of four robotic arms (one camera arm and three instrument arms). Plastic drapes are necessary to drape the arms to achieve sterility. The arms move around fixed pivot points. Each arm has a series of multiple positioning joints and a terminal pivot joint at the attachment with the laparoscopic port. The cart is connected by cables to the console. Once in position for surgery (docked), it is locked in place. For right hepatic resections, it is docked generally at the patient's right shoulder and for left hepatic resection at the head or left shoulder (Fig. 6.1b).

6.2.3 Endoscopic Column (Stack) and Camera System

The endoscopic column and camera system (Fig. 6.1a) has all the features of a standard laparo/endoscopic column with a monitor, a carbon dioxide (CO_2) insufflator, a dual high-intensity light source, and a camera unit. The camera system, 10× magnification, has a dual lens system with two three-chip cameras. The spatial separation of these images projected to the surgeon's eyes in the binocular viewer provides true 3D imaging at the console. It is possible to integrate to the operating-field view images of real-time intraoperative ultrasound (US), CT scan, or others in a highly useful multimedia integration process (Fig. 6.2c).

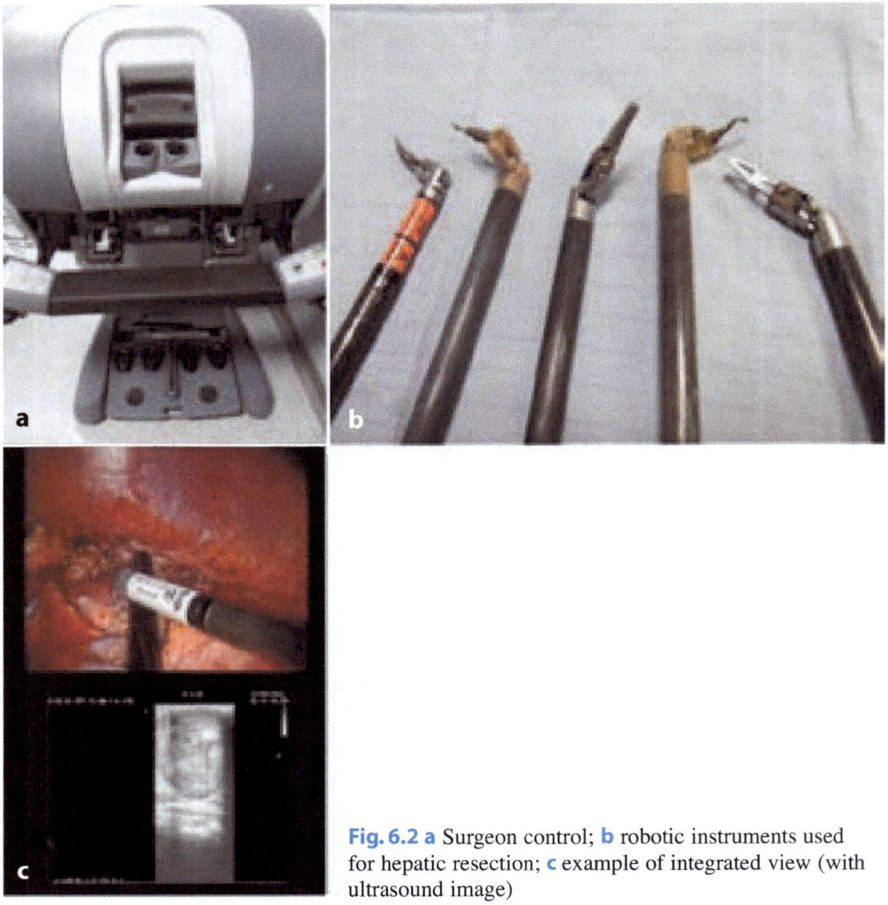

Fig. 6.2 a Surgeon control; b robotic instruments used for hepatic resection; c example of integrated view (with ultrasound image)

6.2.4 EndoWrist Instruments

A wide range of dedicated instruments is available (forceps, hook, bipolar forceps, grasper, etc.) (Fig. 6.2b). Almost all instruments have an articulated tip and have seven degrees of freedom, with the only exception being the ultrasonic dissector. Human tremors are abolished by position sensing. Almost all instruments have only ten lives.

6.2.5 Setup Procedure

The robotic startup sequence includes a self-test that takes approximately 1 min. The arms must be covered by sterile drapes. Once the camera and endoscope are connected, they need to be calibrated. Now the patient is placed in the position desired for the specific operation with the robotic arms attached. The surgeon takes position at the console and the ready button is pressed. An infrared sensor at the head pad engages the instruments and the camera.

6.3 Applications and Hepatobiliary Surgery

More than 1,000 systems have now been set up across the globe, a majority of them being in the USA. The da Vinci robot is now being used in various fields such as:

- Urology: radical prostatectomy, pyeloplasty, cystectomy, nephrectomy, ureteral reimplantation, donor nephron surgery [12]
- Gynecology: hysterectomy, myomectomy, sacrocolpopexy [13]
- General and hepatobiliary surgery: cholecystectomy, Nissen fundoplication, Heller myotomy, gastric bypass, adrenalectomy, splenectomy and bowel resection, liver resection, pancreatic surgery [14–16]
- Cardiothoracic surgery: internal mammary artery and blood vessel mobilization and cardiac tissue ablation; mitral valve repair; endoscopic atrial septal defect closure; cardiac revascularization [17]
- Ear nose and throat: transoral resection of tumors of the upper aerodigestive tract (tonsil, tongue base, larynx), transaxillary thyroidectomy [18].

6.4 Conclusions

The da Vinci Surgical System allows improvement over conventional laparoscopy; it is an electromechanical actuator transmitting movements of the surgeon's hands to the tip of instruments, which have seven degrees of freedom and can articulate up to 90°. The EndoWrist technology eliminates the fulcrum effect and the surgeon's natural hand tremors. This technology provides high stereoscopic definition, a steady view, and movement scaling into micromotions. The console allows the surgeon to operate from a seated and ergonomic position, with eyes and hands positioned in line with the instruments. The articulated instruments allow operation with just a short learning curve for complex laparoscopic procedures (involving dissection or reconstruction steps).

However, robotic surgical systems cannot perform by themselves. They rely on a human operator for all input and are designed to replicate the movement of the surgeon's hands. With regard to clinical evidence, there are many publications on robotic surgery, the majority of which are nonrandomized prospective studies and case series (level of evidence II). Since 1998, >4,000 publications have appeared in various clinical journals, with about one half of these in urology, as well as in cardiothoracic surgery, general surgery, gynecologic surgery; pediatric surgery, ear nose and throat, and others [19]. Urologists and general surgeons are the frontrunners in the use of robotic surgery. One of the most important limitations of this technology is its high cost, and more studies are needed to demonstrate cost-effectiveness in the different fields.

References

1. Ballantyne GH (2002) The pitfalls of laparoscopic surgery: challenges for robotic and telerobotic surgery. Surg Laparosc Endosc Percutan Tech 12:1–5
2. Christensen AG (2007) Karel Čapek. In: The Literary Encyclopedia. First published 16 November 2007 [http://www.litencyc.com/php/speople.php?rec=true&UID=728. Accessed 04 June 2012]
3. Kwoh YS, Hou J, Jonckheere EA Hayall S (1988) A robot with improved absolute positioning accuracy for CT guided stereotactic brain surgery. IEEE Trans Biomed Eng 35(2)153–161
4. Unger SW, Unger HM, Bass RT (1994) AESOP robotic arm. Surg Endosc 8(9):1131
5. McConnell PI, Schneeberger EW, Michler RE (2003) History and development of robotic cardiac surgery. Problems in General Surgery 20(2):20–30
6. Marescaux J, Leroy J, Gagner M, Rubino F, Mutter D, Vix M et al (2001) Transatlantic robot-assisted telesurgery. Nature 27;413(6854):379–380
7. Melvin WS, Krause KR, Needleman BJ, Ellison EC (2003) Robotic resection of pancreatic neuroendocrine tumor. J Laparoendosc Adv Surg Tech 13(1):33–36
8. Giulianotti PC, Sbrana F, Bianco FM, Elli EF, Shah G, Addeo P et al (2010) Robot-assisted laparoscopic pancreatic surgery: single-surgeon experience. Surg Endosc 24:1646–1657. doi:10.1007/s00464-009-0825-4
9. Casciola L, Patriti A, Ceccarelli G, Bartoli A, Ceribelli C, Spaziani A (2011) Robot-assisted parenchymal-sparing liver surgery including lesions located in the posterosuperior segments. Surg Endosc 25(12):3815–3824 [Epub 2011 Jun 8]
10. Giulianotti PC, Tzvetanov I, Jeon H, Bianco F, Spaggiari M, Oberholzer J et al (2012) Robot-assisted right lobe donor hepatectomy.Transpl Int 25(1):e5–e9. doi:10.1111/j.1432-2277.2011.01373.x [Epub 2011 Oct 27]
11. da Vinci Surgery (2012). http://www.davincisurgery.com. Accessed 5 June 2012
12. Skolarus TA, Zhang Y, Hollenbeck BK (2010) Robotic surgery in urologic oncology: gathering the evidence. Expert Rev Pharmacoecon Outcomes Res 10(4):421–432
13. Ahmed K, Khan MS, Vats A, Nagpal K, Priest O, Patel V et al (2009) Current status of robotic assisted pelvic surgery and future developments. Int J Surg 7:431–440
14. Giulianotti PC, Coratti A, Angelini M, Sbrana F, Cecconi S, Baestracci T et al (2003) Robotics in general surgery: personal experience in a large community hospital. Arch Surg 138:777–784
15. Ceccarelli G, Patriti A, Biancafarina A, Spaziani A, Bartoli A, Bellochi R, Casciola L (2009) Intraoperative and postoperative outcome of robot-assisted and traditional laparoscopic Nissen fundoplication. Eur Surg Res 43(2):198–203 [Epub 2009 Jun 10]
16. Patriti A, Ceccarelli G, Ceribelli C et al (2011) Robot-assisted laparoscopic management of cardia carcinoma according to Siewert recommendations. Int J Med Robot 7(2):170–177. doi:10.1002/rcs.385 [Epub 2011 Mar 17]
17. Kypson AP, Randolph Chitwood W Jr (2004) Robotic applications in cardiac surgery. Int J Adv Robotic Syst 1(2):87–92
18. Weinstein GS, O'Malley BW Jr, Desai SC, Quon H (2009) Transoral robotic surgery: does the end justify the means? Curr Opin Otolaryngol Head Neck Surg 17:126–131
19. Intuitive Surgical (2012). http://www.intuitivesurgical.com/. Accessed 5 June 2012
20. Himpens J, Lemans G, Cadiere GB (1998) Telesurgical laparoscopic cholecystectomy. Surg Endosc 12:1091

Part II

Surgery: Principles and Management

Anesthesia

7

Ada De Felice, Giuseppe De Simone, Marcella Marracino
and Ciro Esposito

7.1 Introduction

Liver surgery is associated with a high rate of perioperative morbidity in patients with chronic liver disease and with underlying risk factors such as advanced age and pulmonary and vascular comorbidities, especially when a major hepatic resection is planned [1–4]. Safety and feasibility of laparoscopic liver resection (LLR) have been demonstrated by several studies, and nowadays, the open and laparoscopic approach can be considered equivalent in terms of operative risk [5–7]. Factors contributing to make LLR safe are advances in surgical techniques, more accurate knowledge of liver anatomy and physiology, development of dedicated anesthesiological protocols, and a close collaboration between the anesthesiological and surgical staff. The latter is of paramount importance in order to reduce operative time and intraoperative blood loss with the associated morbidity. Anesthesiologists' duties range from offering skillful technical help and assistance in the operating room, to patient selection and to postoperative monitoring.

7.2 General Considerations

Preoperative evaluation of patient candidates for LLR is necessary to determine perioperative risks related to the disease and the operation and to plan preoperative workup, intraoperative strategy, and postoperative care. LLR has comparable postoperative morbidity and mortality to open hepatic procedures.

A. De Felice (✉)
Anesthesia and Intensive Care Unit- Transplants Department, A. Cardarelli Hospital,
Naples, Italy
e-mail: ada.defelice@gmail.com

F. Calise and G. Casciola (eds.), *Minimally Invasive Surgery of the Liver,*
Updates in Surgery
DOI: 10.1007/978-88-470-2664-3_7,

Some surgical and anesthesiological implications decrease patient eligibility to laparoscopic surgery until exclusion: severe chronic obstructive pulmonary disease (COPD), pulmonary fibrosis, severe cardiopathy or renal hypoperfusion, acute glaucoma, ventricular–peritoneal shunt. The incidence of perioperative complications is associated with many factors, such as high American Society of Anesthesiologists Physical Status Classification (ASA PSC) class, emergency, intraoperative surgical and anesthesiological management, and availability of postoperative intensive care unit. On many issues there is no consensus from experts. Thus, it is good practice to organize a multidisciplinary staff for examining single cases or to integrate different prognostic scores (e.g., Child–Pugh, New York Heart Association classification, Glasgow Prognostic Score) to prevent postoperative complications. When the risk of surgery-related morbidity and mortality is high (advanced age, presence of cardiovascular risk factors, multisystem disease, poor functional status), the thoroughness of the preoperative assessment should be intensified and go beyond the standard history, taking into account physical examination, basic laboratories, and electrocardiogram. Actually, studies suggest performing cardiopulmonary exercise testing (CPET) to evaluate the cardiopulmonary reserve required to respond to metabolic stress of the perioperative period [8–11].

7.3 Anesthesia and Laparoscopic Liver Surgery

Managing hemorrhage and the risk of gas embolism are the major issues in LLR. Improvements in surgical techniques allow hepatic resection to be performed generally with minimal blood loss. This condition was obtained with the introduction of local hemostatic agents and new instruments for liver transection. The cardiopulmonary interventions to reduce blood loss during liver resection include low central venous pressure (LCVP) and hypoventilation. Central venous pressure is controlled <5 cm H_2O with a combination of anesthesia and early intraoperative fluid restriction to reduce venous hemorrhage during parenchymal transection. The anesthetic technique, designed to maintain LCVP and reduce transfusion requirements, also preserves renal function and minimizes short- and long-term adverse effects of transfusions, such as transmission of infection, allergic reactions, hemolysis, anaphylaxis, and transfusion-related acute lung injury (TRALI) [12–15]. Massive hemorrhage is one of the most important adverse events to avoid during LLR. Blood loss is an independent predictor of mortality, and it compromises, in common with postoperative complications, the long-term outcome of oncologic patients. Massive transfusion is commonly defined as replacement of one blood volume over a period of 24 h or transfusion of at least four red blood cell concentrates within 1 h. Massively transfused patients will show evidence of coagulopathy in a high percentage of cases. Preoperative evaluation, including the patient's and his or her relatives' bleeding history and routine coagulation testing aid making a correct diagnosis of inherited and acquired bleeding disorders and set up a possible preoperative preparation (modification of anticoagulant drugs, antifibrinolytic agents, erythropoietin, preadmission blood collection). In the operating room, periodic visual assessment of the surgical field is strongly suggested to detect as soon as possible bleeding due to surgical trauma or coagulopathy. Furthermore, rotational thromboelastometry or modified thrombelastography (point-of-care coagulation monitoring devices) are superior to routine laboratory tests to guide intraoperative hemostasis management. Therapeutic approaches are based on hypothermia, acidosis, and hypocalcemia corrections and on the use of blood products (red cell concentrates, platelets, plasma), coagulation factor concentrates (fibrinogen, prothrombin complex, von Willebrand factor), and pharmacological agents (antifibrinolytic drugs, desmopressin) [16–20].

Laparoscopy and intraperitoneal carbon dioxide (CO_2) insufflation may have detrimental cardiovascular effects. Changes in cardiovascular function due to the high abdominal pressure (12–14 mmHg) are characterized by an immediate decrease in cardiac index and an increase in mean arterial blood pressure and systemic vascular resistance. In the next few minutes, there is partial restoration of cardiac index and resistance, but blood pressure and heart rate do not change during the entire LLR. The pattern is the result of interaction between increased abdominal pressure, neurohumoral responses, and CO_2 absorption. Decailliot [21] has reported an evaluation on hemodynamic consequences of pneumoperitoneum (PP) associated to portal triad clamping (PTC) using trans esophageal echocardiography (TEE) to provide information about left ventricular cavity dimensions, wall thickness and wall motion. Interestingly, analysis of echocardiographic data demonstrated the occurrence of regional wall motion abnormalities (RWMAs) in 50% of patients during PTC with PP, as compared with only 10% of patients during PTC without PP. Although a decrease in preload is the main important change during PTC in open liver surgery, this study also demonstrated a decrease in left ventricular (LV) function, which is likely to be a consequence of decreased LV preload and increased LV afterload during laparoscopic liver resections. In patients without cardiac disease, PTC during a laparoscopic procedure could be performed safely although hemodynamic modifications are reported, while hemodynamic consequences in patients with preexisting cardiac disease and altered LV function need further investigation. The changes in cardiopulmonary function during laparoscopic upper abdominal surgery suggest judicious invasive monitoring and careful interpretation in ASA III-IV patients. Pulmonary function changes are characterized by reduced compliance without large alterations in partial arterial oxygen pressure (PaO_2), but tissue oxygenation can be adversely affected due to reduced O_2 delivery. Difficulty in maintaining normocarbia is due to the abdominal distention reducing pulmonary compliance and to CO_2 absorption [22–23]. End-tidal CO_2 tension is not a reliable index of partial arterial carbon dioxide pressure ($PaCO_2$), particularly in ASA III-IV patients. CO_2 is highly soluble in blood and fairly innocuous to the peritoneum. Small amounts absorbed into the circulation cause slight increases in arterial and alveolar CO_2 and in central venous pressure. When CO_2 enters the venous circulation through iatrogenically opened vascular channels, catastrophic and potentially fatal hemodynamic and respiratory failure may result. Gas embolism may occur each time the vein's internal pressure is lower than the external pressure, not only during a laparoscopic procedure when CO_2 is inflated into the peritoneal cavity, but also during open surgery, such as major liver resections, neurosurgery, and vascular or cardiac surgery. Although symptomatic CO_2 embolism is a rare condition, it is recognized as a potentially fatal complication of laparoscopic surgery. The risk of gas embolism may increase when positive-pressure CO_2 pneumoperitoneum is associated to LCVP used to minimize hemorrhage during liver resection. For these reasons, maximal caution should be exerted when laparoscopic surgery is performed close to large veins. Reinsufflation, inducing gas entry through the injured vessel, might be another risk factor for CO_2 embolism. The risk to the patient may be minimized by the surgical team's awareness of CO_2 embolism, continuous intraoperative monitoring of end-tidal CO_2, and hemogas analysis.

7.4 Conclusions

Laparoscopic surgery offers many benefits to the patient, including shorter hospital stay, shorter recovery time, less postoperative pain, and faster return to normal diet and usual activities. Nevertheless, ascites, postresectional liver failure, bile leakage, intra-abdominal hemorrhage, and intra-abdominal abscess are common complications of both open and laparoscopic liver surgery. The extent of resection and the degree of baseline functional impairment are the main independent risk factors for postoperative complications, including liver failure. Adverse events impairing various organs (kidneys, respiratory apparatus, cardiocirculatory system, nervous system) can occur even after 30 days from surgery. Complications after major surgery are a leading cause of morbidity and mortality. The etiology of postoperative complications is complex, but impaired cardiovascular flow and poor cardiorespiratory reserve appear to be key factors. Primary objects of anesthesiological assistance are an optimal analgesia and the control of hemodynamic parameters. Moreover, in postoperative period the pattern of lung function following laparoscopy is characterized by a transient reduction in lung volumes and capacities with a restrictive breathing pattern and the loss of the abdominal contribution to breathing. These changes are qualitatively similar to but of a lesser magnitude than those following "open" abdominal surgery. Non-invasive ventilation may be useful to avoid atelectasia and consequently hypoxemia. The role of the anesthesiologists is to support the surgical team in early detection of patients who need to be strictly monitored and assisted. A multidisciplinary effort must be made through the entire chain – from the outpatient clinic through discharge from hospital – with the utmost exertion of all team members in order to address the specific needs of the patient [24–27].

References

1. Poon RT, Fan ST, Lo CM et al (2004) Improving perioperative outcome expands the role of hepatectomy in management of benign and malignant hepatobiliary diseases: analysis of 1222 consecutive patients from a prospective database. Ann Surg 240:698–708
2. Imamura H, Seyama Y, Kokudo N et al (2003) One thousand fifty-six hepatectomies without mortality in 8 years. Arch Surg 138(11):1198–1206
3. Kusano T, Sasaki A, Kai S et al (2009) Predictors and prognostic significance of operative complications in patients with hepatocellular carcinoma who underwent hepatic resection. Eur J Surg Oncol 35(11):1179–1185
4. Reddy SK, Tsung A, Geller DA (2011) Laparoscopic liver resection. World J Surg 35(7):1478–1486 Review
5. Buell JF, Cherqui D, Geller DA et al (2009) The international position on laparoscopic liver surgery: The Louisville Statement, 2008. World Consensus Conference on Laparoscopic Surgery. Ann Surg 250(5):825–830
6. Welsh FK, Tekkis PP, John TG, Rees M (2010) Open liver resection for colorectal metastases: better short- and long-term outcomes in patients potentially suitable for laparoscopic liver resection HPB (Oxford) 12(3):188–194
7. Nguyen KT, Gamblin TC, Geller DA (2009) World review of laparoscopic liver resection-2,804 patients. Ann Surg 250(5):831–841 Review
8. Garcia-Miguel FJ, Serrano-Aguilar PG, Lopez-Bastida J (2003) Preoperative assessment. Lancet 362:1749–1757
9. Khuri SF, Henderson WG, DePalma RG et al (2005) Determinants of long-term survival after major surgery and the adverse effect of postoperative complications. Ann Surg 242:326–341
10. Smith TB, Stonell C, Purkayastha S, Paraskevas P (2009) Cardiopulmonary exercise testing as a risk assessment method in non cardio-pulmonary surgery: a systematic review. Anaesthesia 64(8):883–893 Review
11. Fujiwara Y, Shiba H, Furukawa K et al (2010) Glasgow prognostic score is related to blood transfusion requirements and postoperative complications in hepatic resection for hepatocellular carcinoma. Anticancer Res 30:5129–5136
12. Melendez JA, Arsian V, Fischer ME et al (1998) Perioperative outcomes of major hepatic resections under low central venous pressure anesthesia: blood loss, blood transfusion, and the risk of postoperative renal dysfunction. J Am Coll Surg 187:620–625
13. Jones RM, Moulton CE, Hardy KJ (1998) Central venous pressure and its effect on blood loss during liver resection. Br J Surg 85:1058–1060
14. Gurusamy KS, Li J, Sharma D, Davidson BR (2009) Cardiopulmonary interventions to decrease blood loss and blood transfusion requirements for liver resection. Cochrane Database Syst Rev 7(4):CD007338
15. Bux J, Sachs UJ (2007) The pathogenesis of transfusion-related acute lung injury (TRALI). Br J Haematol 136(6):788–799
16. Cockbain AJ, Masudi T, Lodge JP, Toogood GJ, Prasad KR (2010) Predictors of blood transfusion requirement in elective liver resection. HPB (Oxford) 12(1):50–55
17. Sima CS, Jarnagin WR, Fong Y et al (2009) Predicting the risk of perioperative transfusion for patients undergoing elective hepatectomy. Ann Surg 250(6):914–921
18. American Society of Anesthesiologists Task Force on Perioperative Blood Transfusion and Adjuvant Therapies (2006) Practice guidelines for perioperative blood transfusion and adjuvant therapies: an updated report by the American Society of Anesthesiologists Task Force on Perioperative Blood Transfusion and Adjuvant Therapies. Anesthesiology 105(1):198–208
19. Mannucci PM, Levi M (2007) Prevention and treatment of major blood loss. N Engl J Med 31;356(22):2301–2311
20. Liumbruno GM, Bennardello F, Lattanzio A, Piccoli P, Rossetti G; Italian Society of Transfusion Medicine and Immunohaematology (SIMTI) Working Party (2011) Recommendations for the transfusion management of patients in the peri-operative period. II. The intra-operative period. Blood Transfus 9(2):189–217

21. Decailliot F, Streich B, Heurtematte Y et al (2005) Hemodynamic Effects of Portal Triad Clamping With and Without Pneumoperitoneum: An Echocardiographic Study. Anesth Analg 100:617-622
22. Putensen-Himmer G, Putensen C, Lammer H, Lingnau W, Aigner F, Benzer H (1992) Comparison of postoperative respiratory function after laparoscopy or open laparotomy for cholecystectomy. Anesthesiology 177:675-680
23. Cunningham A, Brull S (1993) Laparoscopic cholecystectomy: anesthetic implications. Anesth Analg 76:1120–1125
24. Rothe CF (1993) Mean circulatory filling pressure: its meaning and measurement. J Appl Physiol 74(2):499–509 Review
25. Jhanji S, Lee C, Watson D, Hinds C, Pearse RM (2009) Microvascular flow and tissue oxygenation after major abdominal surgery: association with post-operative complications. Intensive Care Med 35(4):671–677
26. Virani S, Michaelson JS, Hutter MM et al (2007) Morbidity and mortality after liver resection: results of the patient safety in surgery study. J Am Coll Surg 204(6):1284–1292
27. Mirnezami R, Mirnezami AH, Chandrakumaran K et al (2011) Short- and long-term outcomes after laparoscopic and open hepatic resection: systematic review and meta-analysis HPB (Oxford) 13(5):295–308

Costs and Benefits. A Triad in Comparison: Open, Laparoscopic, and Robotic Surgery

8

Graziano Ceccarelli, Alberto Patriti, Raffaele Bellochi, Alessandro Spaziani and Luciano Casciola

8.1 Introduction

Technologic innovation in health care is one of the most important causes of growth in health-care costs. In recent years, financial restrictions have led to careful evaluation of the cost-effectiveness of every new medical procedure [1]. Minimally invasive surgery (laparoscopic and robotic) has improved patient outcomes in different general surgery subspecialties. Liver surgery represents one of the most recent applications of these technologies [2]. The cost-effectiveness of these procedures compared with open procedures is still a matter of debate [3]. Procedure costs include mainly direct costs (operating room time, instruments, and medications) and costs related to hospital stay. In laparoscopic surgery, instrument cost (technological component) is the most important component and is affected by the type (reusable or disposable) and number of instruments used in each procedure. Laparoscopic and robotic procedures are, of course, more expensive compared with open procedures if direct costs only are considered. However, in overall cost evaluation, other parameters must be included, such as costs of complications and patient recovery time and reintegration into social and professional lives. In this evaluation, minimally invasive surgery has demonstrated advantages in cost-effectiveness in the major randomized controlled studies [1, 3, 4].

G. Ceccarelli (✉)
Department of Surgery, Division of General, Minimally Invasive and Robotic Surgery, ASL3 Umbria, San Matteo degli Infermi Hospital, Spoleto, Italy
e-mail:g.cecca2003@libero.it

F. Calise and G. Casciola (eds.), *Minimally Invasive Surgery of the Liver,* Updates in Surgery
DOI: 10.1007/978-88-470-2664-3_8, © Springer-Verlag Italia 2013

8.2 Cost-effectiveness of Laparoscopic Liver Resections

Koffron et al. reported an important shift in the rate of laparoscopic (LLR) and open liver resection in their institution. LLR represented 10% of cases in 2002 and 80% in 2007, showing that laparoscopy decreased operative times (99% vs 182 min), blood loss (102 vs 325 ml), transfusion requirement (2 of 300 vs 8 of 100), length of stay (1.9 vs 5.4 days), overall operative complications (9.3 vs 22%), and local malignancy recurrence rate (2% vs 3%) [5]. A reduction in hospital stay, blood loss, and postoperative pain and morbidity was clearly confirmed by several studies, suggesting that LLR may be a suitable alternative to open surgery [6, 7].

Laparoscopic left lateral sectionectomy (LLS) is the most standardized and anatomically performed LLR. Comparing LLS with the open procedure, decreased intraoperative blood loss and shorter postoperative stay were demonstrated [7, 8]. Vanounou et al., in 2010, published a study comparing 44 laparoscopic and 29 open LLS. In the laparoscopic group, they observed a shorter median length of stay of 2 days ($P = 0.001$) and a reduction in postoperative morbidity ($P = 0.001$). The economic impact of the laparoscopic compared with the open approach was evaluated using the deviation-based cost modeling (DBCM). Laparoscopy resulted in cost savings of US$1,412, demonstrating it to be cost effective. The cost savings observed was due mainly to reduced length of hospital stay and secondarily by a lower complication rate (less need for ancillary services such as laboratory, radiology, and pharmacy) [9]. Edwin et al. reported that total hospital costs for treating patients undergoing laparoscopic left hepatic resection is lower than for open counterparts. This fact is explained mainly by decreased postoperative hospital stay, and faster return of patients to their work would contribute additional values to the society [10]. Polignano et al., comparing laparoscopic versus open liver segmentectomy in a prospective, case-matched fashion, demonstrated a significant reduction in overall hospital costs using laparoscopy. By examining the average unit costs for theater time, disposable instruments, high-dependency-unit (HDU) stay, ward stay, and overall costs, the authors demonstrated that, although theater-time costs did not differ, the laparoscopic approach allowed rapid recovery and reduced HDU and ward costs. The laparoscopic approach was 2,571 euros more cost efficient than the corresponding open approach [3].

Taken as a whole, it appears that any laparoscopic approach may advance patient care and improve patient outcomes compared with traditional open hepatectomy. In addition to its clinical benefit, the laparoscopic approach seems to also offer an economic benefit [5].

8.3 Costs in Robotic Surgery

Robotic technology has increased in recent years. The number of robot-assisted procedures worldwide has nearly tripled, from 80,000 cases in 2007 to

250,000 in 2009 [11]. Cost studies exist for about 20 types of procedures and demonstrate that the average additional cost for the robotic approach is about US$1,600, increasing to more than US$3,000 if the amortized cost of the robot is included. Fixed costs depend greatly upon the number of cases necessary to amortize the life span of the robotic system (breakeven-point analysis). van Dam et al. calculated a direct cost for each patient of 3,920 euros, 1,960 euros, 1,306 euros, and 980 euros if, respectively, 100, 200, 300, and 400 robotic procedures are performed each year (amortization over 7 years). So, for even a few surgical procedures, the costs can be competitive to similar open surgical procedures [12, 13].

8.3.1 Direct and Indirect Costs Associated With Robotic Surgery

8.3.1.1 Equipment Costs

The da Vinci Si robotic device costs more than US$1,800,000, with a yearly service contract of about US$180,000 (about 10%). All robotic instruments ($700–1,000 per case) have a limited, preprogrammed number of uses (generally ten). Another important aspect is the lack of market competition; the only robotic system (da Vinci) is produced by a single company [14].

8.3.1.2 Operative Time

Operative time includes the surgical procedure and the time to prepare the surgical theater and to setup and dock the robotic system. This extra time may be calculated at about 15–40 min, but with experienced surgical teams, it may be lowered to 5–10 min. Operative time decreases, of course, with surgeon and surgical team experience. The learning curve to attain the necessary expertise also has costs. In high-volume centers, the learning curve can be rapidly reached and costs minimized [15].

8.3.1.3 Hospital Stay/Complications

Minimally invasive procedures decrease hospital stay compared with open surgery, so overall costs of hospitalization are decreased [16]. However, for most procedures, there are no demonstrated advantages of the robotic over the conventional laparoscopic approach in terms of hospital stay. On the other hand intra- and postoperative morbidity has added costs linked to longer hospital stay and the patient's inability to work. Many studies demonstrate these advantages of minimally invasive surgery compared with open surgery, but only a few studies were designed to compare robotic to laparoscopy [17, 18].

8.3.1.4 Costs of Patient Inability to Work

It is more difficult to calculate the savings for society, health systems, and insurance when assessing sick leave. Regardless, minimally invasive procedures allow patients to resume their normal professional activities sooner than after undergoing open surgery.

8.3.1.5 Cost of Oncologic Aspects

When we treat oncologic conditions such as liver tumors, other aspects must be considered. In particular, these are estimated blood loss (and transfusions), lymph node harvesting, recurrence, and survival.

8.3.2 Cost-effectiveness in Robotic Liver Surgery

Only a few studies deal with this aspect. They suggest that compared with laparoscopy, robotic surgery is easier to set up, has the advantages of more easily identifying and dealing with vascular and biliary structures, and allows very precise dissection and safe parenchymal division, as a result reducing the risk of complications. The robotic instrumentation in general adds US$500 per case to the laparoscopic equipment cost. These costs can be reduced with heavy use of the robot by other surgical specialties. In some studies, perioperative outcomes were similar between robotic and conventional liver resection groups, with the robotic procedures taking about 25 min longer. We believe that this additional time is related to the learning curve of the surgical team. For selected liver lesions, the robotic approach provides better perioperative outcomes compared with laparoscopy, with better visualization and dexterity. The robotic approach merits further attention because of its potential to mimic open liver resection [19–21]

8.4 Models to Compare Surgical Costs

To compare costs of different procedures, comparing direct costs only is not sufficient. One of the most used cost-model tools is deviation-based cost modeling, which consists of defining deviations from the expected hospital course and comparing the clinical and economic impact of complications in different procedures, particularly in relation to length of stay. Postoperative complications (in particular, their severity) are an important factor conditioning patients' overall hospital course and costs [9, 22]. The principal advantage of using this model is its availability to any surgical procedure at any institution. Data are analyzed using the appropriate statistical analysis, permitting a rigorous comparison of two or more techniques (e.g., laparoscopy vs open vs robotic procedures) [23]. Another important cost-effective model is the Markov model. Cost-effectiveness was measured in terms of incremental cost per life-year gained and incremental cost per quality-adjusted life-years (QALY) for a time horizon up to 25 years [24].

8.5 Innovation Governance and Health Technology Assessment (HTA)

Every new health technology has the problem of "innovation governance," and every health service around the world needs to rely on sufficient information. Every new technology can stimulate enthusiasm, competition, and desire to pioneer; so every heath system needs to avoid availability of these innovations to health-care providers with limited experience. One method adopted to evaluate cost-effectiveness of new technology is a multistep evaluation process (a multidisciplinary panel) to determine:

1. Definition of the technology's evidence profile and all relevant clinical outcomes
2. Systematic review of scientific literature
3. Definition of the acceptable level of uncertainty for investing research resources
4. Analysis of local context
5. Identification of clinical indications with promising clinical return [25].

In Great Britain, the National Institute for Health and Clinical Excellence (NICE), a special health authority of the English National Health Service (NHS), publishes guidelines in three areas: the use of health technologies (medicines, treatments, and procedures); clinical practice; and health promotion. These appraisals are based primarily on evaluating efficacy and cost-effectiveness. NICE tries to provide standards of care. It helps improve the quality of the service and the evidence of clinical practice of the NHS. NICE uses the Health Technology Assessment (HTA) program that was set up in 1993. This program aims to ensure that high-quality research information on costs, effectiveness, and broader impact of health technologies is available for physicians and managers of the NHS. Through its Technology Assessment Report (TAR), the HTA program is able to commission bespoke reports, not only for NICE, but also for other policy customers.

8.6 Conclusions

The use of every new technology begins a process of formally conducted research. In particular, a cost-effective evaluation needs to collect acceptable levels of evidence proving the clinical effectiveness of the new procedures. The safety and advantages in outcomes of minimally invasive hepatic surgery have been demonstrated in many operative procedures, even in those performed for malignancies [26]. Studies of overall costs (direct and indirect) comparing open and laparoscopic procedures demonstrate interesting results in better cost-effectiveness of laparoscopy, above all as the consequence of reduced postoperative hospital stay, blood loss, and complications, demonstrating a reduction in overall costs [5–7]. Robotic technology was developed to overcome these limitations, although it is expensive due to the high costs of purchase, maintenance, and need for dedicated instruments. However, the learning curve for robotic surgery is shorter compared with that for laparoscopy, and the conversion rate lower [18, 27, 28]. The more precise movements and dexterity of robotic technology demonstrates fewer postoperative complications in other procedures also, such as gastric bypass and cardiothoracic surgery.

To reduce laparoscopic – and especially robotic – surgery costs, it may be useful to concentrate activity (high-volume centers), create specialized laparoscopic or robotic units, train dedicated theater staff, reduce the number of disposable instruments per operation, reduce setup time, and reduce the learning curve with the help of expert surgeons at the beginning of the training period. When addressing cost savings of robotic surgery, it is necessary to increase multidisciplinary use and overall annual use (at least 300 procedures are necessary to amortized the purchase cost). Reduced equipment costs, development of new types of instruments – as occurs with almost all electronic devices – competition between major equipment manufacturing companies, multicenter prospective randomized trials incorporating economic aspects, and evidence-based benefits are all necessary to justify the use of minimally invasive surgery. More studies concerning oncologic outcomes of minimally invasive surgery (lymph-node harvesting; recurrence and survival rates) are also necessary.

In 2006, Lanzafame wrote: "...we must use technology responsibly and honestly, evaluating our outcomes. It is our responsibility to educate our patients, ourselves, and our colleagues about the benefits and limits of minimally invasive surgery" [29].

Acknowledgments A special thank to Dr. Giuliano Metastasio, Dr. Maria Bonadonna and Ms. Erminia Ciammarughi.

References

1. Künzli BM, Friess H, Shrikhande SV (2010) Is laparoscopic colorectal cancer surgery equal to open surgery? An evidence based perspective. World J Gastrointest Surg 272(4):101–108
2. Nguyen KT, Gamblin TC, Geller DA (2009) World review of laparoscopic liver resection – 3,804 patients. Ann Surg 250(5):831–841
3. Polignano FM, Quyn AJ, de Figueiredo RS, Henderson NA, Kulli C, Tait IS (2008) Laparoscopic vs open liver segmentectomy: prospective, case-matched, intention-to-treat analysis of clinical outcomes and cost effectiveness. Surg Endosc 22(12):2564–2270
4. Koffron AJ, Kung RD, Auffenberg GB, Abecassis MM (2007) Laparoscopic liver surgery for everyone: the hybrid method. Surgery 142(4):463–468

5. Koffron AJ, Auffenberg G, Kung R, Abecassis M (2007) Evaluation of 300 minimally invasive liver resections at a single institution: less is more. Ann Surg 246(3):385–392
6. Lee KF, Cheung YS, Chong CN, Tsang YY, Ng WW, Ling E et al (2007) Laparoscopic vs open hepatectomy for liver tumours: a case control study. Hong Kong Med J 13:442–448
7. Aldrighetti L, Pulitanò C, Catena M, Arru M, Guzzetti E, Casati M et al (2008) A prospective evaluation of laparoscopic vs open left lateral hepatic sectionectomy. J Gastrointest Surg 12(3):457–462
8. Abu Hilal M, McPhail MJ, Zeidan B, Zeidan S, Hallam MJ, Armstrong T et al (2008) Laparoscopic vs open left lateral hepatic sectionectomy: a comparative study. Eur J Surg Oncol 34(12):1285–1288
9. Vanounou T, Steel JL, Nguyen KT, Tsung A, Marsh JW, Geller DA et al (2010) Comparing the clinical and economic impact of laparoscopic vs open liver resection. Ann Surg Oncol 17(4):998–1009
10. Edwin B, Nordin A, Kazaryan AM (2011) Laparoscopic liver surgery: new frontiers. Scand J Surg 100:54–65
11. Barbash GI, Glied SA (2010) New technology and health care costs – the case of robot-assisted surgery. N Engl J Med 363:701–704
12. van Dam P, Hauspy J, Verkinderen L, Trinh B, Van Looy L, Dirix L (2011) Do costs of robotic surgery matter? In: Darwish A (ed) Advanced endoscopic gynecology. InTech, Rijeka
13. van Dam P, Hauspy J, Verkinderen L, Trinh XB, van Dam PJ, Van Looy L et al (2011) Are costs of robot-assisted surgery warranted for gynecological procedures? Obstet Gynecol Int 973830 [Epub Sept 18]
14. Lotan Y (2010) Economics of robotics in urology. Curr Opin Urol 20(1):92–97
15. Leddy LS, Lendvay TS, Satava RM (2010) Robotic surgery: applications and cost effectiveness. Open Access Surgery 3:99–107
16. Lotan Y, Cadeddu JA (2006) Financial aspects of laparoscopy. In: Gill I (ed) Textbook of laparoscopic urology. Informa Healthcare USA, New York
17. Bell MC, Torgerson J, Seshadri-Kreaden U, Suttle AW, Hunt S (2008) Comparison of outcomes and cost for endometrial cancer staging via traditional laparotomy, standard laparoscopy and robotic techniques. Gynecol Oncol 111(3):407–411
18. Ceccarelli G, Patriti A, Biancafarina A, Spaziani A, Bartoli A, Bellochi R, Casciola L (2009) Intraoperative and postoperative outcome of robot-assisted and traditional laparoscopic Nissen fundoplication. Eur Surg Res 43(2):198–203 [Epub 2009 Jun 10]
19. Choi SB, Park JS, Kim JK, Hyung WJ, Kim KS, Yoon DS et al (2008) Early experiences of robotic-assisted laparoscopic liver resection. Yonsei Med J 49(4):632–638
20. Dindo D, Demartines N, Clavien PA (2004) Classification of surgical complications: a new proposal with evaluation in a cohort of 6336 patients and results of a survey. Ann Surg 240:205–213
21. Casciola L, Patriti A, Ceccarelli G, Bartoli A, Ceribelli C, Spaziani A (2011) Robot-assisted parenchymal-sparing liver surgery including lesions located in the posterosuperior segments. Surg Endosc 25(12):3815–3824 [Epub 2011 Jun 8]
22. Vanounou T, Pratt W, Fischer JE, Vollmer CM Jr, Callery MP (2007) Deviation-based cost modeling: a novel model to evaluate the clinical and economic impact of clinical pathways. J Am Coll Surg 204(4):570–579
23. Murray A, Lourenco T, de Verteuil R, Hernandez R, Fraser C, McKinley A et al (2006) Clinical effectiveness and cost-effectiveness of laparoscopic surgery for colorectal cancer: systematic reviews and economic evaluation. Health Technol Assess 10(45):1–141
24. Ballini L, Minozzi S, Negro A, Pirini G, Grilli R (2010) A method for addressing research gaps in HTA, developed whilst evaluating robotic-assisted surgery: a proposal. Health Res Policy Syst 20;8:27
25. Simillis C, Constantinides VA, Tekkis PP, Darzi A, Lovegrove R, Jiao L et al (2007) Laparoscopic vs open hepatic resections for benign and malignant neoplasms-a meta-analysis. Surgery 141:203–211
26. Buell JF, Thomas MT, Rudich S et al (2008) Experience with more than 500 minimally invasive hepatic procedures. Ann Surg 248(3):475–486

27. Dev H, Sharma NL, Dawson SN, Neal DE, Shah N (2012) Detailed analysis of operating time learning curves in robotic prostatectomy by a novice surgeon. BJU Int 109:1074–1080 [Epub 2011 Oct 28]
28. Secin FP (2011) The learning curve of robotic assisted laparoscopic radical prostatectomy: what is the evidence? Arch Esp Urol 64(8):830–838
29. Lanzafame RJ (2006) Issues in the acquisition, development, and use of technology in health care. JSLS 10:401–408

Transection Devices

9

Alberto Patriti

9.1 Introduction

The risk of massive bleeding during liver transection and postoperative biliary leaks are due to the complex biliary and vascular anatomy of the liver. Hemorrhage was once the leading cause of death in liver resection, and the now reduced hospital mortality rate of ≤5% can be attributed to better intraoperative bleeding control. Hemorrhage and perioperative blood transfusion not only increase the risk of operative morbidity and mortality but jeopardize long-term survival after resection of liver malignancies because of the associated immunosuppression, leading to a higher risk of tumor recurrence [1]. Bleeding control is the result of the evolution of different aspects of liver surgery and anesthesia. Technological advances led to the development of specific instruments for liver transection; intraoperative ultrasound allows better delineation of the transection plane; and a better understanding of physiology and anatomy improved control of inflow and outflow. Inflow occlusion and low central venous pressure (CVP) anesthesia have been widely used to reduce bleeding from inflow vessels and backflow in the transection surface. Inflow occlusion (Pringle maneuver) has been used since the early twentieth century to prevent bleeding during transection, which is performed by crushing the liver parenchyma with the fingers or forceps (Kelly-clamp crushing), and the concomitant low CVP induced by anesthesia further minimizes blood loss by preventing retrograde bleeding from the hepatic veins. Assuming that inflow occlusion and low CVP cause significant damage due to ischemia and reperfusion, there has been a growing interest in using new devices that facil-

A. Patriti (✉)
Department of Surgery, Division of General, Minimally Invasive and Robotic Surgery,
ASL3 Umbria, San Matteo degli Infermi Hospital, Spoleto, Italy
e-mail: albertopatriti@gmail.com

F. Calise and G. Casciola (eds.), *Minimally Invasive Surgery of the Liver*,
Updates in Surgery
DOI: 10.1007/978-88-470-2664-3_9, © Springer-Verlag Italia 2013

itate bloodless transection, obviating the need for inflow occlusion. In the laparoscopic setting, these factors, associated with the struggle to perform an intermittent Pringle maneuver and clamp crushing, have led to a wide diffusion of a variety of transection devices, mostly derived from those routinely used in open surgery. This chapter provides a description of the main transection device features and considerations on the Pringle maneuver associated with clamp crushing in the laparoscopic setting.

9.2 Main Features of Transection Devices

The ideal surgical instrument for liver transection must effectively cut through the parenchyma while simultaneously recognizing and sealing any vessels or bile ducts. In reality, because no such single instrument yet exists, transection is achieved by a combination of instruments and techniques, which first facilitate division of the liver tissue, with subsequent vessel identification and sealing. The two basic actions of liver transection can be achieved by a variety of energy types [2].

9.2.1 Mechanical Energy

Manual fragmentation of the liver parenchyma is the more primeval transection device. The "finger fracture technique" under cycles of inflow occlusion was first introduced by Lin et al. in 1958. This technique was subsequently improved through the use of surgical instruments, such as a small Kelly clamp for blunt dissection. Nowadays, clamp crushing remains one of the most widely used techniques for liver transection in open surgery. A more evolved device employing mechanical energy is the water-jet dissector. The Hydro-Jet® (ERBE, Tuebingen, Germany) employs a pressurized jet of water to fragment the liver parenchyma tissue and expose the vascular and ductal structures. In all systems employing mechanical energy, vessels and bile duct must be sealed by other devices, clipped, or ligated before division. The endoscopic stapler can be considered the only method employing mechanical energy that crushes and divides the liver parenchyma, securing blood and bile vessels at the same time. A straight clamp is used to fracture the hepatic parenchyma, and subsequently, this portion of the liver is transected with a vascular stapler.

9.2.2 Alternating Current

High-frequency alternating current can be delivered in either a monopolar or a bipolar fashion. Most electrosurgical devices work in a radiowave frequency range of 500,000 Hz to 3 MHz. The monopolar device is composed of a generator, an electrode of application, and an electrode for the returning current to complete the circuit. The patient's body becomes part of the circuit when the system is activated. As the effectiveness of energy conversion into heat is inversely related to the area of contact, the application electrode is designed to be small to efficiently generate heat, and the returning electrode is designed to be large to disperse energy and prevent burn injury to the patient. Generated heat is dependent on three other factors in addition to the size of the contact area: power setting/frequency of the current; length of activation time; whether the waveform released from the generator is continuous or intermittent. Unipolar devices can be used to incise tissue when activated with a constant waveform and to coagulate when activated with an intermittent waveform. In the cutting mode, much heat is generated relatively quickly over the target area, with minimum lateral thermal spread. As a result, the device cuts through tissue without coagulating underlying vessels. In contrast with the coagulation mode, the electrocautery generates less heat on a slower frequency, with potential for large lateral thermal spread. This results in tissue dehydration and vessel thrombosis. A blind waveform can be chosen to take advantage of both cutting and coagulation mode. A large grounding pad must be placed securely on the patient for the unipolar electrocautery device to function properly and prevent thermal burn injury to the patient at the current reentry electrode site. Bipolar electrocautery establishes a short circuit between the tips of the instrument, whether a tissue grasper or forceps, without requiring a grounding pad. The tissue grasped between the tips of the instrument completes the circuit. In generating heat that only affects the tissue within the short circuit, it provides precise thermal coagulation. Bipolar electrocautery is more effective than the monopolar instrument for coagulating vessels because it adds the mechanical advantage of tissue compression between the tips of the instrument to the thermal coagulation. Bipolar electrocautery is particularly useful for conducting a procedure in which lateral thermal injury or arcing phenomenon need to be avoided. The argon-beam coagulator is a special form of monopolar electrocautery. The device creates a monopolar electric circuit between a handheld probe and target tissue by establishing a steady flow of electrons through a channel of electrically activated ionized argon gas. This high-flow argon gas conducts electrical current to the target tissue, where it generates thermal coagulation. The depth of thermal penetration of tissue varies from fractions of a millimeter to a maximum of 6 mm, depending on three factors: (1) power setting, (2) distance between probe and target, and (3) length of application. It is most commonly used to control oozing on the cutting liver surface.

9.2.3 Radiofrequency Energy

Radiofrequency (RF) is a rate of oscillation in the range of about 3 kHz to 300 GHz, which corresponds to the frequency of radio waves and alternating currents carrying radio signals. RF usually refers to electrical rather than mechanical oscillations; therefore, RF devices work basically as a monopolar and bipolar electrocautery. A special feature of RF current is the "skin effect": the current does not penetrate deeply into electrical conductors but flows along their surfaces. The rapid alternating directional movements of ions in the tissue surrounding the probe result in the release of kinetic energy, determining high temperatures. It can raise the temperature of target tissue to >100°C and cause protein denaturation, desiccation, and coagulation necrosis, with a built-in sensor terminating the transmission of the current automatically at a particular set point, preventing overheating and unwanted collateral damage. To date, RF devices are largely used for resection and ablation of liver tumors. The most popular RF-based devices for parenchymal transection are LigaSure, TissueLink, and radiofrequency-assisted devices. LigaSure (Valley Lab, Tyco Healthcare, Boulder, CO, USA) is designed to seal small vessels. By a combination of compression pressure and bipolar RF energy, it causes shrinkage of collagen and elastin in the vessel wall, and it is effective in sealing small vessels up to 7 mm in diameter. Gyrus Plasma Trisector (PK) (Gyrus Group PLC, Gyrus International, Ltd., UK) is a novel RF system based on bipolar technology and is available for open and laparoscopic surgery. The PK system uses high-powered pulsed bipolar energy to produce a plasma kinetic field around the working elements and is designed to operate at temperatures that allow effective tissue dissection but to cause minimal collateral damage and adherence to tissue. The TissueLink (Medical, Inc., Dover, NH, USA) is new technology using saline-linked RF energy. In this instrument, saline runs to the tip of the electrode to couple RF energy to the liver surface and achieve coagulation. The radiofrequency-assisted devices are designed to be inserted along the transection plane serially 1- to 2-cm apart, and RF energy is applied for 1–2 min to create overlapping cylinders of coagulated tissue, followed by transection of the coagulated liver using a simple scalpel. Devices in this category are the Cool-tip RF electrode (Radionics Inc., Burlington, MA, USA) and the Laparoscopic Habib™ 4X (LH4X; RITA Medical Systems, Inc. California, CA, USA). The LH4X consists of a 2 × 2 array of needles arranged in a rectangle and uses bipolar RF energy to create an area of necrosis that can be cut with scissors or scalpel.

9.2.4 High-frequency Sound Waves

Ultrasound effects on living tissue depend on its frequency. At a low power level, it causes no tissue damage and is mainly used for diagnostic purposes. With a high frequency setting, ultrasound can be used to dissect, cut, and coagulate. There are several high-frequency ultrasonic devices available for surgical practice. The Harmonic Scalpel (Ethicon Endo-Surgery, Cincinnati, OH, USA) and the LOTUS (Laparoscopic Operation by Torsional Ultrasound; S.R.A. Developments Ltd., Devon, UK) use ultrasonically activated shears to seal small vessels between the vibrating blades. The blade longitudinal

(Harmonic Scalpel) or tortional (LOTUS) vibration with a frequency of 55.5 kHz can dissect liver parenchyma easily. The coagulation effect is caused by protein denaturation, which occurs as a result of destruction of the hydrogen bonds in proteins and generation of heat in the vibrating tissue. Blood vessels up to 2–3 mm in diameter are coagulated on contact with the vibrating blade. The tissue-cutting effect derives from a saw mechanism in the direction of the vibrating blade.

The Cavitron Ultrasonic Surgical Aspirator (CUSA) (Tyco Healthcare, Mansfield, MA, USA) fragments and aspirates liver parenchyma with ultrasonic energy, thus exposing vascular and ductal structures that can be ligated, clipped, or sealed with other transection devices. CUSA uses lower frequency ultrasound energy and works basically as an ultrasound probe combined with an aspirator. The main feature of CUSA is the ability to fragment and aspirate tissue of low collagen and high water content (hepatocytes), leaving intact anatomical structures with high collagen content, such as blood and biliary vessels.

9.3 What is the Best Transection Device?

Even though some of these devices have gained wide acceptance for hepatectomy, their efficacy has been tested in only a few randomized studies conducted in the open surgical setting. In a recent meta-analysis, data from 556 patients undergoing elective liver resection and randomized in seven trials were reviewed [3]. In that systematic review, there were no significant differences in mortality and complication rates (including bile leak) of liver resection, irrespective of the method used for parenchymal transection. Markers of liver parenchymal injury or liver dysfunction were also similar, and there was no difference in intensive care unit or hospital stay between groups. On the other hand, the clamp crushing coupled with the Pringle maneuver appeared to have the lowest blood loss and lowest transfusion requirements compared with the other techniques. Clamp crushing was quicker than CUSA, Hydro-Jet, and RFDS. Ikeda et al. randomized 120 patients to undergo clamp crushing or liver resection with the LigaSure. In both groups, intermittent pedicle clamping was applied, and the technique of parenchymal transection differed only for the method of securing blood and bile vessels. After liver capsule cauterization, the liver parenchyma was fractured by clamp crushing, and vascular structures, including portal triads and hepatic veins, were sealed with LigaSure in one group and ligated in the other. The two groups did not differ in terms of transection speed and postoperative morbidity [4]. Therefore, there is no evidence of superiority of any technique over clamp crushing and the Pringle maneuver for open liver resections. Clamp crushing is hardly reproducible laparoscopically and several concerns limited the wide diffusion of the Pringle maneuver among laparoscopic liver surgeons.

Encircling the liver pedicle is technically challenging with the rigid laparoscopic tools, and the risk of injury to the inferior vena cava and structures within the liver pedicle is considered potentially life threatening. Moreover, an effective intermittent clamping is difficult to achieve due to the continuous changes of the visual field between the section line and liver pedicle. Finally, some preclinical data show that a high abdominal pressure could decrease liver backflow, enhancing ischemic damage induced by inflow occlusion [5, 6].

These are several reasons the use of the other devices has been adopted in laparoscopic liver resection instead of clamp crushing. Nevertheless, to date, there are no randomized studies demonstrating the superiority of one technique or device over the others for liver transection in laparoscopic surgery. A consensus among authors regards the use of vascular staplers to secure and divide the portal pedicles and hepatic veins [7–9]. Similarly, transection of the superficial liver layer (2 cm beneath the glissonian sheet), with the absence of large vessels and bile ducts, is unanimously considered safe with all devices, monopolar and ultrasonically activated shears included. When a deeper transection is required, two methods can be followed: (1) indiscriminately coagulate all tissues and vessels along the transection plane, or (2) destroy the liver parenchyma to expose the inner vascular and biliary structures to be clipped or sealed separately and then divided.

RF-based devices work in the first manner. The LigaSure Atlas crushes the parenchyma between the instrument jaws and then coagulates any vessel and bile duct to at least 7 mm in diameter. The reported advantage of this device is the minimal adjacent tissue damage due to thermal spread [10]. In fact, when compared with other laparoscopic devices used for hepatic parenchymal dissection, the LigaSure device demonstrates a lower mean temperature in the surrounding parenchyma. In a study by Kim et al., the mean temperature in the liver was 121.3 ± 9.7°C and 76 ± 2.9°C for the Harmonic scalpel and LigaSure, respectively [11]. The LH4X is another bipolar RF device. It produces coagulative necrosis along the line of intended parenchymal transection without vascular clamping of either portal triads or major vessels. The area of necrosis is then cut with laparoscopic scissors [12]. Limitations of RF-based devices are possible vascular injuries when the transection plane is close to a major liver vessel, and the wide area of necrosis at the surgical margin, which can increase the risk of postoperative septic complications and make identifying marginal recurrences difficult during follow-up [13].

Rather than creating massive coagulation of the transection plane, ultrasonic and water-jet devices fragment the liver parenchyma, leaving intact arteries, veins, and bile ducts crossing the line of division, which can be sealed or clipped and divided using the ultrasonic coagulating cutter, electrocautery, or an RF-based device [7, 8, 14]. Therefore, the choice of transection techniques in laparoscopic and open surgery is a matter of surgeon preference, as there are no data from prospective randomized trials that compared different techniques. Frequently, liver transection is performed using more than one device. Therefore, it will be difficult to design controlled studies to prospectively compare all the available devices [15].

9.4 Clamp-crushing Technique in Laparoscopic Liver Surgery

Whereas clamp crushing associated with intermittent Pringle maneuver is considered the safer and more accurate method of parenchymal transection in open liver surgery, its use in the laparoscopic practice seems to be neglected [3]. As discussed above, there are many technical and theoretical reasons explaining this preference of laparoscopic surgeons. Despite the extensive use of high-tech devices, the principle behind clamp crushing – to fragment the parenchyma while preserving vascular structures – has been reproduced by some devices, such as water-jet devices and CUSA. Moreover, parenchymal fragmentation under intermittent inflow occlusion has been described in laparoscopic series. Cuschieri, in 2005, claimed that to transect the liver he uses a long-jawed crushing laparoscopic forceps under intermittent Pringle maneuvers carried out with a laparoscopic vascular clamp introduced through a port on the right flank [16]. Lee et al., in their series of 100 laparoscopic hepatectomies, used clamp crushing in selected cases [21].

Renewed interest in this old-school technique is paralleled by ingenious and easy techniques to achieve a safe and reproducible inflow occlusion. Four groups independently describe a novel way to clamp the liver pedicle extracorporeally [17–19]. The availability of a tourniquet encircling the liver pedicle that can be managed from the outside has the advantage of guaranteeing intermittent and effective inflow occlusions, as in open surgery. This feature is especially appreciable when major bleeding occurs, as the surgeon or the assistant can rapidly occlude the inflow, avoiding the struggle of looking for the liver pedicle in a bloody field and changing the focus of attention from the bleeding site.

With the advent of robotics, clamp crushing has become the standard for parenchymal transection in our institution. Using the EndoWrist bipolar Precise forceps (Intuitive Surgical Systems, Sunnyvale, CA, USA), the parenchyma can be easily fragmented, exposing the inner vessels as in open surgery. The on-table surgeon uses the device to perform intermittent inflow occlusion, thereby allowing the console surgeon to focus attention only on the transection line [20]. The wristed instruments gave back to the laparoscopic surgeon the possibility of performing curved and angled resections in all liver segments, an ability that was lost with the rigid laparoscopic tools. The effect of the poor ergonomics of laparoscopic devices has detrimental effects on the outcome of liver resection.

Straight resections are easier, forcing the surgeon to favor major hepatectomies for lesions located in the posterior liver segments. Therefore, it is the author's opinion that robotic clamp crushing may improve parenchymal preservation, even for deeply located lesions, thus widening the indications for a minimally invasive approach even to lesions in the posterolateral segments and located close to major liver vessels.

References

1. Kooby DA, Stockman J, Ben-Porat Let al (2003) Influence of transfusions on perioperative and long-term outcome in patients following hepatic resection for colorectal metastases. Ann Surg 237(6):860–869. doi:10.1097/01.SLA.0000072371.95588.DA
2. Poon RT (2007) Current techniques of liver transection. HPB (Oxford) 9(3):166–173. doi:10.1080/13651820701216182
3. Gurusamy KS, Pamecha V, Sharma D, Davidson BR (2009) Techniques for liver parenchymal transection in liver resection. Cochrane Database Syst Rev (1):CD006880. doi:10.1002/14651858.CD006880.pub2
4. Ikeda M, Hasegawa K, Sano K et al (2009) The vessel sealing system (LigaSure) in hepatic resection: a randomized controlled trial. Ann Surg 250(2):199–203. doi:10.1097/SLA.0b013e3181a334f9
5. Nsadi B, Gilson N, Pire E et al (2011) Consequences of pneumoperitoneum on liver ischemia during laparoscopic portal triad clamping in a swine model. J Surg Res 166(1):e35–43
6. Takagi S (1998) Hepatic and portal vein blood flow during carbon dioxide pneumoperitoneum for laparoscopic hepatectomy. Surg Endosc 12(5):427–431
7. Aldrighetti L, Pulitano C, Arru M et al (2008) Ultrasonic-mediated laparoscopic liver transection. Am J Surg 195(2):270–272. doi:10.1016/j.amjsurg.2007.02.022
8. Troisi RI, Van Huysse J, Berrevoet F et al (2011) Evolution of laparoscopic left lateral sectionectomy without the Pringle maneuver: through resection of benign and malignant tumors to living liver donation. Surg Endosc 25(1):79–87. doi:10.1007/s00464-010-1133-8
9. Kaneko H, Otsuka Y, Tsuchiya M et al (2008) Application of devices for safe laparoscopic hepatectomy. HPB (Oxford) 10(4):219–224. doi:10.1080/13651820802166831
10. Slakey DP (2008) Laparoscopic liver resection using a bipolar vessel-sealing device: LigaSure. HPB (Oxford) 10(4):253–255. doi:10.1080/13651820802166880
11. Kim FJ, Chammas MF Jr, Gewehr E et al (2008) Temperature safety profile of laparoscopic devices: Harmonic ACE (ACE), LigaSure V (LV), and plasma trisector (PT). Surg Endosc 22(6):1464–1469. doi:10.1007/s00464-007-9650-9
12. Pai M, Navarra G, Ayav A et al (2008) Laparoscopic Habib 4X: a bipolar radiofrequency device for bloodless laparoscopic liver resection. HPB (Oxford) 10(4):261–264. doi:10.1080/13651820802167862
13. McGahan JP, Khatri VP (2008) Imaging findings after liver resection by using radiofrequency parenchymal coagulation devices: initial experiences. Radiology 247(3):896–902. doi:10.1148/radiol.2473070949
14. Vollmer CM, Dixon E, Sahajpal A et al (2006) Water-jet dissection for parenchymal division during hepatectomy. HPB (Oxford) 8(5):377–385. doi:10.1080/13651820600839449
15. Nguyen KT, Gamblin TC, Geller DA (2009) World review of laparoscopic liver resection-2,804 patients. Ann Surg 250(5):831–841. doi:10.1097/SLA.0b013e3181b0c4df
16. Cuschieri A (2005) Laparoscopic liver resections. J Minim Access Surg 1(3):99–109. doi:10.4103/0972-9941.18993
17. Patriti A, Ceccarelli G, Bartoli A, Casciola L (2011) Extracorporeal Pringle maneuver in robot-assisted liver surgery. Surg Laparosc Endosc Percutan Tech 21(5):e242–244. doi:10.1097/SLE.0b013e31822d7fb4
18. Saif R, Jacob M, Robinson S et al (2011) Laparoscopic Pringle's manoeuvre for liver resection?: how I do it. Minim Invasive Ther Allied Technol 20(6):365–368
19. Rotellar F, Pardo F, Bueno A et al (2011) Extracorporeal tourniquet method for intermittent hepatic pedicle clamping during laparoscopic liver surgery: an easy, cheap, and effective technique. Langenbecks Arch Surg. doi:10.1007/s00423-011-0887-3
20. Casciola L, Patriti A, Ceccarelli G et al (2011) Robot-assisted parenchymal-sparing liver surgery including lesions located in the posterosuperior segments. Surg Endosc 25(12):3815–3824. doi:10.1007/s00464-011-1796-9
21. Lee MR, Kim YH, Roh YH et al (2011) Lessons learned from 100 initial cases of laparoscopic liver surgery. J Korean Surg Soc 80:334-341. Epub 2011, May 6

The Louisville Consensus Conference: Conclusions and Perspectives

10

Giulio Belli, Luigi Cioffi, Andrea Belli and Corrado Fantini

10.1 Introduction

After an initial period of scepticism, especially concerning technical and oncological problems, laparoscopic liver surgery (LLS) has become a feasible and safe technique. Over the past decade, the minimally invasive approach has been used increasingly in to manage hepatic diseases, showing that this technique in liver surgery, despite the technical challenges, reduces operative blood loss and results in fewer early postoperative complications, less postoperative analgesic drug consumption, and shorter hospital stay, with an oncologic clearance and a survival rate similar to that of open surgery [1–5]. Therefore, the place of laparoscopy in liver surgery is increasing, and many types of liver resections, including major hepatectomies, are now performed by laparoscopy in specialized centers [6–9]. Nevertheless, no international consensus on laparoscopic surgical management of liver lesions has been published, and no worldwide criteria exist for the indications for minimally invasive liver resection. Thus, there are no evidence-based criteria assisting the surgeon with management strategies for the laparoscopic treatment of liver tumors. For example, how large should the lesion be? Where should the lesion be located? Should we modify the well-accepted surgical indication only because we can perform a liver resection using a minimally invasive approach? Should we perform laparoscopic liver resection (LLR) only for benign lesions, or can we resect even malignant lesions by laparoscopy?

In view of these factors, a project was launched to prepare an international position on LLS. A working group consisting of 45 experts in hepatobiliary surgery (Fig. 10.1) on 7–8 November 2008 was invited to participate in a con-

G. Belli (✉)
General and HPB Surgery, Loreto Nuovo Hospital, Naples, Italy
e-mail: giubelli@unina.it

F. Calise and G. Casciola (eds.), *Minimally Invasive Surgery of the Liver,*
Updates in Surgery
DOI: 10.1007/978-88-470-2664-3_10, © Springer-Verlag Italia 2013

Fig. 10.1 Louisville Consensus Conference working group consisting of 45 experts in hepatobiliary surgery from five continents
Joseph F. Buell[1], Daniel Cherqui[2], David Geller[3], Nicholas O'Rourke[4], David Iannitti[5], Ibrahim Dagher[6], Alan J. Koffron[7], Mark Thomas[8], Brice Gayet[9], Ho Seong Han[10], Go Wakabayashi[11], Giulio Belli[12], Hironori Kaneko[13], Chen-Guo Ker[14], Olivier Scatton[15], Alexis Laurent[2], Eddie K. Abdalla[16], Prosanto Chaudhury[17], Erik Dutson[18], Clark Gamblin[3], Michael D'Angelica[19], David Nagorney[20], Giuliano Testa[21], Daniel Labow[22], Derrik Manas[23], Ronnie T Poon[24], Heidi Nelson[20], Robert Martin[1], Bryan Clary[25], C. Wright Pinson[26], John Martinie[5], Jean-Nicolas Vauthey[16], Robert Goldstein[27], Sasan Roayaie[22], David Barlett[3], Joseph Espat[28], Michael Abecassis[29], Myrddin Rees[30], Yuman Fong[19], Kelly M. McMasters[1], Christoph Broelsch[31], Ron Busuttil[19], Jacques Belghiti[32], Steven Strasberg[33], and Ravi S. Chari[22].

sensus conference convened in Louisville, KY, USA; more than 300 attendees were present from five continents. Specific areas of discussion included indications for surgery, patient selection, surgical techniques, complications, patient safety, and surgeon training. The final objective was to summarize the current world position on LLS [10].

During the conference, it was established that only three terms should be used to describe LLR – pure laparoscopy, hand-assisted laparoscopy, and the hybrid technique – in order to standardize terminology. Concerning the role of major laparoscopic hepatectomy, all experts agreed that major LLR have been performed in highly specialized centers with safety and efficacy equalling open surgery, and major LLS should proceed only when a reported degree of safety is published that is equivalent to open liver surgery.

Patient selection was considered the key point of discussion. Although most types of liver resections can be performed laparoscopically, the technique

[1]University of Louisville; [2]Henri Mondor Hospital; [3]University of Pittsburgh; [4]Royal Brisbane Hospital; [5]Carolinas Medical Center; [6]Antoine Beclere Hospital; [7]William Beaumont; [8]Cincinnati, Ohio; [9]Institut Mutualiste; [10]SNU, Seoul Korea; [11] Iwate Medical University; [12]Loreto Nuovo Hospital; [13]Toho University; [14]Kaohsiung Medical University Hospital; [15]Hospital Cochin; [16]The University of Texas, M.D. Anderson Cancer Center; [17]McGill University; [18]University of California-Los Angeles; [19]Memorial Sloan Kettering; [20]Mayo Clinic; [21]University of Chicago; [22]Mount Sinai Medical Center; [23]NHS, Newcastle, England; [24]Queen Mary Hospital; [25]Duke University; [26]Vanderbilt University Medical Center; [27]Baylor, Houston; [28]Roger Williams Medical Center; [29]Northwestern University; [30]North Hampshire Hospital; [31]Essen, Germany; [32]Hospital Beaujon; [33]Washington University

should be limited to experienced surgeons already facile with more limited laparoscopic resections. It is important to stress that left lateral sectionectomy was found by all participants to be the most straightforward moderately sized laparoscopic procedure, and there was agreement that in experienced hands, laparoscopy should be the standard approach for this particular operation [11]. An important aspect debated was when and how a patient should be converted to an open procedure. There was general acceptance that conversion should not be viewed as a complication and should be performed for lack of case progress and/or patient safety. In case of bleeding, efforts should be made to control hemorrhage laparoscopically before converting to hand assist or laparotomy, as significant time and blood loss can occur during the process of conversion. Another important and widely discussed topic was indications for resection of asymptomatic benign hepatic lesions. All participants agreed that indications should not be widened. On the other hand, traditional indications should be respected, even if the laparoscopic approach is less invasive that an open one [12, 13]. Unroofing simple hepatic cysts should not be considered a liver resection and should not be included in the analyses of LLR [24–30]. The role of LLS for primary and metastatic liver tumors was widely discussed. The major concern with LLR for colorectal metastases included a potential increase in positive resection margins and failure to detect occult lesions, although there are no trials as yet that clearly demonstrate that laparoscopic hepatic resections have equivalent long-term outcomes to open hepatic resection. There was agreement that surgeons with experience in both open and laparoscopic approaches to liver resection are permitted to perform LLS for metastases [14]. Hepatocellular carcinoma has been considered a good indication for LLS, even in patients with a cirrhotic liver. There was general agreement that laparoscopic resection of small hepatocellular cancers in a cirrhotic liver is feasible and safe in centers with experienced surgeons, with reduced morbidity compared with open resection, especially with reduced occurrence of postoperative ascites. Thus, follow-up data suggest that the long-term oncological outcome has not been compromised by the laparoscopic approach compared with open resection [17–21].

The role of LLS in a transplant setting was considered differently. The consideration was that laparoscopic live-donor hepatectomy is the most controversial application for LLS and should only proceed in the confines of a worldwide registry [22, 23]. All experts agreed that registries are effective, and in some cases more effective, than randomized control trials in detecting and recording uncommon but severe negative events [24]. All experts also felt that there may be a role for a prospective randomized trial, but study population and length of time to perform the trial may make this impracticable. There was consensus that understanding the role and safety of LLS would be advanced through a cooperative patient registry.

Finally, there was agreement that LLS should be initiated only in centers in which the combined expertise in laparoscopic and hepatic surgery exists. National and international societies, as well as governing boards, should become involved in the goal of establishing training standards and credentialing to ensure a high and consistent clinical outcome.

10.2 Perspectives

Only 3 years have passed since the Louisville Consensus Conference, but many changes have already occurred in clinical practice. LLS is an evolving field, and a new (r)evolution is likely to be realized in the near future. In fact, even if at the moment several centers worldwide are progressively introducing the laparoscopic approach in liver surgery, some pioneering centers (many of those involved in the Louisville Consensus Conference) are already pushing the "traditional" indications for laparoscopy. Progressively more patients are being treated with a laparoscopic approach, even those affected by liver lesions located in the posterolateral segments [25, 26], which were generally considered as a contraindication to laparoscopy. Robotic surgery is gaining popularity and possibly will help standardize LLS, making it safer [27]. Also, the position on major laparoscopic hepatectomy is slowly moving toward new perspectives. The International Consensus Group for Laparoscopic Liver Surgery is, in fact, evaluating the potential role of laparoscopic left hemihepatectomy as a standard of care [28]. In conclusion, LLS, after an initially slow diffusion, is now a strong reality and is progressively offered more and more as a therapeutic option in many centers worldwide. It is likely that the introduction of new dedicated technologies, together with the advent of a new generation of hepatic surgeons trained in advanced laparoscopy, will result in a dramatic increase in the percentage of patients treated for a liver lesion by a laparoscopic approach.

References

1. Gagner M, Rheault M, Dubuc J (1992) Laparoscopic partial hepatectomy for liver tumor (abstract). Surg Endosc 6:99
2. Azagra JS, Goergen M, Gilbart E et al (1996) Laparoscopic anatomical (hepatic) left lateral segmentectomy-technical aspects. Surg Endosc 10:758–761
3. Kaneko H, Takagi S, Shiba T (1996) Laparoscopic partial hepatectomy and left lateral segmentectomy: technique and results of a clinical series. Surgery 120:468–475
4. Huscher CG, Lirici MM, Chiodini S (1998) Laparoscopic liver resections. Semin Laparosc Surg 5:204–210
5. Cherqui D, Husson E, Hammoud R et al (2000) Laparoscopic liver resections: a feasibility study in 30 patients. Ann Surg 232:753–762
6. Koffron AJ, Auffenberg G, Kung R et al (2007) Evaluation of 300 minimally invasive liver resections at a single institution: less is more. Ann Surg 246:385–392
7. O'Rourke N, Fielding G (2004) Laparoscopic right hepatectomy: surgical technique. J Gastrointest Surg 8:213–216
8. Gigot JF, Glineur D, Azagra JS et al (2002) Laparoscopic liver resection for malignant liver tumors: preliminary results of a multicenter European study. Ann Surg 236:90–97
9. Vibert E, Perniceni T, Levard H et al (2006) Laparoscopic liver resection. Br J Surg 93:67–72

10. Buell JF, Cherqui D, Geller DA et al (2009) World Consensus Conference on Laparoscopic Surgery. Ann Surg 250:825–830
11. Lesurtel M, Cherqui D, Laurent A et al (2003) Laparoscopic versus open left lateral hepatic lobectomy: a case-control study. J Am Coll Surg 196:236–242
12. Koffron A, Geller D, Gamblin TC et al (2006) Laparoscopic liver surgery: shifting the management of liver tumors. Hepatology 44:1694–1700
13. Buell JF, Koffron AJ, Thomas MJ et al (2005) Laparoscopic liver resection. J Am Coll Surg 200:472– 480
14. O'Rourke N, Shaw I, Nathanson L et al (2004) Laparoscopic resection of hepatic colorectal metastases. HPB (Oxford) 6:230–235
15. Kaneko H, Otsuka Y, Tsuchiya M et al (2008) Application of devices for safe laparoscopic hepatectomy. HPB (Oxford) 10:219–224
16. Aloia TA, Vauthey JN, Loyer EM et al (2006) Solitary colorectal liver metastasis: resection determines outcome. Arch Surg 141:460–466
17. Laurent A, Cherqui D, Lesurtel M et al (2003) Laparoscopic liver resection for subcapsular hepatocellular carcinoma complicating chronic liver disease. Arch Surg 138:763–769
18. Belli G, Fantini C, D'Agostino A et al (2007) Laparoscopic versus open liver resection for hepatocellular carcinoma in patients with histologically proven cirrhosis: short- and middleterm results. Surg Endosc 21:2004–2011
19. Dagher I, Proske JM, Carloni A et al (2007) Laparoscopic liver resection: results for 70 patients. Surg Endosc 21:619–624
20. Kaneko H (2005) Laparoscopic hepatectomy: indications and outcomes. J Hepatobiliary PancreatSurg 12:438–443
21. Chen HY, Juan CC, Ker CG (2008) Laparoscopic liver surgery for patients with hepatocellular carcinoma. Ann Surg Oncol 15:800–806
22. Cherqui D, Soubrane O, Husson E et al (2002) Laparoscopic living donor hepatectomy for liver transplantation in children. Lancet 359:392–399
23. Soubrane O, Cherqui D, Scatton O et al (2006) Laparoscopic left lateral sectionectomy in living donors: safety and reproducibility of the technique in a single center. Ann Surg 244:815–820
24. Barone JE, Lincer RM (1991) A prospective analysis of 1518 laparoscopic cholecystectomies: the Southern Surgeons Club. N Engl J Med 324:1073–1078
25. Cho JY, Han HS, Yoon YS et al (2008) Feasibility of laparoscopic liver resection for tumors located in the posterosuperior segments of the liver, with a special reference to overcoming current limitations on tumor location. Surgery 144:32–38
26. Kazaryan A, Rosok BI, Marangos Pavlik I et al (2011) Comparative evaluation of laparoscopic liver resection for posterosuperior and anterolateral segments. Surg Endosc 25:3881–3889
27. Casciola L, Patriti A, Ceccarelli G et al (2011) Robot-assisted parenchymal-sparing liver surgery including lesions located in the posterosuperior segments. Surg End 25:3815–3824
28. Belli G, Gayet B, Han HS et al (2012) International Consensus Group for Laparoscopic Liver Surgery Laparoscopic Hemi-Hepatectomy: a consideration for acceptance as standard ofcare. New Orleans, LA, 10th World Congress of IHPBA, Paris, July 1-5, 2012

Indications to Surgery: Laparoscopic or Robotic Approach

11

Luciano Casciola and Alberto Patriti

11.1 Introduction

Minimally invasive liver surgery is growing worldwide. To date, more than 3,000 cases of laparoscopic liver resections (LLR) have been published, and the interest in robot-assisted resections (RAR) is rising [1–9]. Indications for LLR are similar to those of open liver resections, as recently stated in the Louisville Statement Consensus Conference [2]. LLR was initially indicated for benign and peripherally located lesions, suggesting a concern about the safety and oncologic effectiveness of these procedures. In the last few years, the number of laparoscopic complex liver resections has increased, and even major hepatectomies and segmentectomies of the posterosuperior (PS) segments have been successfully performed [10, 11]. Recently, robotics was introduced into general surgery with the aim of overcoming some of the limitations associated with traditional laparoscopy, thus providing greater manueverability with a set of articulated instruments and a tridimensional vision [12, 13]. The first application of robotics in liver surgery dates back to 2008, when Choi et al. published their first series of four left lateral sectionectomies [14]. After this report, a few case series were published showing the feasibility of robot-assisted minor and major liver resections [5–8, 15–17] and – to date – there is only one prospective comparative study [17]: Berber at al. compared nine patients undergoing RAR matched with 23 patients undergoing LLR. However, the study has some limitations, including small sample size. One inclusion criteria was tumor location in the peripheral liver segments, precluding arriving at definitive conclusions on possible advantages of RAR in situations considered demanding with a laparoscopic approach. Some tech-

A. Patriti (✉)
Department of Surgery, Division of General, Minimally Invasive and Robotic Surgery,
ASL3 Umbria, San Matteo degli Infermi Hospital, Spoleto, Italy
e-mail: albertopatriti@gmail.com

F. Calise and G. Casciola (eds.), *Minimally Invasive Surgery of the Liver*,
Updates in Surgery
DOI: 10.1007/978-88-470-2664-3_11, © Springer-Verlag Italia 2013

nical aspects preclude investigating the real role of the da Vinci robotic system in liver resection. The on-table assistant carried out parenchymal division using the Harmonic Scalpel, making it impossible to analyze differences related to the use of the robotic endowristed instruments. Therefore, as of yet, we have no high level of evidence in favor of RAR over LLR, and the associated considerations of the authors regarding this issue come from published case series and their personal experiences.

11.2 Limitations of Laparoscopic Liver Surgery

Theoretically, the only limitation to laparoscopic liver surgery is surgeon ability and preferences. In fact, reports show that the limitations of LLR described in the Louisville Statement Consensus Conference can be overcome. The first limitation is restricted lesions located in PS segments (1, 7, 8, 4a). Most laparoscopic liver surgeons prefer a right hepatectomy to remove a lesion in segments 7 and 8, but Cho et al. showed that this was possible even with a parenchyma-preserving resection [11]. Nevertheless, LLR of lesions located in the PS segments are associated with significantly longer operative time and intraoperative transfusion compared with anterolateral segment resections (segments 2, 3, 4b, 5, 6) [11]. Tumors that are either large (>5 cm), central, multiple, bilateral, or with connections with the hilum, major hepatic veins, or inferior vena cava are at the moment considered not good candidates for a laparoscopic approach. Also, for these particular cases, reports show that in experience hands, LLR can be safely performed: Yoon et al. reported that LLR for multiple hepatocellular carcinoma lesions does not increase operative time, intraoperative transfusion rate, length of postoperative hospital stay, or postoperative complications [18]. A major LLR can be an option for lesions >5 cm or with connections with one portal pedicle or one hepatic vein, and studies report comparable early and oncologic results between laparoscopic and open major resections [19]. Conversely, these situations are a real limitation for surgeons who do not consider a major hepatectomy an effective option when the laparoscopic approach would change the surgical strategy from that adopted in open surgery. Finally, biliary reconstruction after an extended right hepatectomy is an objective limitation of laparoscopic surgery that does not allow a rage of movements fine enough to safely complete a bilioenteric anastomosis.

11.3 Possible Advantages of Robot-assisted Surgery

To date, only a few reports have focused on robotics in liver surgery. The number of patients involved is rather small, and most surgeries are carried out on a retrospective basis [5, 7, 8, 12, 19]. Potential advantages of robotic assistance arising from these studies may include facilitating complex reconstructions (i.e., biliary and vascular anastomoses) [5, 6, 20] and parenchyma-preserving resections of lesions located in PS segments [8]. Giulianotti et al. demonstrated that robot-assisted major hepatectomies are safe and feasible, even when a biliary reconstruction is required, as in the case of hilar cholangiocarcinoma [5, 6]. However, there are no studies comparing the outcome of robot-assisted and major laparoscopic hepatectomies. In a study at our institution, where every attempt is made to perform parenchyma-preserving surgery, we show the possibility allowed by the robot to preserve liver parenchyma, even in cases of tumor location in the PS segments or close to a major liver vessel. In 23 cases, ten patients (47.8%) had liver nodules in the PS segments; in three cases, the tumour was connected with a portal branch, in two cases with a hepatic vein, and in one case with both vascular structures. No major hepatectomies were carried out in the manner of the surgical plan we would have followed in the open setting [8].

11.4 Conclusions

Even if randomized controlled studies are still absent, RAR could be an attractive option for surgeons who wish to perform a minimally invasive parenchyma-preserving surgery, even in cases of lesions close to the main liver vessels or located in segments 7, 8, and 1. As there are no randomized studies demonstrating the superiority of RAR over LLR in major hepatectomies and antero-lateral-segment resections, the two approaches can be considered analogous. There is some evidence from Giulianotti et al.'s study that applications of robotics in major hepatectomies could improve two phases of liver resection: hilar and hepatocaval confluence dissection. This aspect provides the basis for prospective studies on the da Vinci system application for liver resections requiring meticulous vascular dissection and reconstruction [5]. Nevertheless, if a RAR program is planned, even resections in the anterior segments should be considered in the first phase of the learning curve in order for the surgeon and all staff members to gain expertise and subsequently safely approach PS segments and complex major hepatectomies using robotic surgery techniques.

References

1. Ardito F, Tayar C, Laurent A et al (2007) Laparoscopic liver resection for benign disease. Arch Surg 142(12):1188–1193. doi:142/12/1188 [pii] 10.1001/archsurg.142.12.1188
2. Buell JF, Cherqui D, Geller DA et al (2009) The international position on laparoscopic liver surgery: The Louisville Statement, 2008. Ann Surg 250(5):825–830
3. Buell JF, Thomas MT, Rudich S et al (2008) Experience with more than 500 minimally invasive hepatic procedures. Ann Surg 248 (3):475–486. doi:10.1097/SLA.0b013e318185e647 00000658-200809000-00014 [pii
4. Nguyen KT, Laurent A, Dagher I et al (2009) Minimally invasive liver resection for metastatic colorectal cancer: a multi-institutional, international report of safety, feasibility, and early outcomes. Ann Surg 250(5):842–848. doi:10.1097/SLA.0b013e3181bc789c

5. Giulianotti PC, Coratti A, Sbrana F et al (2011) Robotic liver surgery: results for 70 resections. Surgery 149(1):29–39. doi:10.1016/j.surg.2010.04.002
6. Giulianotti PC, Sbrana F, Coratti A et al (2011) Totally robotic right hepatectomy: surgical technique and outcomes. Arch Surg 146(7):844–850. doi:10.1001/archsurg.2011.145
7. Patriti A, Ceccarelli G, Bartoli A et al (2009) Laparoscopic and robot-assisted one-stage resection of colorectal cancer with synchronous liver metastases: a pilot study. J Hepatobiliary Pancreat Surg 16(4):450–457. doi:10.1007/s00534-009-0073-y
8. Casciola L, Patriti A, Ceccarelli G et al (2011) Robot-assisted parenchymal-sparing liver surgery including lesions located in the posterosuperior segments. Surg Endosc 25(12):3815–3824. doi:10.1007/s00464-011-1796-9
9. Descottes B, Glineur D, Lachachi F et al (2003) Laparoscopic liver resection of benign liver tumors. Surg Endosc 17(1):23–30. doi:10.1007/s00464-002-9047-8
10. Nguyen KT, Gamblin TC, Geller DA (2009) World review of laparoscopic liver resection-2,804 patients. Ann Surg 250(5):831–841. doi:10.1097/SLA.0b013e3181b0c4df
11. Cho JY, Han HS, Yoon YS, Shin SH (2008) Feasibility of laparoscopic liver resection for tumors located in the posterosuperior segments of the liver, with a special reference to overcoming current limitations on tumor location. Surgery 144(1):32–38. doi:10.1016/j.surg.2008.03.020
12. Giulianotti PC, Coratti A, Angelini M et al (2003) Robotics in general surgery: personal experience in a large community hospital. Arch Surg 138(7):777–784. doi:10.1001/archsurg.138.7.777
13. Giulianotti PC, Addeo P, Buchs NC et al (2011) Robotic extended pancreatectomy with vascular resection for locally advanced pancreatic tumors. Pancreas 40(8):1264–1270. doi:10.1097/MPA.0b013e318220e3a4
14. Choi SB, Park JS, Kim JK et al (2008) Early experiences of robotic-assisted laparoscopic liver resection. Yonsei Med J 49(4):632–638. doi:10.3349/ymj.2008.49.4.632
15. Giulianotti PC, Tzvetanov I, Jeon H et al (2012) Robot-assisted right lobe donor hepatectomy. Transpl Int 25(1):e5–9. doi:10.1111/j.1432-2277.2011.01373.x
16. Ji WB, Wang HG, Zhao ZM et al (2011) Robotic-assisted laparoscopic anatomic hepatectomy in China: initial experience. Ann Surg 253(2):342–348. doi:10.1097/SLA.0b013e3181ff4601
17. Berber E, Akyildiz HY, Aucejo F et al (2010) Robotic versus laparoscopic resection of liver tumours. HPB (Oxford) 12(8):583–586. doi:10.1111/j.1477-2574.2010.00234.x
18. Yoon YS, Han HS, Cho JY et al (2012) Laparoscopic approach for treatment of multiple hepatocellular carcinomas. Surg Endosc. doi:10.1007/s00464-012-2304-6
19. Topal H, Tiek J, Aerts R, Topal B (2012) Outcome of laparoscopic major liver resection for colorectal metastases. Surg Endosc. doi:10.1007/s00464-012-2209-4
20. Giulianotti PC, Sbrana F, Bianco FM, Addeo P (2010) Robot-assisted laparoscopic extended right hepatectomy with biliary reconstruction. J Laparoendosc Adv Surg Techn A 20(2):159–163. doi:10.1089/lap.2009.0383

Diagnostic Laparoscopy and Allied Technologies

12

Roberto Santambrogio and Enrico Opocher

12.1 Introduction

The concept of using ultrasound (US) through a laparoscopic access was first described by Yamakawa et al. in 1958 [1], but only since the end of the 1980s were laparoscopic US probes introduced in the clinical practice. Available data indicate that laparoscopy with laparoscopic US (LUS) provides information similar to that obtained by intraoperative US and can identify lesions that are too small to be visible by preoperative imaging techniques. Furthermore, LUS also allows performance of US-guided biopsy or interstitial therapies such as ethanol injection, cryoablation, or radiofrequency thermal ablation in the same session.

12.2 Indications

12.2.1 Hepatocellular Carcinoma

LUS avoids unnecessary laparotomy in 12–39% of cases (Table 12.1) [2–10]. However, it is still not known whether a routine or a selective laparoscopy approach should be adopted, and selection criteria are not well defined. On the other hand, with the increasing use of the laparoscopic treatment approach to hepatocellular carcinoma (HCC), the use of LUS will not be limited to avoiding unnecessary laparotomy for patients with inoperable HCC: LUS staging will allow us to choose the optimal treatment strategy for patients with operable or inoperable HCC [9]. In our experience [10], in patients with a single

R. Santambrogio (✉)
Hepatobiliary, Pancreatic and Digestive Surgery,
San Paolo Hospital, University of Milan, Milan, Italy
e-mail: rsantambrogio@mclink.it

F. Calise and G. Casciola (eds.), *Minimally Invasive Surgery of the Liver*,
Updates in Surgery
DOI: 10.1007/978-88-470-2664-3_12, © Springer-Verlag Italia 2013

Table 12.1 Series of laparoscopic ultrasound staging for hepatocellular carcinoma (HCC) and metastases (Meta)

Authors	Year	No. patients	HCC	Information
John et al [2]	'94	43	15%	42% better staging
Tandan et al [3]	'97	31	10%	10% no LPT
Ido et al [4]	'99	186	100%	12% better staging
Lo et al [5]	'00	198	100%	16% no LPT
D'Angelica et al [6]	'03	401	8%	20% no LPT
De Castro et al [7]	'04	76	39%	23% no LPT
Klegar et al [8]	'05	20	100%	42% better staging
Lai et al [9]	'08	119	100%	39% no LPT
Montorsi et al [10]	'01	132	100%	25% better staging 14% no LPT
Authors	**Year**	**No. patients**	**Meta**	**Information**
Rahusen et al [11]	'99	50	100%	38% no LPT
Jarnagin et al [12]	'01	103	100%	CRS ≤2: 12% better staging CRS >2: 42% better staging
Koea et al [13]	'04	57	100%	5% no LPT
Thaler et al [14]	'05	136	100%	25% no LPT
Khan et al [15]	'07	210	100%	8% no LPT
Mann et al [16]	'07	200	100%	CRS ≤2: 6% better staging CRS >2: 38% better staging
Pilkington et al [17]	'07	77	100%	21% no LPT
Li Destri et al [18]	'08	43	100%	CRS ≤2: 12% no LPT CRS >2: 39% no LPT
Shah et al [19]	'10	79	100%	CRS ≤2: 7% no LPT CRS >2: 24% no LPT

LPT, laparotomy; *CRS*, clinical scoring system [12].

HCC nodule at preoperative imaging, there was a 19% risk of finding a new lesion compared with 29% for patients with more than one nodule. Even if no exhaustive criteria exist, we do not usually perform LUS staging if patients who are candidates for hepatic resection have a single nodule on preoperative imaging studies. In these cases, if a new nodule is detected by intraoperative US, it can be treated with radiofrequency or alcohol injection at that time.

12.2.2 Liver Metastasis

Publications on the role of LUS in potentially resectable colorectal liver metastasis (CRLM) suggest a limited value, mainly because of the low sensitivity and heterogeneity caused by varied study selection criteria and institutional practice, evolution of techniques and technologies, and broadening of criteria used for resection (Table 12.1) [11–19]. A more selective approach is therefore justified, with large, prospective, multicenter studies evaluating the role of LUS in patients with resectable CRLM, accompanied by well-defined risk stratification and appropriate algorithms incorporating intention to treat using all available modalities rather than using pure surgical evaluation alone [20].

12.3 Laparoscopic Ultrasound

12.3.1 Materials

12.3.1.1 Ultrasound Equipment

We used a US machine connected to an LUS probe with a flexible tip, 10 mm in diameter and 50 cm in length. A 5- to 7.5-MHz linear-array transducer was sidemounted near the tip of the shaft. The length of the transducer surface was 38 mm, which produced an image footprint of approximately 4 cm in length and 6 cm in depth. Recently, we also used a microconvex probe that permits application of intravenously administered US contrast agents during LUS of the liver. The addition of contrast enhancement during intraoperative US improves image conspicuity and allows correct diagnosis of new malignant nodules and ablation efficacy in hepatic tumors and oncologic margin outcomes.

12.3.1.2 Tru-Cut Biopsy Needle

It is useful to use a cutting needle with an automatic trigger mechanism to enable holding the probe with one hand and the needle with the other. Because of the presence of the pneumoperitoneum, which separates the surface of the liver from the abdominal wall, the use of longer needles (25–27 cm) could be necessary for lesions localized in the posterior segments or in the highest part of the liver (segments 4a and 8).

12.3.2 Laparoscopic Technique

Usually it is sufficient to use two 10-mm trocar accesses. However, patient positioning is dependent on the location of the hepatic lesions to be treated (Fig. 12.1). Generally, patient position on the operating table is supine with the left arm extended: the surgeon stands either on the right side or between the patient's legs (Fig. 12.1a, b). Patients with tumors in segments 6 and 7 can be placed in either an oblique position with the right side elevated up to 45° or left decubitus position with the right arm elevated and across the chest and the surgeon at either the right or the left side of the patient (Fig. 12.1c).

The location of LUS probe introduction is limited by trocar locations: the umbilical port can be made for laparoscopic exploration; the second trocar site for LUS can be selected depending upon both preoperative imaging evaluation and intraoperative conditions as determined by laparoscopy.

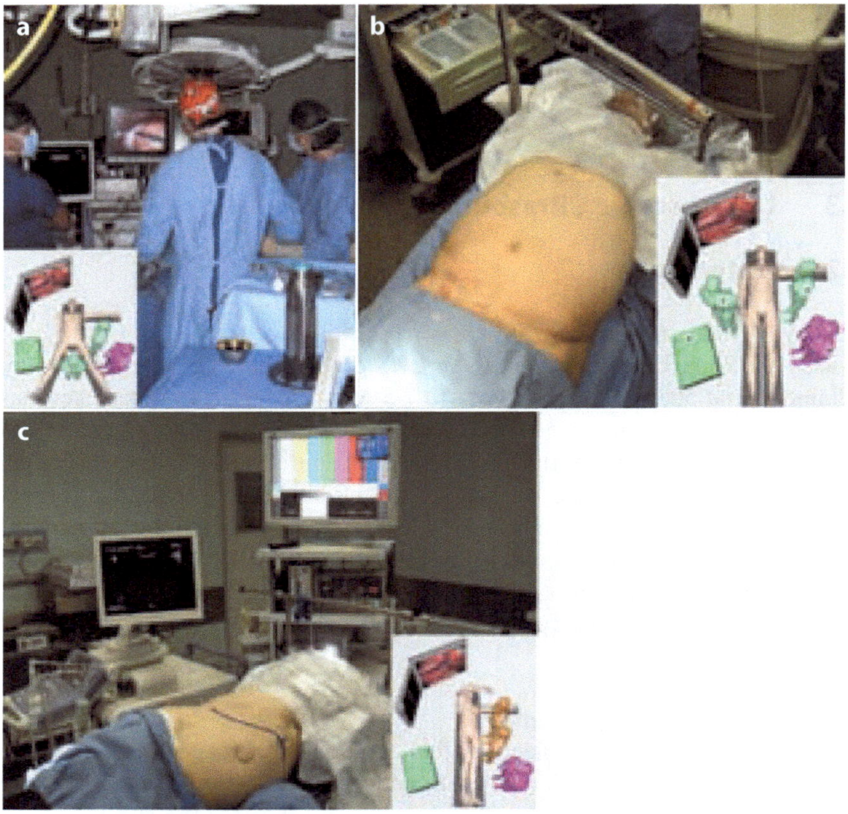

Fig. 12.1a–c Patient and operating room positions for laparoscopic radiofrequency ablation (LRFA). **a** Supine position with patient's legs abducted (surgeon stands between the legs) for lesions in segments 2, 3, and 4; **b** supine position with patient's legs adducted (surgeon stands on the patient's right side) for lesions in segments 4, 5, and 8; **c** oblique position with the patient's right side elevated up to 45°, or left decubitus position with the right arm elevated and across the chest (surgeon stands at the left side of the patient) for lesions in segments 6 and 7

Exploring the liver parenchyma can usually be performed with a direct-contact technique thanks to the natural humidity of the liver surface, which allows good acoustic contact with the transducer. However, the dome of the liver may be difficult to examine due to the lack of adequate contact between the probe and the convex liver surface. This can be overcome by instilling saline solution and scanning the highest part of the organ through the fluid (water-immersion method). Furthermore, in some instances, it is useful to decrease the pneumoperitoneum to 6–8 mmHg, favoring a correct angle of transducer contact with the liver surface. The entire liver is initially screened, and the size of each tumor is measured by LUS and described according to the Couinaud classification of liver anatomy [21].

12.4 Laparoscopic Radiofrequency Ablation

12.4.1 Materials

12.4.1.1 Radiofrequency (RFA)/Microwave (MW) Machines and Electrode/Antenna

We used a dual-ablation system that has both an MW and an RFA energy generator in the same hardware. For RFA technology, we prefer a a 17-gauge, internally cooled electrode with an exposed tip length of 3 cm and shaft length of 250 mm. For MW technology, we use a 14-gauge interstitial antenna with a shaft length of 270 mm. MW offers all the benefits of RFA energy for thermal ablation but is not as dependent on tissue properties and has the ability to heat faster in a larger volume. Thus, MW is less susceptible to perfusion or heat sinks and may be able to penetrate deeper into low-conductivity materials.

12.4.2 Laparoscopic Technique

After lesions have been identified, the therapeutic electrode can be accurately inserted into the tumor. When dealing with lesions localized in segment 1 or in the posterior segments of an enlarged liver, a longer laparoscopic electrode could be necessary (27 cm). The electrode must pass the abdominal wall and the pneumoperitoneum space prior to reaching the liver surface, meeting a fulcrum that increases the difficulty of moving the tip of the electrode. This problem can be overcome with the aid of either a 2-mm trocar or a 14-gauge cannula needle placed through the abdominal wall and the electrode then placed through this sheath.

A LUS-guided interventional procedure [22] can be successfully performed if the following ideal working conditions are fulfilled: (1) the lesion is well visible – the US probe must be oriented on the liver surface to display the largest diameter of the entire lesion; (2) the electrode must be positioned near the LUS probe transducer in order to introduce it slightly oblique to the transducer and with an acute angle to the axis of the LUS probe. In fact, after inserting the electrode into the liver parenchyma, slight rotation of the probe can identify the mark of the electrode and guide its tip into the lesion. If electrode access is too acute with respect to the liver surface, it is well possible that the electrode remains superficial and parallel to the long axis of the probe without reaching the lesion. For lesions located in posterior segments, it is necessary to insert the electrode on the liver surface further than the lesion: in this case, the transducer cannot visualize contemporarily the tumor and the electrode tip. On the other hand, because any LUS-guided interventional procedure is totally freehand, a puncture adapter has been proposed: incorporation of the biopsy channel into the shaft of the US probe permits accurate electrode placement only in lesions seated in some areas of the liver [23]. Laparoscopic RFA has a

Table 12.2 Results of laparoscopic radiofrequency ablation

Author	Publication year	No. patients	Follow-up (months)	Total necrosis	Intrahepatic recurrence	Local recurrences
Ido et al [24]	1997	15	15.9	100%	13%	7%
Ito et al [25]	1999	14	NA	100%	28.6%	28.6%
Seki et al [26]	2000	24	23.5	92%	45.8%	12.5%
Podnos et al [27]	2001	12	7.4	100%	8%	8%
Noguchi et al [28]	2003	51	NA	98%	53%	NA
Hsieh et al [29]	2004	40	12.5	NA	47.5%	NA
Casaccia et al [30]	2008	24	NA	90%	21%	NA
Ballem et al [31]	2008	104	NA	NA	30%[a]	NA
Sakaguchi et al[b] [32]	2009	391	NA	NA	NA	8.2%
Simo et al [33]	2011	39	3	97%	15%	NA
Personal	2012	354	26.3	91%	56%	23%

[a]Disease-free recurrence; [b]multicenter study.

limited diffusion due to the technical difficulties of the procedure, but it seems to produce promising results (Table 12.2) [24–33].

12.4.3 Technical Variants: Intrahepatic Vascular Occlusion

This approach determines an ischemic area surrounding the lesion, thus increasing the necrosis volume [34]. This effect could reduce the risk of immediate therapy failure (partial ablation) and of local recurrences. In order to obtain a selective intrahepatic portal venous occlusion, the primary vessel of the lesion is identified by color Doppler imaging. using US guidance, we direct the electrode toward this area, with direct puncture of the nearby blood vessel; the ablation cycle lasts either 2–4 min using RFA or 60–90 seconds using MWA. We perform another evaluation with color Doppler imaging to confirm a coagulative ablation of the vascular area: a discolored area on the liver surface can be also visualized (Fig. 12.2). Then, the lesion is treated with the insertion of the electrode in the usual way. This approach decreases local recurrence after RFA ablation, with similar results to surgical resection [35].

12.4.4 Technical Variants: Cooling Technique

If the tumor is located in the hepatic hilar region, RFA may cause bile duct stenosis due to physical or heat damage. It is possible to prevent this biliary damage by inserting preoperatively an endoscopic nasobiliary drainage tube, and percutaneous RFA is then performed with intraductal perfusion of cold 5%

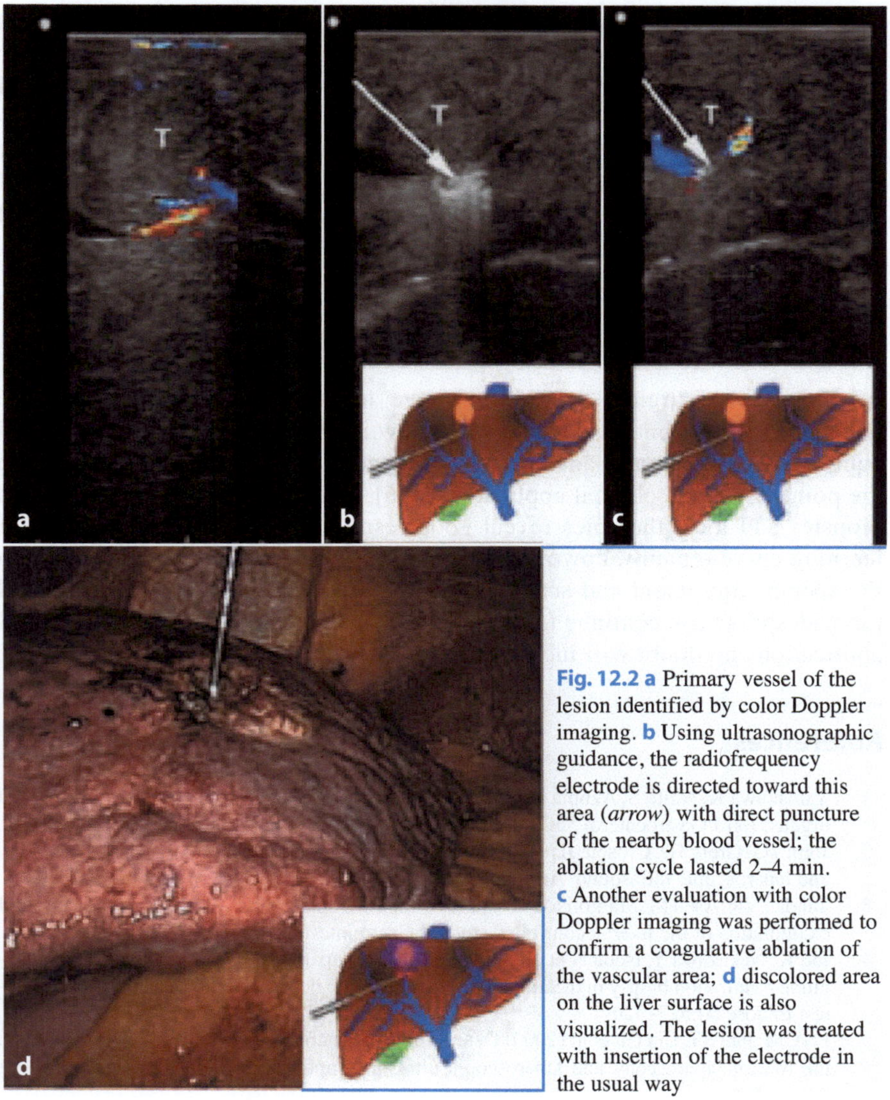

Fig. 12.2 **a** Primary vessel of the lesion identified by color Doppler imaging. **b** Using ultrasonographic guidance, the radiofrequency electrode is directed toward this area (*arrow*) with direct puncture of the nearby blood vessel; the ablation cycle lasted 2–4 min. **c** Another evaluation with color Doppler imaging was performed to confirm a coagulative ablation of the vascular area; **d** discolored area on the liver surface is also visualized. The lesion was treated with insertion of the electrode in the usual way

glucose isotonic solution. It is possible to perform the same cooling technique through percutaneous transhepatic biliary drainage, but this is a more demanding technique because intrahepatic biliary ducts are usually thin. Another technique is intraductal cooling of the main bile ducts during laparoscopic RFA. The tube can be inserted in the main biliary duct either through the cystic duct (cholecystectomy must be performed) or through a direct incision of the main bile duct (choledochotomy). However, these procedures could be very difficult in cirrhotic patients.

We prefer a less invasive approach: we insert some gauze in the peritoneal cavity and place them around the hepatoduodenal ligament. Then, an irrigator with cold saline solution infuses them: cold vascular flow in the portal and arterial systems protects bile ducts from the thermal effects of RFA by dissipating the heat generated in the ablated area. The possible "heat-sink" effect of central-bile-duct cooling might also affect the efficacy of RFA in terms of local recurrences, but in our experience, neither biliary damage nor partial ablation or local HCC recurrence was observed.

12.5 Conclusions

As more open surgical procedures move to laparoscopic approaches, the demand for LUS continues to increase. New improvements in technology and equipment have already advanced LUS from its experimental beginnings to the point of routine clinical applications [36]. The applications in LUS-guided biopsies and local therapies reveal good results in recent years. An operator learning curve remains, however, and mastering LUS requires familiarity with the special equipment and scanning techniques. However, as technology and the widespread use continue to advance, the full range and importance of LUS applications no doubt will increase.

References

1. Yamakawa K, Naito S, Azuma K et al (1958) Laparoscopic diagnosis of the intraabdominal organs. Jpn J Gastroenterol 55:741–747
2. John TG, Greig JD, Crosbie JL et al (1994) Superior staging of liver tumors with laparoscopy and laparoscopic ultrasound. Ann Surg 6:711–719
3. Tandan VR, Asch M, Margolis M et al (1997) Laparoscopic vs open intraoperative ultrasound examination of the liver: a controlled study. J Gastrointest Surg 1:146–151
4. Ido K, Nakazawa Y, Isoda N et al (1999) The role of laparoscopic US and laparoscopic US-guided aspiration biopsy in the diagnosis of multicentric hepatocellular carcinoma. Gastrointest Endosc 50:523–526
5. Lo CM, Fan ST, Liu CL et al (2000) Determining resectability for hepatocellular carcinoma: the role of laparoscopy and laparoscopic ultrasonography. J Hepatobiliary Pancreat Surg 7:260–264
6. D'Angelica M, Fong Y, Weber S et al (2003) The role of staging laparoscopy in hepatobiliary malignancy: prospective analysis of 401 cases. Ann Surg Oncol 10:183–189
7. De Castro SMM, Tilleman EHBM, Busch ORC et al (2004) Diagnostic laparoscopy for primary and secondary liver malignancies: impact of improved imaging and changed criteria for resection. Ann Surg Oncol 11:522–529
8. Klegar EK, Marcus SG, Newman E, Hiotis SP (2005) Diagnostic laparoscopy in the evaluation of the viral hepatitis patient with potentially resectable hepatocellular carcinoma. HPB (Oxford) 7:204–207
9. Lai ECH, Tang CN, Ha JPY et al (2008) The evolving influence of laparoscopy and laparoscopic ultrasonography on patients with hepatocellular carcinoma. Am J Surg 196:736–740
10. Montorsi M, Santambrogio R, Bianchi P et al (2001) Laparoscopy with laparoscopic ultrasound (L-LUS) for the pre-treatment staging of hepatocellular carcinoma: a prospective study. J Gastrointest Surg 5:312–315

11. Rahusen FD, Cuesta MA, Borgstein PJ et al (1999) Selection of patients for resection of colorectal metastases to the liver using diagnostic laparoscopy and laparoscopic ultrasonography. Ann Surg 230:31–37
12. Jarnagin WR, Conlon K, Bodniewicz J et al (2001) A clinical scoring system predicts the yield of diagnostic laparoscopy in patients with potentially resectable hepatic colorectal metastases. Cancer 91:1121–1128
13. Koea J, Rodgers M, Thompson P et al (2004) Laparoscopy in the management of colorectal cancer metastatic to the liver. ANZ J Surg 74:1056–1059
14. Thaler K, Kanneganti S, Khajanchee Y et al (2005) The evolving role of staging laparoscopy in the treatment of colorectal hepatic metastasis. Arch Surg 140:727–734
15. Khan AZ, Karanjia ND (2007) The impact of staging laparoscopic prior to hepatic resection for colorectal metastases. EJSO 33:1010–1013
16. Mann CD, Neal CP, Metcalfe MS et al (2007) Clinical risk score predicts yield of staging laparoscopy in patients with colorectal liver metastases. Br J Surg 94:855–859
17. Pilkington SA, Rees M, Peppercorn D, John TJ (2007) Laparoscopic staging in selected patients with colorectal liver metastases as a prelude to liver resection. HPB (Oxford) 9:58–63
18. Li Destri G, Di Benedetto F, Torrisi B et al (2008) Metachronous liver metastases and resectability: Fong's score and laparoscopic evaluation. HPB (Oxford) 10:1–17
19. Shah AJ, Phull J, Finch-Jones MD (2010) Clinical risk score can be used to select patients for staging laparoscopy and laparoscopic ultrasound for colorectal liver metastases. World J Surg 34:2141–2145
20. Hariharan D, Constantinides V, Kocher H, Tekkis PP (2012) The role of laparoscopy and laparoscopic ultrasound in the preoperative staging of patients with resectable colorectal liver metastases: a meta-analysis. Am J Surg [Epub ahead of print, January 12, PMID: 22244586]
21. Santambrogio R, Opocher E, Pisani Ceretti A et al (2007) Impact of intraoperative ultrasonography in laparoscopic liver surgery. Surg Endosc 21:181–188
22. Santambrogio R, Bianchi P, Pasta A et al (2002) Ultrasound-guided interventional procedures of the liver during laparoscopy: technical considerations. Surg Endosc 16:349–354
23. Hassan H, Vilmann P, Sharma V, Holm J (2009) Initial experience with a new laparoscopic ultrasound probe for guided biopsy in the staging of upper gastrointestinal cancer. Surg Endosc 23:1552–1558
24. Ido K, Isoda N, Kawamoto C et al (1997) Laparoscopic microwave coagulation therapy for solitary hepatocellular carcinoma performed under laparoscopic ultrasonography. Gastrointest Endosc 45:415–420
25. Ito T, Niiyama G, Kawanaka M et al (1999) Laparoscopic microwave coagulation for the treatment of hepatocellular carcinoma. Dig Endosc 11:137–143
26. Seki S, Sakaguchi H, Kadoya H et al (2000) Laparoscopic microwave coagulation therapy for hepatocellular carcinoma. Endoscopy 32:591–597
27. Podnos YD, Henry G, Ortiz JA et al (2001) Laparoscopic ultrasound with radiofrequency ablation in cirrhotic patients with hepatocellular carcinoma: technique and technical considerations. Am Surg 67:1181–1184
28. Noguchi O, Izumi N, Inoue K et al (2003) Laparoscopic ablation therapy for hepatocellular carcinoma. Clinical significance of a newly developed laparoscopic sector ultrasonic probe. Dig Endosc 15:179–184
29. Hsieh CB, Chang HM, Chen TW et al (2004) Comparison of transcatheter arterial chemoembolization, laparoscopic radiofrequency, and conservative treatment for decompensated cirrhotic patients with hepatocellular carcinoma. World J Gastroenterol 10:505–508
30. Casaccia M, Adorno E, Nardi I et al (2008) Laparoscopic US-guided radiofrequency ablation of unresectable hepatocellular carcinoma in liver cirrhosis: feasibility and clinical outcome. J Laparoendosc Adv Surg Tech A 18:797–801
31. Ballem N, Berber E, Pitt T, Siperstein A (2008) Laparoscopic radiofrequency ablation of unresectable hepatocellular carcinoma: long-term follow-up. HPB (Oxford) 10:315–320
32. Sakaguchi H, Seki S, Tsuji K et al (2009) Endoscopic thermal ablation therapies for hepatocellular carcinoma: a multi-center study. Hepatol Res 39:47–52

33. Simo KA, Sereika SE, Newton KN, Gerber DA (2011) Laparoscopic-assisted microwave ablation for hepatocellular carcinoma: safety and efficacy in comparison with radiofrequency ablation. J Surg Oncol 104:822–829
34. Santambrogio R, Costa M, Barabino M, Opocher E (2008) Laparoscopic radiofrequency of hepatocellular carcinoma using ultrasound-guided selective intrahepatic vascular occlusion. Surg Endosc 22-2051–2055
35. Santambrogio R, Opocher E, Zuin M (2009) Surgical resection versus laparoscopic radiofrequency ablation in patients with hepatocellular carcinoma and Child-Pugh class A liver cirrhosis. Ann Surg Oncol 16:3289–3298
36. Vapenstad C, Rethy A, Lango T et al (2010) Laparoscopic ultrasound: a survey of its current and future use, requirements and integration with navigation technology. Surg Endosc 24:2944–2953

Part III

Surgery: General Aspects

Intraoperative Ultrasound

13

Alessandro Ferrero, Luca Viganò, Roberto Lo Tesoriere,
Nadia Russolillo and Lorenzo Capussotti

13.1 Introduction

In recent years, liver surgery has become more popular and safer, and morbidity and mortality rates following surgery have significantly decreased. Anesthesia and technical refinements, together with technological innovations, are the main factors responsible for improvements in these results. Intraoperative ultrasonography (IUS) is considered an indispensable tool in both neoplastic disease staging and operative decision making and surgical guidance during liver resection. Laparoscopic ultrasound (LUS) use was first reported in 1981 by Fukuda and Nakano [1]. In the 1990s, LUS gained popularity with the introduction of laparoscopic surgery. Although LUS is reported to increase surgical safety [2], it remains far from being routinely used: in a recent international survey, LUS use was reported in only 67% of procedures related to the liver [3]. Nevertheless, we strongly recommend the routine use of LUS not only for better disease staging at the beginning of the operation but during every step of the resection in order to guide the hepatectomy and to make it safer for the patient and easier for the surgeon.

To perform adequate exploration of the liver, a compact, mobile, real-time B-mode medium- to top-level machine is mandatory, with the availability of color-flow and power-flow imaging to evaluate blood flow within the liver. The most advanced laparoscopic US transducers on the market can be introduced through a 10- to 12-mm port and usually have a 5- to 10-MHz frequency range; the distal end of the probe, both linear and convex shaped, can be moved in both the left/right and up/down directions to better adhere to the

A. Ferrero (✉)
Department of HPB and Digestive Surgery, Ospedale Mauriziano Umberto I
Turin, Italy
e-mail: aferrero@mauriziano.it

F. Calise and G. Casciola (eds.), *Minimally Invasive Surgery of the Liver*,
Updates in Surgery
DOI: 10.1007/978-88-470-2664-3_13,

liver surface. A built-in biopsy channel to make it easier to obtain the desired samples and the possibility of supporting contrast imaging are highly desirable.

13.2 Exploration Techniques

General anesthesia is typically used. The primary trocar for the 30° laparoscope is inserted with an open approach in the umbilicus or periumbilically, depending upon previous abdominal scars. Two 10- to 12-mm trocars for the probe are then inserted in the right and left upper quadrants between the midaxillary line and the anterior axillary line (Fig. 13.1). Through these ports, all areas of the liver should be viewable.

US study of the liver should represent the first step of any liver surgery and does not require any previous liver mobilization. LUS exploration should be performed before any other maneuver on the liver, because surgical dissection during pneumoperitoneum causes gas to spread into the anatomic planes, causing artifacts. The round and falciform ligaments are not sectioned, and lysis of the adhesions is performed to allow adequate access to both hemilivers. Two standardized explorations are then performed: the first to study liver anatomy (caval confluence, portal bifurcation, left segmental pedicles, right segmental pedicles, hepatic pedicle) and the second to study liver parenchyma and its lesions.

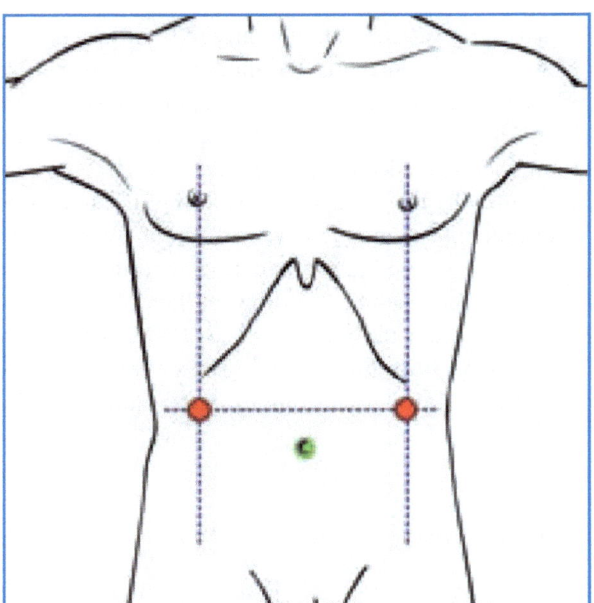

Fig. 13.1 Trocar position. *Green* laparoscope port; *red* operative ports (ultrasound probe, grasper, biopsy forceps)

Liver anatomy can be precisely identified during LUS. Hepatocaval confluence is usually visualized with the probe introduced through the right port and positioned on S4a. By moving the probe between the diaphragm and the liver dome, the right hepatic vein is then identified. Portal bifurcation and right hepatic pedicles are visualized by positioning the transducer on S4b and moving it to the right side. Left portal pedicles and left hepatic vein are better identified from the left upper quadrant. The hepatic pedicle can be visualized from the left port both by positioning the probe on the left liver and directly on the hepatoduodenal ligament.

13.3 Laparoscopic Ultrasonography and Staging

IUS and direct inspection of the abdominal cavity are used to assess the resectability of liver tumors. However, open exploration is associated with morbidity and increased length of hospital stay. For these reasons, a minimally invasive approach, based on laparoscopic exploration, was proposed for intraoperative staging in patients with hepatic malignant tumors. LUS overcomes the 2D capability of laparoscopy by providing a third arm that could change the surgical strategy by detecting additional, deeply located, hepatic tumors not seen at laparoscopic inspection. Nevertheless, its value as a staging technique is still debated as a consequence of improvement in preoperative staging techniques throughout the years.

13.3.1 General Data

Few data in the literature analyzed the role of LUS for staging malignant liver disease (Table 13.1). Clarke et al. [4], in their prospective study comparing LUS with conventional preoperative imaging modalities in patients undergoing resection for colorectal liver metastases, found 25–35% additional lesions intraoperatively. Moreover, 40% of liver tumors detected by LUS had not been visible or palpable at surgery. Although this study is interesting, it was performed when preoperative staging techniques were less effective. John et al. [5] analyzed a cohort of 50 consecutive patients with potentially resectable liver tumors. Laparoscopic exploration demonstrated factors precluding curative resection in 23 patients (46%). LUS identified liver tumors not visible during laparoscopy in 14 patients (33%) and provided additional staging information in 18 of 43 patients (42%). Foroutani and colleagues [6] performed computed tomography (CT) scans and LUS in 55 patients with a total of 222 liver lesions, including primary and metastatic tumors. Triphasic spiral CT scans were obtained less than 1 week before surgery. Liver LUS was performed with a linear-array 7.5-MHz side-viewing laparoscopic transducer. LUS detected all 201 tumors seen on preoperative CT and an additional 21 tumors (9.5%) in 11 patients (20%). Lesions missed by CT were broken down

Table 13.1 Studies assessing the role of LUS in staging of malignant liver disease

Study/Year	Accrual period	Diagnosis	# of patients (SL/LUS)	Unresectable Disease at SL # of patients	LUS finding
D'Angelica (2003) [13]	1997-2001	Hepatobiliary cancer	401/153	84 (20.9%)	8 (9.5%)
Connor (2005)	1992-2003	HC	84/84	34 (40%)	14 (41%)
Mann (2007) [8]	2000-2004	Colorectal Metastases	200/159	39 (19.5%)	20 (51%)
Lai (2008) [10]	2001-2007	HCC	119/119	46 (38.6%)	7 (15%)
Author's data (2012)	2000-2011	GC/HC/IHC	131/71	35 (26.7%)	5 (29%*)

SL, staging laparoscopy; *LUS*, laparoscopic ultrasound; *HCC*, hepatocellular carcinoma; *HC*, hilar cholangiocarcinoma; *IHC*, intrahepatic cholangiocarcinoma
* 5 patients out of 17 considered unresectable after SL+LUS.

by size: more than a quarter were <1 cm, and none was >3 cm. CT was more likely to miss tumors close to the falciform ligament of the liver and along the liver's periphery.

13.3.2 Colorectal Liver Metastases

Many studies published in the 1990s reported that patients with colorectal liver metastases had unresectable disease at laparoscopy in 22–48% of cases. This high rate may be due to the inadequacy of preoperative imaging studies used to determine resectability. The results of more recent studies show that the majority of patients considered to have resectable disease based on good-quality preoperative imaging do not have occult unresectable disease, and the yield of laparoscopy was lower than previously reported (10–20%). In 2001 [7] Jarnagin at al. used a scoring system, CRS, (Clinical Risk Score, derived from the nodal status of the primary disease, disease-free interval, number of hepatic metastases, largest diameter of hepatic tumor and the carcinoembryonic antigen level) to identify high risk patients most likely to benefit from staging laparoscopy (SL). The authors reported that with a high CRS, the chances of unresectable disease were 42%, whereas, with a low score, resection was precluded in only 12% of patients. In that study, LUS was performed only in 60% of patients and identified unresectable disease that was not appreciated by inspection alone in three of 14 patients (21%). Subsequently, Mann et al. [8] validated the role of the CRS in 200 patients and confirmed that the potential benefit of SL augments progressively with increasing CRS.

In conclusion, data from the literature support the theory that SL + LUS should be used only in selected patients with a high risk of having unresectable disease.

13.3.3 Hepatocellular Carcinoma

The amount of available literature pertaining to SL in hepatocellular carcinoma (HCC) is limited. As peritoneal disease is uncommon with HCC, SL may be useful. Nevertheless, the rate of contraindications to surgery found during laparoscopy ranged from 16% to 37%. These results could be explained by intraoperative findings by LUS. During liver laparoscopy, LUS exploration improves the yield of detecting lesions. Montorsi et al. reported that LUS identified the presence of new HCC nodules in 20.5% of patients [9]. Lai et al. [10] reported the results of 122 consecutive patients with potentially operable HCC who underwent preoperative SL and LUS; 46 patients were found to have an inoperable tumor, 44 after laparoscopic staging and 2 after laparatomy. In particular, 37 patients presented inadequate liver remnants or severe cirrhosis, two patients had peritoneal metastasis, and LUS detected vascular invasion and bilobar liver lesions in seven patients. According to these results, Weitz et al. [11] proposed SL only in patients with clinically evident liver cirrhosis and major vascular invasion or bilobar tumors.

13.3.4 Biliary Cancer

Biliary tract tumors can be divided into two main categories: gallbladder cancers and cholangiocarcinomas. The two groups differ in patterns of spread and prognosis. Gallbladder cancer tends to grow more rapidly and has earlier dissemination, which makes SL a useful tool in this setting. In contrast, cholangiocarcinomas tend to be more locally invasive, decreasing the yield of SL. In the Goere et al. experience [12], overall SL yield was higher for gallbladder cancer (GC; 62%), followed by intrahepatic cholangiocarcinoma (IHC; 36%), then hilar cholangiocarcinoma (HC; 25%). Nevertheless, all patients with vascular or nodal involvement and two with liver metastases were missed by laparoscopy, probably because the authors did not associate LUS. A study by D'Angelica et al. [13] of 410 patients with resectable hepatobiliary malignancy at preoperative imaging shows that SL was completed in 73% of patients. Moreover, in 84 (55%) of the 153 evaluated patients, SL identified disease that precluded resection. The authors confirmed that the highest yield was for biliary cancers (GC 50%, IHC 25%, HC 20%). LUS identified clinically important additional disease in 14.9% of patients and was responsible for approximately 10% of the findings of unresectability.

In our center from 2000 to 2011, 131 of 240 potentially resectable patients (54.5%) with proximal biliary cancers underwent SL based on the following criteria: suspicion of GC with carbohydrate antigen (CA) 19.9 >100 U/ml, borderline resectable IHC or HC. After 2008, all patients with HC underwent SL.

LUS was systematically performed after 2006. Overall, 71 patients (54.1%) underwent LUS. At SL, 35 cases (26.7%) were considered unresectable for carcinomatosis (28), new liver nodules (10), and vascular invasion (9). LUS found additional liver nodules in six IHC, five GC, and two HC patients and vascular invasion in four IHC, eight GC, and three HC cases. Finally, LUS identified unresectable disease in five patients (7%) otherwise considered resectable by SL only. All but one patient, who had a laparoscopic hepatectomy, underwent laparotomy (n = 95), with evidence in 11 (11.5%) of carcinomatosis (n = 8), new liver nodules (n = 3), or vascular invasion (n = 2). Of note, five of these patients had an incomplete SL. The overall accuracy of SL was 76%, increasing to 89.4% with LUS. The overall yield of SL + LUS was 23.9%, with highest rates for GC (65%). Thus, in our study, LUS increased SL accuracy for detecting unresectable disease in a subset of patients with proximal biliary cancer.

In conclusion, available data indicate that laparoscopy with LUS provides information similar to that obtained by intraoperative open US and that it is able to identify small intrahepatic lesions not diagnosed by conventional preoperative techniques.

13.4 Performance of Laparoscopic Ultrasonography Compared with Intraoperative Open Ultrasonography

The reliability of LUS should be validated by comparing its performances with that of open IUS. In the literature, this issue is poorly analyzed. In 1996, Cozzi et al. demonstrated similar sensitivity of LUS and open IUS in detecting hypoechoic artificial lesions in five fresh pig livers [14]. In 1997, Tandan et al. reported a small series of patients (27 cases) scheduled for liver resection undergoing both LUS and open IUS [15]: LUS had good sensitivity and specificity in comparison with open IUS (93% and 100%, respectively). Some additional data derive from the series of SL: open IUS rarely disclosed additional liver nodules whenever LUS was previously performed.

In this context, the authors of this chapter performed a prospective study to definitively compare LUS and open IUS. Between September 2009 and March 2011, all 230 patients scheduled for liver resection at the Mauriziano Umberto I hospital were considered. Inclusion criteria were diagnosis of primary or metastatic liver tumor. Exclusion criteria were diagnosis of hilar or gallbladder cholangiocarcinoma, more than ten lesions at preoperative imaging, liver reresection, liver resection completed by laparoscopy, adhesions precluding complete LUS, and unresectability at exploration. Study protocol scheduled laparoscopy with LUS and then laparotomy, open IUS, and hepatectomy. Contrast-enhanced IUS (CE-IUS) was performed in discordant cases. Reference standards were final pathology and 6-month follow-up. Sixty-five patients were assessed (12 with HCC; 12 with peripheral cholangiocarcinoma; 37 with colorectal metastases, and four with noncolorectal metastases).

Median number of preoperative imaging studies was three: percutaneous US and CT scan in 100% of cases, magnetic resonance imaging (MRI) in 67%, and positron emission tomography (PET)-CT in 54%. One hundred and nineteen lesions were diagnosed at preoperative imaging: median number per patient was one (1–9); median diameter was 18 (3–160) mm.

In comparison with preoperative imaging, LUS detected 22 additional lesions (+18.5%) in 14 patients (21.5%). The median diameter of new nodules was 7 (2–15) mm. In comparison with LUS, open IUS detected two additional lesions in two patients. Conversely, open IUS did not confirm (even after CE-IUS) four lesions detected by LUS. Overall, 20 new lesions (+16.8%) in ten patients (15.4%) were detected by open IUS. In comparison with open IUS, per-lesion LUS sensitivity was 98.6% and per-patient accuracy was 93.8%. Agreement between the two procedures for newly detected malignant nodules was 99.3% in per-lesion and 100% in per-patient analyses. Similar results were observed for detecting vascular and biliary infiltrations: in comparison with preoperative imaging, new data were added by both LUS and open IUS, with an agreement between the two procedures of 99.3% (Fig. 13.2). Liver resection

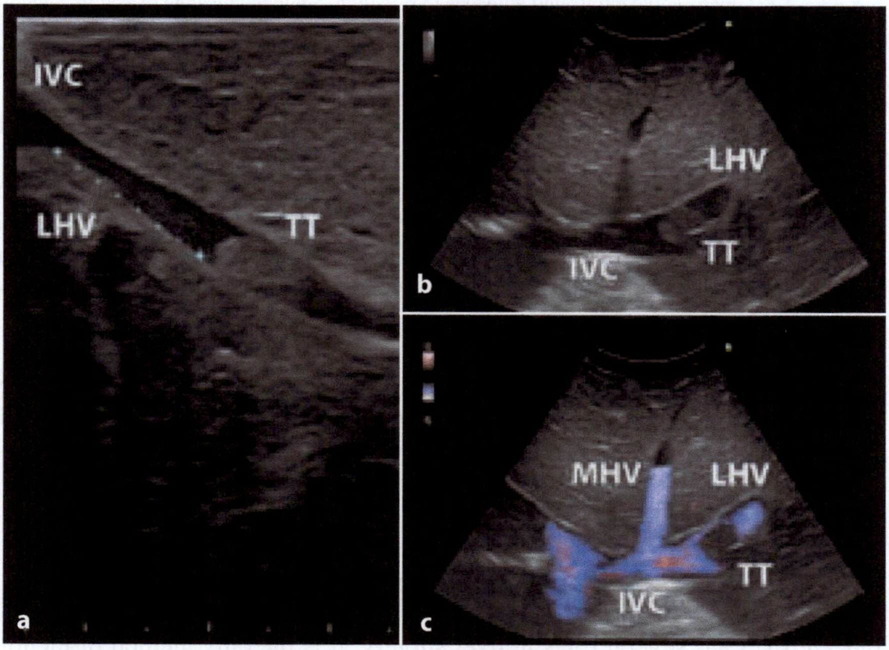

Fig. 13.2a–c Laparoscopic ultrasound (LUS) intraoperative staging. An 82-year old man affected by segments 2–3 hepatocellular carcinoma with cirrhosis scheduled for laparoscopic left lateral sectionectomy. LUS detected a tumor thrombus in the left hepatic vein (**a**) that was not evident at preoperative imaging. The procedure was converted to laparotomy. Open intraoperative ultrasound (IUS) confirmed vascular infiltration (**b**), easily detectable by means of e-flow US (**c**). *IVC*, inferior vena cava; *LHV*, left hepatic vein; *MHV*, middle hepatic vein; *TT*, tumor thrombus

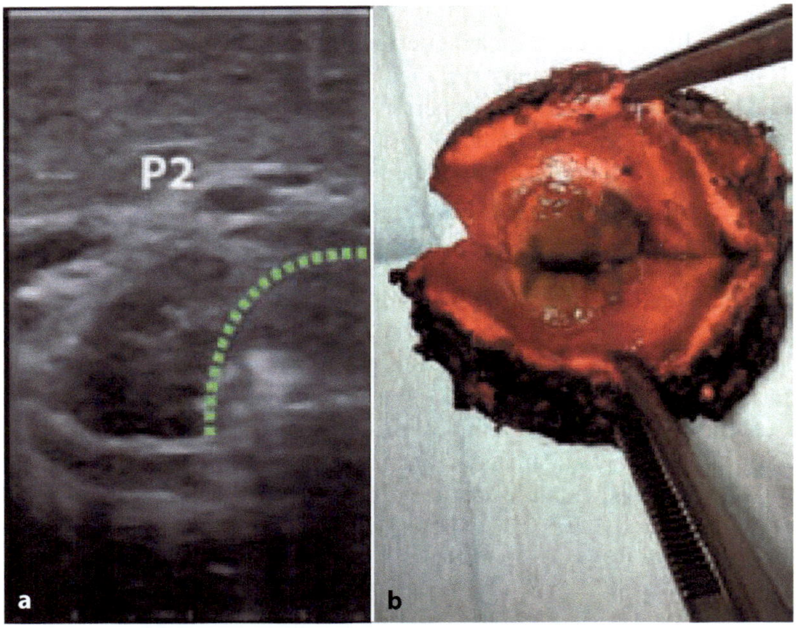

Fig. 13.3 a, b Laparoscopic ultrasound guidance to resection. Segment 2 wedge resection. Section plane runs along segment 2 pedicle (P2) (a); adequate tumor specimen margin (b)

was finally modified in 13 (20%) patients, and agreement between LUS and open IUS in surgical strategy modification was 92.3%.

Final pathology confirmed 14 newly detected malignant nodules (+11.8%) in eight patients (12.3%). Six nodules (five detected by both LUS and open IUS and one by open IUS only) were benign. After a 6-month follow-up, ten new hepatic malignant nodules were identified in six (9.2%) patients. Per-lesion sensitivity of preoperative imaging, LUS, and open IUS were 83.1%, 92.3%, and 93.0%, respectively.

In conclusion, the reliability of LUS for staging liver diseases and planning surgical treatment was demonstrated by its performances similar to those of open IUS. These data strengthen the safety concept of laparoscopic liver surgery, which is based on its capability to reproduce open surgical procedures.

13.5 Laparoscopic Ultrasonography and Liver Resection

Indeed, LUS has several roles in liver surgery: it is the only method of identifying the precise liver anatomy and recognizing the presence of vasculobiliary anomalies. Furthermore, it plays an important role in intraoperative staging of the disease and identifying the exact site of the lesion and eventual new nodules within the liver.

Moreover, probably the most important role of LUS is to assist the surgeon during resection. Laparoscopic surgery presents some inherent limitations, such as the lack of tactile sensation, impairing operative planning: US can compensate for these limitations by allowing the surgeon to see beyond the surface of the liver. During the first step of the operation, relationships between tumor and intrahepatic vasculobiliary pedicles can be precisely visualized, and resection lines are designed on the liver surface with monopolar coagulation. During transection, the resection plane is repeatedly checked by US to maintain a safe margin, providing the surgeon with immediate feedback of possible necessary changes (Fig. 13.3). Any glissonian pedicle or vein encountered during liver transection can be checked and recognized by LUS before ligature and section. Color-Doppler US allows the user to visualize blood flow and assess flow in and near the area of interest, thus avoiding injury to important vessels during dissection. The possibility of guiding interstitial treatments associated with hepatectomy, such as radiofrequency or microwave ablation, gives an added value to the procedure, mainly in HCC management.

LUS has some drawbacks and limitations. Hand–eye coordination of the probe visualized through the video laparoscope can be difficult. Orientation and the following image interpretation can be complicated. Furthermore, the field of view is limited due to transducer size. Some US-specific drawbacks add to the difficulties of interpretation, such as shadowing, multiple reflections, variable contrast depending on liver parenchyma status, and the fact that image quality also may be somewhat operator dependent. Nevertheless, despite these difficulties, LUS remains an indispensable tool for surgeons dealing with laparoscopic liver resections.

References

1. Fukuda MMF, Nakano Y (1982) Studies on echolaparoscopy. Scan J Gastroenterol 17:186
2. Machi J, Johnson JO, Deziel DJ et al (2009) The routine use of laparoscopic ultrasound decreases bile duct injury: a multicenter study. Surg Endosc 23:384–388
3. Vapenstad C, Rethy A, Langø T et al (2010) Laparoscopic ultrasound: a survey of its current and future use, requirements, and integration with navigation technology. Surg Endosc 24:2944–2953
4. Clarke MP, Kane RA, Steele G et al (1989) Prospective comparison of preoperative imaging and intraoperative ultrasonography in the detection of liver tumors. Surgical 106:849–855
5. John TG, Greig JD, Crosbie JL et al (1994) Superior staging of liver tumors with laparoscopy and laparoscopic ultrasound. Ann Surg 220:711–719
6. Foroutani A, Gerland A, Berber E (2000) Laparoscopic ultrasound vs. triphasic computed tomography for detecting liver tumors. Arch Surg 135:933–938
7. Jarnagin WR, Conlon K, Bodniewicz J, Dougherty E, DeMatteo RP, Blumgart LH, Fong Y (2001) A clinical scoring system predicts the yield of diagnostic laparoscopy in patients with potentially resectable hepatic colorectal metastases. Cancer 91:1121–1128
8. Mann CD, Neal CP, Metcalfe MS et al (2007) Clinical Risk Score predicts yield of staging laparoscopy in patients with colorectal liver metastases. Br J Surg 94:855–859

9. Montorsi M, Santambrogio R, Bianchi P et al (2001) Laparoscopy with laparoscopic ultrasound for pretreatment staging of hepatocellular carcinoma: a prospective study. J Gastrointest Surg 5:312–315
10. Lai EC, Tang CN, Ha JP et al (2008) The evolving influence of laparoscopy and laparoscopic ultrasonography on patients with hepatocellular carcinoma. Am J Surg 196:736–740
11. Weitz J, D'Angelica M, Jarnagin W et al (2004) Selective use of diagnostic laparoscopy prior to planned hepatectomy for patients with hepatocellular carcinoma. Surgery 135:273–281
12. Goere D, Wagholikar GD, Pessaux P et al (2006) Utility of staging laparoscopy in subsets of biliary cancers: laparoscopy is a powerful diagnostic tool in patients with intrahepatic and gallbladder carcinoma. Surg Endosc 20:721–725
13. D'Angelica M, Fong Y, Weber S et al (2003) The role of staging laparoscopy in hepatobiliary malignancy: prospective analysis of 401 cases. Ann Surg Oncol 10:183–189
14. Cozzi PJ, McCall JL, Jorgensen JO, Morris DL (1996) Laparoscopic vs. open ultrasound of the liver: an in vitro study. HPB Surg 10(2):87–89
15. Tandan VR, Asch M, Margolis M et al (1997) Laparoscopic vs. open intraoperative ultrasound examination of the liver: a controlled study. J Gastrointest Surg 1(2):146–150

A New Anatomical Vision: Liver Surgery on the Screen

14

Fulvio Calise, Giuseppe Surfaro, Antonio Giuliani,
Carla Migliaccio and Antonio Ceriello

14.1 Introduction

Shall we see through a camera better than with our eyes? The question, still appropriate at the present time, dates back to the seventeenth century. It seems that Vermeer, and then Canaletto some decades later, introduced in their paintings the use of the *camera obscura*, or optic camera (Figs. 14.1, 14.2). The simplest arrangement, with a lens fastened in the pinhole, projected an inversed and reversed image on a vertical screen opposite the aperture. A variation on this employed a translucent screen, allowing the viewer to see the image from the other side, thereby correcting the left-to-right reversal. These two types of cameras projected the image directly and could be combined in one device. The optical *camera obscura* ushered in a new approach to optics, opening up new views of the visible world and shaping a new understanding of vision itself.

After the Second World War, the operating microscope was introduced in several surgical fields: in other words, a prism allowing splitting of the light beam in order that assistants may also visualize the procedure or to allow photography or video to be taken of the operating field.

In his famous essay "*Homo Videns*," Giovanni Sartori [1] raised his leading questions: Is it true that the tele-vision, namely, to see through a screen, changes our human nature? Is it really true that the screen is an anthropogenic instrument? Probably yes, is the devastating answer to both questions.

Therefore, it seems that for a long time, humans have needed help to better visualize reality. The laparoscopic revolution has made this need a day-by-

F. Calise (✉)
Hepatobiliary Surgery and Liver Transplant Center, A. Cardarelli Hospital
Naples, Italy
e-mail: fulvio.calise@libero.it

F. Calise and G. Casciola (eds.), *Minimally Invasive Surgery of the Liver*,
Updates in Surgery
DOI: 10.1007/978-88-470-2664-3_14, © Springer-Verlag Italia 2013

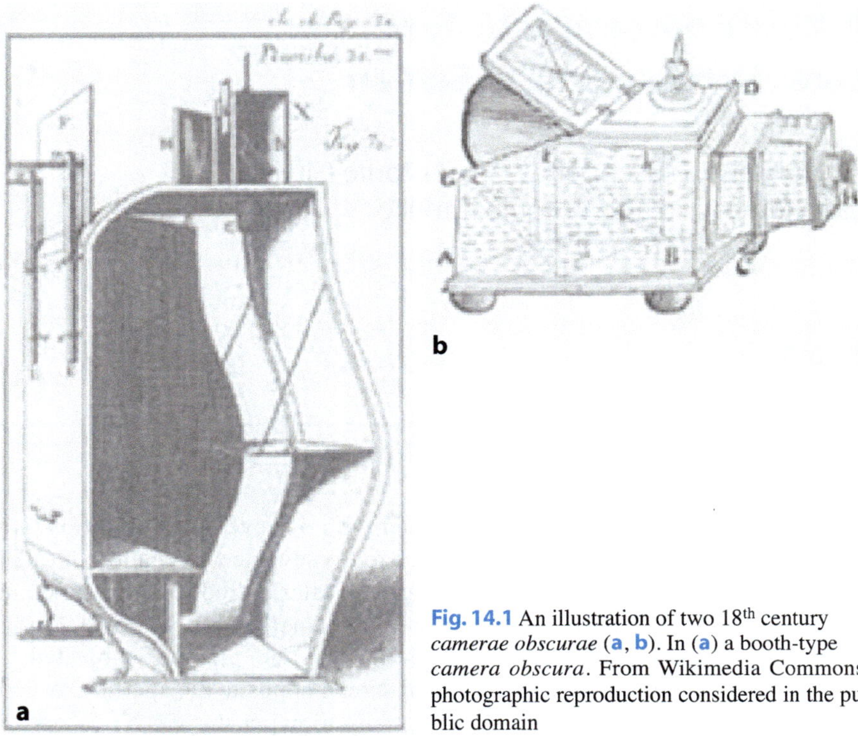

Fig. 14.1 An illustration of two 18th century *camerae obscurae* (**a**, **b**). In (**a**) a booth-type *camera obscura*. From Wikimedia Commons, photographic reproduction considered in the public domain

Fig. 14.2 Four drawings by Canaletto, representing Campo San Giovanni e Paolo in Venice, obtained with a *camera obscura* (Venice, Gallerie dell'Accademia). From Wikimedia Commons, photographic reproduction considered in the public domain

day opportunity and challenge for surgeons to better serve their patients. However, at the same time, the physical essence of the body – to touch, to handle the organs – disappears in favor of vision through a screen: the surgeon no longer looks where his or her hands (i.e., the surgical instruments) are, but looks upward, in another direction. So, the surgeon's operative approach, and consequently the approach to the anatomy of organs, has changed.

In a laparoscopic approach and at the present time, particularly with the use of the robot, we may move our camera (our technological eye) where we want, and rotate it and our forceps a full 360° – an impossible feat for our human eyes and hands. However, at the same time, we lose our touch, which is laparoscopy's burden. When we operate on the stomach, the spleen, and the pancreas, these organs remain in place until almost the end of surgery. This is not so for the liver.

14.2 Surgeon Position

When we open the abdomen to operate, we have a lateral vision of the organs. To approach the basic structures of the liver, a 3D organ, we need to free it from its ligaments and to raise, lower, turn, rotate, and change the position of the operating table. In laparoscopic liver surgery (LLS), the first main change is that the surgeon is no longer on the right side of the patient but between his or her legs. In other words, the liver is in front of the surgeon, who can now immediately explore the four faces of the liver and has an immediate approach to the hepatic hilum. This, alone, shows how feasible the approach is to the left-lateral and anterior liver segments. Finally, we do not directly use our eyes to see what we are operating on, but we look at a screen, which is generally placed at our left, on the right side of the patient. We thus return to our optic camera….

14.3 Trocar Position

Generally in LLS, the best trocar positioning is to describe an arch with its concavity upward in order to obtain correct crossing of the operating instruments, which is the focus of the tools' triangulation. Therefore, again differently from open liver surgery (OLS), in LLS, we are obliged to use axial movements with a fixed point of rotation (except in some cases with the possibility of using an angled instrument) and to work in a straight line to obtain the most appropriate visual field [2].

14.4 Pedicle Approach

Change in technique does not mean change in surgical principles. Therefore, in LLS, as direct manual control of the pedicle is impossible, it becomes necessary to previously prepare the Pringle maneuver. Indeed, in the laparoscopic approach, it is possible to obtain good frontal vision of the liver and perform a safe and rapid maneuver by underpassing the pedicle to obtain ready control in case of bleeding. Laparoscopic visualization provides a frontal view of the pedicle, not bilateral as in OLS. It is then necessary to use a Roticulator Endo Dissect (or other articulating device) that may allow passing through the Winslow foramen, helping compensate for the loss of adequate lateral vision.

14.5 Intrahepatic Approach and Vascular Pedicles Vision

Isolation of vascular branches at the hilum follows exactly the same pattern than in OLS and is even easier to do provided that the surgeon can work with both right and left hands. In fact, hilum dissection is more simply done if we use the HARMONIC scalpel or bipolar forceps with the right and left hands alternatively, according, namely, to portal vein direction. The surgeon must also remember that any section of a major vessel needs ultrasound guidance due to lack of pulse touch.

The intrahepatic approach was first described by Tung and Quang [3] in 1965 and involves hepatic parenchyma dissection along the hepatic fissures and pedicle ligation directly within the liver. Employing this technique fits well with LLS, which allows good vision of the anterior face of the liver, and being problematic, manipulate or rotate the liver. In other words, in major hepatectomy (right and left), the approach to resection mimics the anterior approach described by Belghiti et al. for OLS [4].

The approach to the vena cava for right hepatectomy is frontal and under the liver, being extremely difficult to rotate the liver towards the left. So, the surgeon works as if the operating field was within a tunnel. The right ligaments can be mobilized after completing the resection. The final approach to the right hepatic vein is from within the liver, just as Tung and Quang wrote many years ago: *historia magistra vitae…*

Conversely, left hepatectomy using LLS allows better vision compared with OLS, with optimal access to the left side of the vena cava after dissection of the Arantius ligament and also – but only if strictly necessary and easy to do – isolation of the left hepatic vein on a tape.

Some authors suggest an intrahepatic glissonian approach for laparoscopic right-segmental liver resection and for pure right hepatectomy [5, 6] in order to obtain appropriate vascular control during transection. On this topic, we emphasize the possibility with laparoscopy of good visualization of the portal branches either to isolate them and obtain vascular control or to ligate them in the case of a two-stage hepatectomy.

Finally, although we see better using LLS, we actually see less. The extensive use of a linear vascular stapler (LVS) during LLS is a novelty compared with the OLS approach. The use of LVS is of great help in LLR because it speeds up the procedure and reduces bleeding [7, 8]. Regardless, most of the time, the surgeon does not see what is being stapled and cut. *O tempora o mores…* The intra-abdominal pressure allows this technical "trick": however, the surgeon must pay attention when cutting the biliary branches with a stapler because there the seal is not always optimal and often requires an additional stitch to prevent bile leak.

14.6 Conclusions

Due to advances in operating-room technology, surgeons, when performing LLS surgery, have lost their traditional surgical position at the right side of the patient, have lost direct visualization of what they do, have lost the ability to touch with their hands the operating field, and have lost the predominance of one hand over the other. And yet, this 360° change in operative technique has occurred and is of great advantage – to the surgeon and thus to the patient. Not bad in only one generation. We are rewriting our tablets of Moses, and this is the wonder of our job!

References

1. Sartori G (2000) Homo videns. Laterza, Bari
2. Azagra JS, Goergen M, Gilbart E et al (1996) Laparoscopic anatomical (hepatic) left lateral segmentectomy-technical aspects. Surg Endosc 10:758–761
3. Ton-That-Tung, Nguyen-Duong-Quang (1965) Segmentary hepatectomy by transparenchymatous vascular ligation. Presse Med 73(52):3015–3017
4. Belghiti J, Guevara OA, Noun R et al (2001) Liver hanging maneuver: a safe approach to right hepatectomy without liver mobilization. J Am Coll Surg 193(1):109–111
5. Machado MA, Makdissi FF, Galvão FH et al (2008) Intrahepatic Glissonian approach for laparoscopic right segmental liver resections. Am J Surg 196:38–42
6. Ikeda T, Yonemura Y, Ueda N et al (2011) Pure laparoscopic right hepatectomy in the semiprone position using the intrahepatic glissonian approach and a modified hanging maneuver to minimize intraoperative bleeding. Surg Today 41:1592–1598
7. Kaneko H, Otsuka Y, Takagi S et al (2004) Hepatic resection using stapling devices. Am J Surg 187:280–284
8. Gumbs AA, Gayet B, Gagner M (2008) Laparoscopic liver resection: when to use the laparoscopic stapler device. HPB (Oxford) 10(4):296–303

Encircling the Pedicle for the Pringle Maneuver

15

Vincenzo Scuderi, Antonio Ceriello, Giuseppe Aragiusto, Antonio Giuliani and Fulvio Calise

15.1 Introduction

Although some liver resections may now be safely performed without vascular clamping, blood saving remains an important concern in hepatobiliary surgery, especially in patients with liver tumors, as blood loss seems to directly affect early and late outcome.

Hepatic inflow occlusion has been recommended for reducing blood loss during liver resection. Although various techniques for hepatic vascular control have been presented, Pringle maneuver (PM), the oldest and simplest, is still favored by many surgeons in conventional open surgery.

During laparoscopic liver surgery (LLS), however, PM is not easily performed because of the narrow field of vision and because the laparoscopic forceps is usually too obtuse to encircle the hepatoduodenal ligament (HDL). Nevertheless, the good results using LLS are actually based on the continuous effort to reproduce the same maneuvers used in open surgery. Therefore, prudently loading the hepatic pedicle on a tape is advisable when using a minimally invasive approach. Furthermore, bleeding from the cut hepatic surface during laparoscopic surgery can dramatically complicate the outcome of the procedure by darkening the operative field.

The importance of the ability to clamp the hepatic hilum in the course of LLS emerges in the study of Nguyen et al., who reviewed 2,804 laparoscopic liver resections (LLR) and reported that the main cause of conversion was bleeding (34%; 40/116) [1]. In this chapter, methods proposed for encircling

V. Scuderi (✉)
Hepatobiliary Surgery and Liver Transplant Unit, A. Cardarelli Hospital, Naples, Italy
e-mail: scuderi.enzo@gmail.com

F. Calise and G. Casciola (eds.), *Minimally Invasive Surgery of the Liver,* Updates in Surgery
DOI: 10.1007/978-88-470-2664-3_15,

the hepatic hilum and general principles of liver inflow control are described, outlining correlations between hepatic pedicle clamping and pneumoperitoneum (PP).

15.2 Anatomic and Hemodynamic Remarks Regarding Hepatic Inflow Occlusion

The adult liver, accounting for 2–3% of overall body weight, has a dual vascular inflow (1,500 ml/min, approximately one quarter of the total cardiac output) through the portal vein (75% of total hepatic blood flow) and the hepatic artery (25%) and an outflow drainage through the main and accessory hepatic veins. The main portal vein drains the splanchnic territory; its blood pressure is usually low, with a portocaval gradient <5 mmHg.

In the most common anatomy, the main hepatic artery arises from the celiac trunk; in 20–25% of cases, several types of anatomic variations may be encountered. The most common ones include the right hepatic artery arising from the superior mesenteric artery and running behind the pancreatic head along the right posterolateral side of the portal vein, and the left hepatic artery arising from the left gastric artery and running in the lesser omentum.

Once the lesser omentum is opened, a blunt dissector may be passed through the foramen of Winslow and the pedicle encircled with an umbilical tape; in this manner, the HDL can usually be clamped en masse using atraumatic flexible clamps or a tourniquet. When a right hepatic artery originating from the superior mesenteric artery is present, it runs within the HDL and is therefore included in the PM. By contrast, a left hepatic artery originating from the left gastric artery runs separately and requires individual clamping in order to totally occlude the arterial inflow.

Blood pressure in the hepatic sinusoids depends on hepatic blood flow, hepatic resistance, and pressure in the hepatic veins; the latter is directly linked to the pressure in the inferior vena cava (IVC) and right atrium. Hepatic vein pressure can be estimated during open procedures by monitoring the central venous pressure (CVP). In cases of inflow clamping, intrahepatic blood pressure as well as pressure in the hepatic veins depends exclusively on CVP.

The hemodynamic repercussions of liver inflow clamping have been extensively studied by means of intraoperative hemodynamic assessment (invasive arterial pressure and CVP monitoring) during traditional and minimally invasive liver resections. During laparoscopy, the PP renders CVP monitoring unreliable, and it therefore must be estimated by observing the IVC: when it appears flaccid and fluctuates with heart and lung movements it can be assumed that CVP is sufficiently low and suitable for parenchymal transection.

Persistent bleeding during parenchymal transection during inflow occlusion results either from incomplete inflow occlusion or, more significantly, from backflow bleeding through the hepatic veins. Incomplete inflow occlusion can be avoided by applying the clamp or tourniquet until pulsation in the distal hepatic artery is absent, clamping any accessory hepatic artery and lysing hypervascular adhesions around the liver. Moreover, the PM interrupts the arterial and portal venous inflow to the liver but has no effect on backflow bleeding from branches of the hepatic veins. This is the reason for the importance of low CVP to reduce bleeding during transection: backflow bleeding is reduced when CVP is maintained <5 cm H_2O, although it can increase the risk of air embolism.

Hepatic backflow during inflow occlusion, as well as being responsible of persistent bleeding, provides a liver parenchymal perfusion that, even though suboptimal, is sufficient to guarantee liver tolerance to normothermic ischemia, even under PP conditions. Nsadi et al. [2] show in the swine model that high abdominal pressure during PP decreases liver backflow, enhancing ischemic liver damage due to inflow occlusion. The severity of ischemic damage could be increased by low hepatic oxygenation and by decreased hepatic backflow due to higher abdominal pressure. Evidence of applicability during a minimally invasive approach [3] to intermittent portal clamping (IPC) is therefore also important, just as it is for open hepatic resection [4]. IPC of 15 min and a 5-min clamp-free interval is considered the best option to achieve both hemorrhage control and reduce ischemia–reperfusion liver damage.

Moreover, continuous portal clamping generates splanchnic congestion with intestinal edema, which may lead to postoperative morbidity, especially in cases of bowel resection for colorectal cancer and synchronous liver metastases. Splanchnic congestion from portal clamping tends to be mild with IPC.

15.3 Surgical Strategies for Laparoscopic Inflow Control

Laparoscopic encircling of the HDL can be difficult because of the narrow field due to the 2D view and because the device, with its obtuse curves and sharp and hard tips, provides limited degrees of freedom. This concern may induce some surgeon to avoid passing tape around the HDL under blind manipulation because of the risk of injury to the IVC and other vital structures within porta hepatis. Furthermore, previous surgery (recurrent hepatocellular cancer or colorectal cancer metastases) may cause multiple adhesions, making this surgical procedure hazardous.

In case of severe bleeding during laparoscopic hepatic transection without preparation of the hepatic hilum, it is prudent to immediately convert to an open procedure. Rotellar et al.'s [5] experience is quite significant: after 16 LLR procedures, including major hepatectomies without hilum preparation, the authors report conversion to the open approach due to uncontrollable bleeding from a portal pedicle during resection of a 13-mm lesion in S6. After this event, the authors decided to almost systematically perform inflow control.

The approach to the HDL is not univocal. It was proposed that inflow occlusion with laparoscopic vascular clamp or through an intraperitoneal or extracorporeal tourniquet method be performed. The laparoscopic vascular clamp is reported to be not robust enough to give satisfactory inflow occlusion; furthermore, its hasty and blind placement in an emergency bleeding situation can be dangerous.

A tape placed around the porta hepatis and passed through a 16-F rubber drain to be used as a tourniquet for PM, proposed originally by Cherqui et al. [6], appears safer and more efficient.

Robotic HDL encircling is described as being similar to and easier than that in the open setting. The robotic system overcomes laparoscopic limitations, thanks to the fine movements of the dedicated devices that, with 6 degrees of freedom, can reproduce basic movements of human hands [7].

In case of a traditional laparoscopic approach with inarticulate device, Rotellar et al. [5] proposed passing behind the HDL with a normal laparoscopic forceps; when inserted from the right flank and through the foramen of Winslow, it exits from the opened lesser omentum to grasp the tape. The authors propose two different dispositions for the tourniquet, based on tumor site, to perform an efficient inflow occlusion without interference between tourniquet and operative devices (Figs. 15.1 and 15.2)

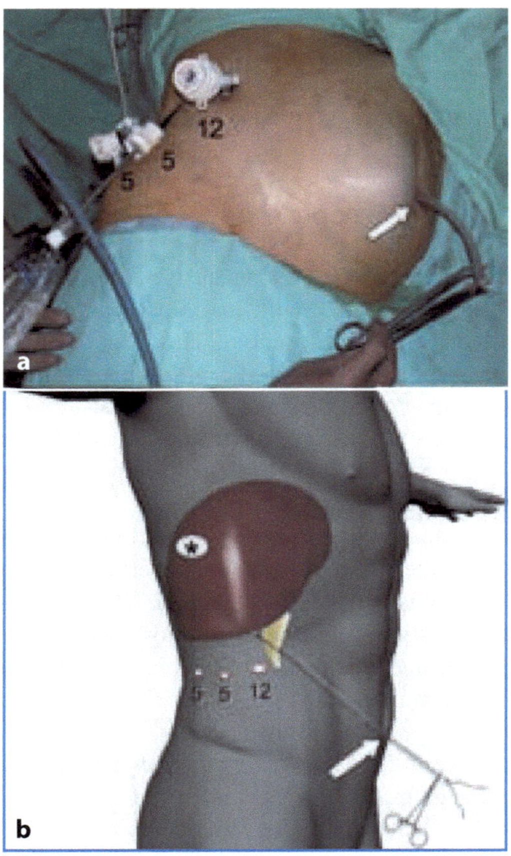

Fig. 15.1 a Patient and trocar position for laparoscopic resection of lesions located posteriorly: segments 6 and 7 and posterior areas of segments 5 and 8.
b Same position with tourniquet in place. *Asterisk* indicates the lesion. *White arrow* indicates the catheter where the tourniquet becomes extracorporeal. Reproduced from [5] with permission

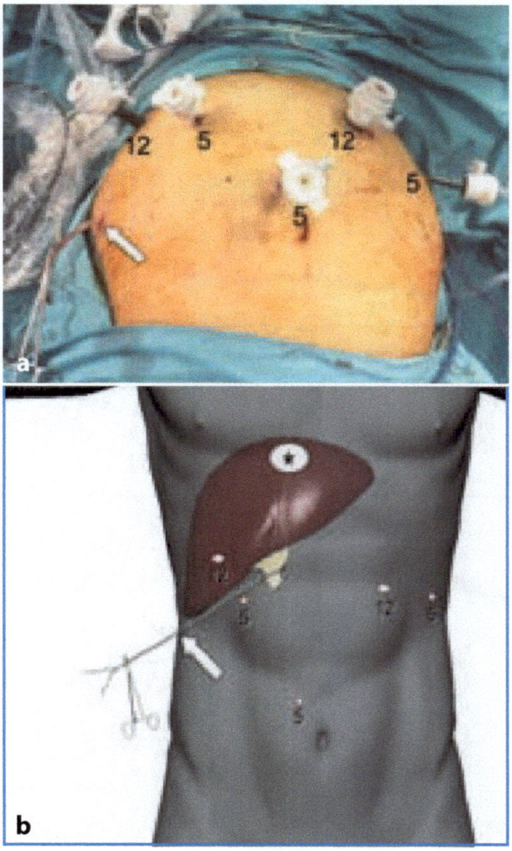

Fig. 15.2 **a** Patient and trocar position for laparoscopic resection of lesions located anteriorly: segments 2, 3, 4 a–b, 5, and 8. **b** Position with the tourniquet in place. *Asterisk* indicates the lesion. *White arrow* indicates the catheter where the tourniquet becomes extracorporeal. Reproduced from [5] with permission

An easier and safer approach appears to be the use of articulate device. Maehara et al. [8] propose the use of a biliary scope (BS) introduced through the opened lesser omentum until the light of its tip is visible on the right side of pedicle; the tape is then grasped with endoscopic forceps, passed inside the BS, and a stopper is positioned along it extracorporeally if clamping is needed.

Belli et al. [9] and Cho et al. [10] independently proposed the use of the Endo Retract Maxi (United Surgical, Tyco Healthcare Group, Norwalk, CT, USA) with a silicon tape (Vesseloops; Argon Medical Devices, TX, USA) preliminarily fixed to this tip (Fig. 15.3). When this device is behind the HDL, its metallic arch is prudently extended, allowing visualization of its tip and the tape at the right side of the pedicle (see video related to this chapter).

With the same purpose, Saif et al. [11] describe the use of the Gold Finger (Ethicon Endo Surgery, Johnson & Johnson, New Brunswick, NJ, USA), a flexible articulating device developed for bariatric surgery. After positioning the tape around the HDL, the two ends of the device are positioned through a port, inside a tube (chest tube or bladder catheter or drain tube), extracorpore-

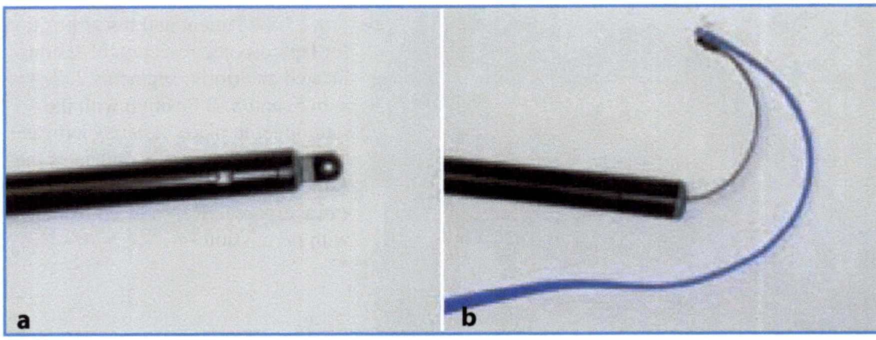

Fig. 15.3 **a** Endo Retract (Covidien) Maxi in closed position. **b** Endo Retract Maxi in activated position. Vessel tape is preliminarily fixed to the tip of the metallic arch

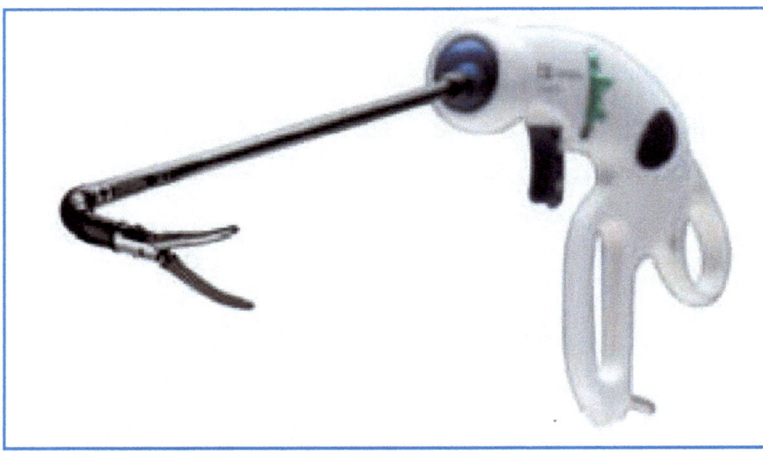

Fig. 15.4 SILS™ Dissector (courtesy of Covidien)

ally. Before starting the parenchyma transection, or in case of bleeding, it is possible to blindly pull the tape with the left hand and push the tube with the right one, without changing the visual field.

In our experience, we always prepare the HDL for eventual inflow occlusion. We pass a silicon tape (Vessel Loops; Argon Medical Devices TX, USA) behind the pedicle from the left or right using the Endo Dissect articulating device (Covidien) (Figs. 15.4 and 15.5); two clips join the extremes of the tape, which remains in the intraperitoneal position. If clamping is necessary, the surgeon can expose the HDL by pulling the tape with the left hand and placing a laparoscopic vascular clamp on the portal pedicle with the right hand from the left flank through an additional trocar.

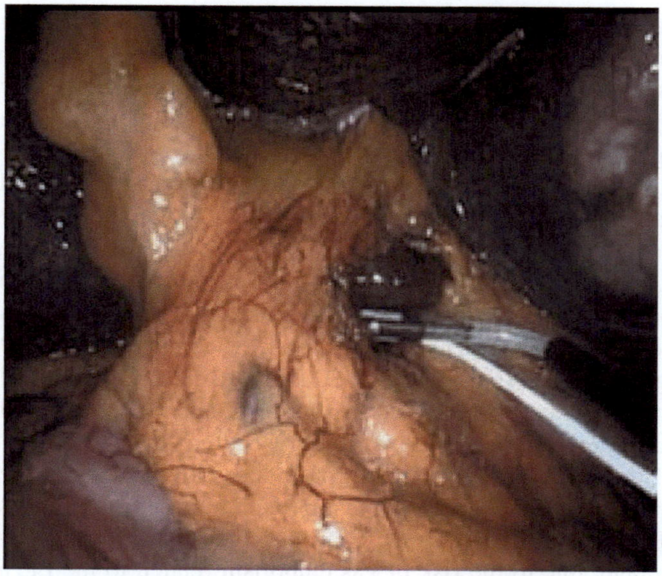

Fig. 15.5 SILS™ Dissector with Vessel Loop (Argon Medical Devices, Athens, TX, USA) around hepatic hilum

Few data are available regarding inflow hepatic control in the event of a laparoscopic redo. Belli et al. [12] report data on 15 laparoscopic redo surgeries (12 resection and three radiofrequency ablation) for recurrent HCC after open or minimally invasive resection; the authors prepared the PM in nine patients, but there was no need to perform it.

Cheung et al. [13] reported a case of laparoscopic hepatectomy for recurrent HCC; PM was not applied during transection.

Shafaee et al. [14] describe a large tri-institutional analysis of 76 repeat hepatectomy for both benign and malign recurrence, but no information is available regarding inflow control during this redo surgery.

15.4 Conclusions

PM has almost no systemic hemodynamic repercussions, although some patients with unstable cardiovascular status can present dangerous arterial hypotension, requiring fluid refilling that increases venous pressure, leading to blood loss from hepatic veins.

Refinements in surgical tools and improvements in anesthesiology management therefore allow surgeons to perform major hepatectomies with an acceptable morbidity rate. Irrespective of this global advancement, part of the morbidity is related to ischemia–reperfusion injury of the small, and often diseased, remnant liver. Regardless, it is remarkable to note that hepatic resection without vascular clamping is possible, feasible, and safe: many reports show that major liver resection can also be safely performed without vascular clamping. In fact, after years of systematic vascular clamping during major hepatic resection, the advances in parenchymal transection, including the use of coagulating and gently dissecting devices, make PM not always necessary.

In LLR, thanks to different options proposed for encircling the pedicle and the improvements in transection devices, widespread and systematic inflow occlusion appears sometimes unjustified.

Topal et al. [15], in a large series, statistically analyzes blood loss between laparoscopic and open liver resections, demonstrating that operative blood loss was significantly reduced in 109 patients undergoing LLR compared with 250 patients undergoing open resection (150 vs. 300 ml). Simillis et al. [16] described reduced blood loss with LLR but also more frequent use of portal clamping during LLR in respect to traditional liver resection.

A CVP <5 cm H_2O during open liver resection reduces bleeding from the transection surface, but during the laparoscopic approach, this setting changes greatly. First, the PP collapses the IVC and CVP determination becomes unreliable. On the other hand, due to the risk of gas embolism related to the PP, a CVP slightly increased at 3–6 cm H_2O is considered protective. This warning should always be present in our minds.

Several groups report significant LLR series also in cirrhotic patients without using the PM [17, 18]. The less frequent use of hepatic inflow occlusion during resection is a parameter evaluated during the progress of the learning curve in minimally invasive liver surgery [19].

The spread of living-donor transplantation has revitalized the debate surrounding ischemia–reperfusion injury. Grafts resected from living donors could be harvested without theoretical negative effects of ischemia induced by hepatic pedicle clamping. Always taking into account that donor safety is undoubtedly the highest priority, it is better for the donor's residual liver that PM be avoided [20]. The technical possibility of performing laparoscopic left-lateral sectionectomy for pediatric living-donor transplantation, without performing PM, is remarkable [21] as the ultimate step of the learning curve.

Laparoscopic PM is widely adopted to decrease blood loss during liver surgery. However, thanks to contemporary transection equipment, advanced operative techniques, and increasing skill of hepatic surgeons in minimally invasive liver surgery, routine use of inflow occlusion is becoming less and less frequent. However, it is important that the hepatic laparoscopic surgeon has the skill to perform, in a few minutes, a practical procedure to occlude hepatic blood inflow.

James Hogarth Pringle (1863–1941)

Fig. 15.6 James Hogarth Pringle (1863–1941), Royal College of Physicians and Surgeons of Glasgow

The name of the surgeon who first performed a particular surgical procedure or described an anatomical structure is often used to identify such landmarks, yet the extraordinary professional achievements of the surgeon generally remain unknown. James Hogarth Pringle (Fig.15.6) was born in Parramatta, Australia, on 26 January 1863, the son of well-know surgeon George Hogarth Pringle of Sydney, Australia [22]. He graduated in medicine at Edinburgh University in Scotland in 1885, following which he made a 3-year European grand tour across Berlin, Vienna, and Hamburg, creating fruitful links with German and Austrian surgeons.

In 1892, he began working with Sir William Macewen at Glasgow Royal Infirmary (GRI), Scotland, where remained for the rest of his career, until 1923.

He was a skilled surgeon with an impressive surgical series and an active participation as a member of the British Association of Surgeons. He and his master, Macewen, were among the first to be sympathetic to the cause of the first women medical students, accepting them into their clinics. He also contributed greatly to the fields of malignant melanoma, head injury, and arterial reconstruction with a saphenous vein graft.

His book *Fractures and Their Treatment*, published in 1910, was a milestone textbook for a generation of British orthopedists. There were two reasons for the extraordinary results of his innovative approach to fracture treatment: The first was his support at one of the first hospital radiology departments in the world, at GRI, which was founded by John Macintyre in 1896. The second was confident application of the rules of asepsis, a concept introduced by Joseph Lister, an assistant surgeon at GRI. It is remarkable, moreover, to remember that Pringle was the first to perform hindquarter amputation in a one-stage procedure for sarcoma of the thigh (a 50-min procedure under spinal anaesthesia!).

Pringle could hardly have imagined that well into the future, the maneuver he proposed to control bleeding of the liver during surgery would still be written and talked about, or that his technique would be considerably debated in relation to a minimally invasive approach to liver surgery. From the end of the nineteenth and the beginning of the twentieth century, there was much scientific turmoil surrounding reasons for the inexorable and abrupt death of animals subjected to clamping of the hepatic pedicle. In his paper

of 1908 (Ann Surg 48:541–549), Pringle supports the effectiveness of temporary hepatic hilum clamping. He reported on eight cases of liver injury: three patients died just after arrival at the hospital, and one patient died 3 days after the injury because he refused surgery. In the remaining four patients, temporary clamping of the hepatic pedicle allowed surgery to be performed in a bloodless operating field by successfully obtaining hemostasis without fatal implications for the patients: "An assistant held the portal vein and the hepatic artery between a finger and thumb and completely arrested all bleeding, and we got on a little further in consequence of being able to wipe out blood and blood clots and to examine the rupture" [23].
In support of his observations, he reported successful experiences with liver resection in rabbits by clamping the hepatic hilum.

References

1. Nguyen KT, Gamblin TC, Geller DA (2009) World review of laparoscopic liver resection-2,804 Patients. Ann Sur 250:831–841
2. Nsadi B, Gilson N, Pire E et al (2011) Consequences of pneumoperitoneum on liver ischemia during laparoscopic portal triad clamping in a swine model. J Surg Res 166(1):35–43
3. Patriti A, Ceribelli C, Ceccarelli G et al (2012) Non-cirrhotic liver tolerance to intermittent inflow occlusion during laparoscopic liver resection. Updates Surg. doi:10.1007/s13304-012-0144-5
4. Torzilli G, Procopio F, Donadon M et al (2012) Safety of intermittent Pringle maneuver cumulative time exceeding 120 minutes in liver resection: a further step in favor of the "Radical but Conservative" policy. Ann Surg 255(2):270–280
5. Rotellar F, Pardo F, Bueno A et al (2012) Extracorporeal tourniquet method for intermittent hepatic pedicle clamping during laparoscopic liver surgery: an easy, cheap, and effective technique. Langenbecks Arch Surg 397:481–485
6. Cherqui D, Husson E, Hammoud R et al (2000) Laparoscopic liver resections: a feasibility study in 30 patients. Ann Surg 232:753–762
7. Patriti A, Ceccarelli G, Bartoli A, Casciola L (2011) Extracorporeal Pringle maneuver in robot-assisted liver surgery. Surg Laparosc Endosc Percutan Tech 21(5):e242–e244
8. Maehara S, Adachi E, Shimada M et al (2007) Clinical usefulness of biliary scope for Pringle's maneuver in laparoscopic hepatectomy. J Am Coll Surg 205:816–818
9. Belli G, Fantini C, D'Agostino A et al (2008) Laparoscopic segment VI liver resection using a left lateral decubitus position: a personal modified technique. J Gastrointest Surg 12:2221–2226
10. Cho A, Yamamoto H, Nagata M et al (2009) Safe and feasible inflow occlusion in laparoscopic liver resection. Surg Endosc 23:906–908
11. Saif R, Jacob M, Robinson S et al (2011) Laparoscopic Pringle's manoeuvre for liver resection: how I do it. Minim Invasive Ther Allied Technol 20:365–368
12. Belli G, Cioffi L, Fantini C et al (2009) Laparoscopic redo surgery for recurrent hepatocellular carcinoma in cirrhotic patients: feasibility, safety and results. Surg Endosc 23:1807–1811
13. Cheung TT, Kwok-chai KNg, Tung-ping Poon R et al (2010) A case of laparoscopic hepatectomy for recurrent hepatocellular carcinoma. World J Gastroenterol 16(4):526–530
14. Shafaee Z, Kazaryan AM, Marvin MR et al (2011) Is laparoscopic repeat hepatectomy feasible? A tri-institutional analysis. J Am Coll Surg 212:171–179
15. Topal B, Fieuws S, Aerts R, et al (2008) Laparoscopic versus open liver resection of hepatic neoplasms: comparative analysis of short-term results. Surg Endosc 22:2208–2213

16. Simillis C, Constantinides VA, Tekkis PP et al (2007) Laparoscopic versus open hepatic resections for benign and malignant neoplasms a meta-analysis. Surgery 141(2):203–211
17. Belli G, Fantini C, D'Agostino A et al (2005) Laparoscopic liver resection without a Pringle maneuver for HCC in cirrhotic patients. Chir Ital 57(1):15–25
18. Pulitanò C, Catena M, Arru M et al (2008) Laparoscopic liver resection without portal clamping: a prospective evaluation. Surg Endosc 22:2196–2200
19. Viganò L, Laurent A, Tayar C et al (2009) The learning curve in laparoscopic liver resection improved feasibility and reproducibility. Ann Surg 250:772–782
20. Imamura H, Takayama T, Sugawara Y et al (2002) Pringle's manoeuvre in living donors. Lancet 360:2049–2050
21. Troisi RI, Van Huysse J, Berrevoet F et al (2011) Evolution of laparoscopic left lateral sectionectomy without the Pringle maneuver: through resection of benign and malignant tumors to living liver donation. Surg Endosc 25:79–87
22. Macintyre I (2008) George (1830-72) and his son James Hogarth (1863-1941) Pringle: unsung surgical pioneers. J Med Biogr 16:155–161
23. Pringle JH (1908) Notes on the arrest of hepatic haemorrhage due to trauma. Ann Surg 48:541–549

Minimally Invasive Combined Surgery: Liver and Colon–Rectum

16

Mario Morino and Federico Famiglietti

16.1 Indications and Background

Minimally invasive combined colorectal and liver resection is indicated for both benign (diverticular disease, colonic adenoma unsuitable for endoscopic resection, benign liver lesions) and malignant (primary colorectal cancer, hepatocellular carcinoma or HCC, colorectal liver metastases) etiologies. However, whereas reports for benign diseases are sporadic [1], most of the studies published show results of the combined minimally invasive approach to primary colorectal cancer (CRC) with synchronous liver metastases (SLM) [2–7].

During the natural history of colorectal cancer, about 50 % of patients develop hepatic metastases, and of these, 15–20 % has synchronous disease [8]. Although synchronous metastases may confer a poorer prognosis than metachronous metastases, complete surgical resection remains the only therapy with potential for cure [9]. In this setting, simultaneous colorectal and hepatic resection for primary CRC with SLM compared with the classical staged approach has been described as safe and feasible in selected patients [10–12]. In particular, a meta-analysis of 14 retrospective studies showed that simultaneous resections lead to shorter hospital stay and lower morbidity rates over the staged approach, with similar long-term outcomes [12]. This approach may also be beneficial in terms of cost and patient quality of life, avoiding two hospitalizations and limiting the effect of two major surgeries on immunological homeostasis. Moreover, from a psychological point of view, patients feel better knowing that all the disease has been removed in a single procedure. Given these findings, laparoscopy should be considered the natu-

M. Morino (✉)
Digestive and Colorectal Surgery, University of Turin, Turin, Italy
e-mail: mario.morino@unito.it

F. Calise and G. Casciola (eds.), *Minimally Invasive Surgery of the Liver,*
Updates in Surgery
DOI: 10.1007/978-88-470-2664-3_16, © Springer-Verlag Italia 2013

ral evolution for such an approach, enhancing the benefits with its well-known short-term advantages over open surgery in both colorectal and liver resections [13, 14].

16.2 Patient Selection

Proper patient selection for this procedure is of paramount importance. When considering a minimally invasive combined resection for stage IV CRC, limitations of the simultaneous approach and those of the laparoscopic approach should be taken into account (Table 16.1). First, patient characteristics and tumor behavior should be considered. Concerning obesity, even though laparoscopic colorectal resection in obese patients is safe and feasible [15], it could be an extremely demanding procedure in certain cases (i.e., rectal cancer) [16]. Conversely, the laparoscopic combined procedure may be advantageous, especially in severely obese patients, in terms of reduced surgical-site infection (SSI) and wound complication rates. So, body mass index (BMI) must be carefully evaluated and related to the surgeon's laparoscopic expertise. Moreover, a detailed history should be obtained, and special attention should be given to performance status and symptoms and signs suggesting advanced disease, such as pain, ascites, jaundice, weight loss, and a palpable mass. Similarly, highly (>30 ng/ml) elevated carcinoembryonic antigen (CEA) serum levels were found to be predictive of the benefit of perioperative chemotherapy before surgery [17]. Obstruction and perforation from the primary colorectal cancer are further poor prognostic signs, also representing an adverse clinical condition for a combined procedure. Conversely, extrahepatic metastases should not be considered a contraindication unless unresectable.

The extent and location of metastatic disease within the liver are other important issues. The use of a simultaneous approach for SLM is highly controversial due to concerns about the extent of hepatectomy from both oncological and technical points of view. Several authors describe low survival and

Table 16.1 Contraindications for minimally invasive combined colorectal and liver resection

Absolute contraindications	Relative contraindications
PS > 2	BMI > 30; (GoR D)
Poor prognosis	More than 3 SLM; (GoR C)
Obstructed or perforated CRC	Need for major liver resection (3 or more segments); (GoR C)
Unresectable extrahepatic metastases	Lesions in segments 1, 7, 8; (GoR C)
R0 unlikely	Extraperitoneal rectal cancer; (GoR C/D)
	Primary CRC > 5 cm; (GoR C/D)

PS, performance status; *BMI*, body mass index; *CRC*, colorectal cancer; *SLM*, synchronous liver metastases; *GoR*, grade of recommendation.

high morbidity rates for patients with more than three SLM [10, 11], suggesting a staged approach in these cases. Moreover, tumor location in the posterior and superior liver segments (segments 1, 5, and 8) and the need for major hepatectomy are well-known limitations for laparoscopic liver surgery (LLS) [18]. Despite recent advances in minimally invasive liver surgery and chemotherapeutic regimens, patients with either extensive liver metastatic involvement or scheduled for resections of the posterior and anterior segments should be referred to centers highly experienced in advanced laparoscopic procedures and liver surgery or submitted to hybrid laparoscopic procedures.

Finally, size and site of the primary tumor are predictive of morbidity in metastatic patients who undergo laparoscopic colectomy for CRC. In fact, a retrospective analysis from our institution comparing laparoscopic and open colorectal resections in patients with symptomatic stage IV colorectal cancer raises concerns about the laparoscopic approach to tumors >5 cm and extraperitoneal rectal cancers due to the increased risk of conversion and postoperative complications [19]. In our opinion, these cases should probably be approached directly by laparotomy.

In conclusion, a tailored approach to CRC patients with SLM should be used. The characteristics of patients and their tumors should be discussed within a multidisciplinary team to plan the most appropriate timing of chemotherapy and surgery.

16.3 Techniques of Minimally Invasive Combined Colorectal and Liver Resection

Due to its recent introduction, the laparoscopic technique of combined hepatic and colorectal resection is not yet well standardized. To date, only case reports [1, 6] and a few case series [2–5, 7] are available in the literature, and technical details differ considerably. Moreover, laparoscopic combined resections for SLM are often included as associate procedures in large series of laparoscopic liver or colonic resections in which details of the combined technique and their specific results are not reported [19, 20]. In general, the minimally invasive approach to patients with CRC and SLM could be roughly categorized into two groups according to whether one of the resections is performed via laparotomy.

16.3.1 Pure Laparoscopic Technique

Patients with peripheral SLM located on left lateral (segments 2 and 3) and anterior right (5 and 6) liver segments can benefit from a totally laparoscopic combined resection. When both resections are deemed feasible by laparoscopy, we routinely start with the hepatic resection in order to avoid both septic contamination of the liver surgical field and the adverse effect of a

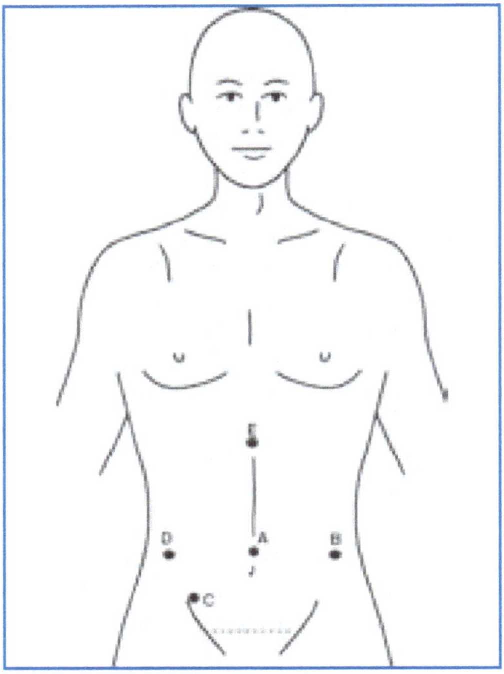

Fig. 16.1 Trocar positioning for left-sided colorectal lesions. During colorectal resection, the operative trocars are C and D; during liver resection, they are D and E for tumors located in the right lobe and B and D for tumors located in the left lobe. *Dotted lines* represent possible sites of minilaparotomy

Pringle maneuver on venous colonic outflow. In fact, liver pedicle clamping causes transient portal hypertension that leads to intestinal edema and consequently to an increased risk of anastomotic leakage. The patient, under general anesthesia, is placed in a supine position with legs abducted. The surgeon stands between the patient's legs and the monitor is positioned on the patient's right shoulder. Usually, the Lloyd–Davis position is preferred for lesions in segments 2–5, whereas the semi-left-lateral decubitus position is used for the lesions in segment 6. Pneumoperitoneum is created by a Veress needle at an intra-abdominal pressure of 12 mmHg. Five trocars are usually placed both for left- (Fig. 16.1) and right-sided lesions (Fig. 16.2). The first phase of the procedure consists of careful exploration of the peritoneal cavity, the liver, and the pelvis. The feasibility of laparoscopic colorectal resection should be verified at the beginning, as it is a crucial point for both successfully completing the combined procedure and preventing complications such as conversion to open surgery. Next, the patient is placed in a reverse Trendelenburg position. A laparoscopic ultrasound should be performed in all cases to confirm resectability and detect any other occult metastases. The hepatic pedicle can be isolated and a tape passed around it; both ends are passed in a tourniquet to enable performance of a Pringle maneuver. The dissection line is demarcated on the liver with monopolar cautery. Liver parenchymal transection is per-

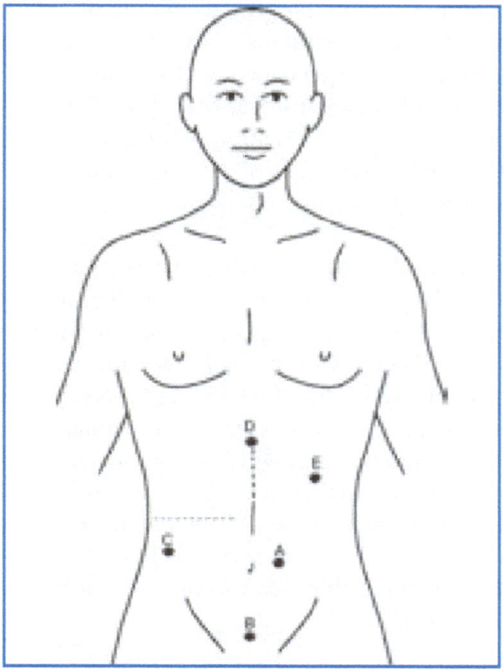

Fig. 16.2 Trocar positioning for right-sided colonic lesions. During colonic resection, operative trocars are B and D; during liver resection, they are C and D for tumors located in the right lobe and C and E for tumors located in the left lobe. *Dotted lines* represent possible sites of minilaparotomy

formed with an ultrasonic scalpel. Intraparenchymal control of the major vessels is achieved with clips or sutures. In left lobectomies (segments 2 and 3), the left hepatic vein is sectioned with a linear vascular endostapler. Sectioning of the segmental bile ducts is performed by extraparenchymal ligation and scissors sectioning in cases of left lobectomy (resections of segments 2 and 3), using the ultrasonic scalpel in other procedures. The specimen is put into a sterile bag and left under the diaphragm for later retrieval. The surgeon then moves to the patient's right side and the monitor is shifted close to the patient's left leg. For resections involving left colon, sigmoid, or rectum, the splenic flexure is mobilized at the beginning of the operation. Next, the patient is placed in a Trendelenburg position with a right tilt. A lateral-to-medial dissection is performed, with a high vascular ligation. The inferior mesenteric artery is divided 1 cm from the aorta after ligation with clips, and the inferior mesenteric vein is divided at the level of Treitz ligament. Complete left-colon mobilization and a distal colonic or rectal transection is performed. A suprapubic minilaparotomy is then performed to extract both colorectal and hepatic specimens, to resect the proximal margin, and to introduce the stapler anvil. Finally, pneumoperitoneum is reestablished, and a double-stapled mechanical anastomosis is fashioned. When the distal clearance of the inferior margin of the tumor is at the level of the surgical anal canal, or in a narrow pelvis where a transverse stapled section is sometimes impossible, the technique of choice is a rectal mucosectomy and a true coloanal anastomosis, executed manually from below and removing the specimen by the anal verge. Right hemicolectomies are completed laparoscopically with a medial-to-lateral dissection, high vascular ligation, and complete right colonic mobilization. An intracorporeal ileocolic stapled anastomosis is fashioned. Irrigation of all ports and extraction sites with povidone–iodine solution is recommended. Suction drains are selectively used.

Sometimes concerns about the feasibility of a totally laparoscopic combined procedure could be raised at the beginning of the operation due to technical or anesthesiological problems. In these cases, it is advisable to attempt laparoscopic colorectal resection first. Once the colorectal resection is completed, three main factors should be evaluated before proceeding to liver resection: (1) an R0 colorectal resection must be likely, (2) technical evaluation of the quality of the anastomosis must be satisfactory, and (3) the expected operative time of the entire combined procedure should not exceed 8 h [3]. If these criteria are met, the patient is turned into the reverse Trendelenburg position and the laparoscopic hepatic resection can be then performed. Conversely, when one of the resections is deemed unfeasible laparoscopically, a hybrid video-assisted technique becomes necessary.

16.3.2 Hybrid Video-Assisted Techniques

In general, hybrid methods are advisable when a complete resection either of the hepatic or colorectal lesion is difficult to achieve laparoscopically.

When a major liver resection (three or more segments) is needed, this can be completed through a right subcostal incision, whereas the laparoscopic CRC resection can be easily performed with a classic four-trocar technique, as described above. The CRC resection should be performed first, but the intestinal anastomosis could be fashioned at the end of the liver parenchymal transection, as a Pringle maneuver is probably necessary in these cases and the intestinal anastomosis may be compromised. Moreover, blood loss and transfusions might occur during such resections, and anastomotic perfusion may be affected. The resected specimens are then brought out through the midline or the right subcostal incision.

Similarly, in patients with low, bulky T4 rectal cancers and SLM, a totally laparoscopic combined resection is not recommended. Four to five trocars are placed along an ideal semicircular line, with the concavity facing the subcostal margin to perform laparoscopic liver resection (LLR), as described previously, whereas open anterior resection is classically performed through a midline incision.

16.4 Outcomes of Minimally Invasive Combined Colorectal and Liver Resection

Outcomes of laparoscopic combined colorectal and liver resections are infrequently reported in the literature due to its recent introduction into clinical practice. Moreover, most studies are on video-assisted hybrid operations, whereas totally laparoscopic combined procedures are often included as associate procedures in large series of LLR or colonic resections, in which their specific results are not reported [19, 20]. The procedure is safe and feasible in highly selected patients. Morbidity rate ranges from 10 % to 50 % [4, 7], median hospital stay varies from 10 to 19 days [4, 7], and operative time is between 300 and 446 min [3, 4]. Only one comparative study between laparoscopic and the open approach to combined colorectal and liver resection has been published so far [7]. The authors found that the laparoscopic-assisted procedure was associated with significantly longer operative time (358 vs 278 min) but lower blood loss (350 vs 500 ml). Morbidity rate was similar between groups and the 3-year survival rate. Of note, 13 of 20 patients underwent a hybrid procedure with open liver resection.

16.5 Summary

The use of laparoscopy in this specific field seems promising. Adequate liver exploration during synchronous resections is crucial. Compared with open surgery, the laparoscopic approach allows minimization of abdominal wall trauma and consequently reduces postoperative pain, the incidence of incisional hernia, and SSIs. Moreover, as recently reported [13, 14], laparoscopy enhances recovery and could lead to an earlier start of chemotherapy in these patients. Adequate patient selection and extensive surgeon experience in hepatic and laparoscopic surgery are essential prerequisites to optimize outcomes. Nevertheless, several issues remain unsolved (resection timing and concerns about anastomotic leakage and septic contamination of the liver surface), and the technique needs to be standardized. Future studies should try to answer these open questions.

References

1. Inagaki H, Kurokawa T, Sakamoto J et al (2003) Laparoscopic left hemihepatectomy combined with right hemicolectomy for liver tumor and hemorrhagic diverticulosis. Surg Endosc 17:158–159
2. Kim SH, Lim SB, Ha YH et al (2008) Laparoscopic-assisted combined colon and liver resection for primary colorectal cancer with synchronous liver metastases: initial experience. World J Surg 32:2701–2706

3. Bretagnol F, Hatwell C, Farges O et al (2008) Benefit of laparoscopy for rectal resection in patients operated simultaneously for synchronous liver metastases: preliminary experience. Surgery 144:436–441
4. Akiyoshi T, Kuroyanagi H, Saiura A et al (2009) Simultaneous resection of colorectal cancer and synchronous liver metastases: initial experience of laparoscopy for colorectal cancer resection. Dig Surg 26:471–475
5. Lee JS, Hong HT, Kim JH et al (2010) Simultaneous laparoscopic resection of primary colorectal cancer and metastatic liver tumor: initial experience of single institute. J Laparoendosc Adv Surg Tech A 20:683–687
6. Casaccia M, Famiglietti F, Andorno E et al (2010) Simultaneous laparoscopic anterior resection and left hepatic lobectomy for stage IV rectal cancer. JSLS 14:414–417
7. Huh JW, Koh YS, Kim HR et al (2011) Comparison of laparoscopic and open colorectal resections for patients undergoing simultaneous R0 resection for liver metastases. Surg Endosc 25:193–198
8. Manfredi S, Lepage C, Hatem C et al (2006) Epidemiology and management of liver metastases from colorectal cancer. Ann Surg 244:254–259
9. Tsai MS, Su YH, Ho MC et al (2007) Clinicopathological features and prognosis in resectable synchronous and metachronous colorectal liver metastasis. Ann Surg Oncol 14:786–794
10. Capussotti L, Viganò L, Ferrero A et al (2007) Timing of resection of liver metastases synchronous to colorectal tumor: proposal of prognosis-based decisional model. Ann Surg Oncol 14:1143–1150
11. Reddy SK, Pawlik TM, Zorzi D et al (2007) Simultaneous resections of colorectal cancer and synchronous liver metastases: a multi-institutional analysis. Ann Surg Oncol 14:3481–3491
12. Chen J, Li Q, Wang C et al (2011) Simultaneous vs. staged resection for synchronous colorectal liver metastases: a metaanalysis. Int J Colorectal Dis 26:191–199
13. Schwenk W, Haase O, Neudecker J et al (2005) Short term benefits for laparoscopic colorectal resection. Cochrane Database Syst Rev 3:CD003145
14. Croome KP, Yamashita MH (2010) Laparoscopic vs open hepatic resection for benign and malignant tumors: an updated meta-analysis. Arch Surg 145:1109–1118
15. Makino T, Shukla PJ, Rubino F et al (2012) The impact of obesity on perioperative outcomes after laparoscopic colorectal resection. Ann Surg 255:228–236
16. Buchanan GN, Malik A, Parvaiz A et al (2008) Laparoscopic resection for colorectal cancer. Br J Surg 95:893–902
17. Sorbye H, Mauer M, Gruenberger T et al (2012) Predictive factors for the benefit of perioperative FOLFOX for resectable liver metastasis in colorectal cancer patients (EORTC Intergroup Trial 40983). Ann Surg 255:534–539
18. Morino M, Morra I, Rosso E et al (2003) Laparoscopic vs open hepatic resection: a comparative study. Surg Endosc 17:1914–1918
19. Allaix ME, Degiuli M, Giraudo G et al (2012) Laparoscopic versus open colorectal resections in patients with symptomatic stage IV colorectal cancer. Surg Endosc 5 [Epub ahead of print, PMID: 22476839]
20. Castaing D, Vibert E, Ricca L et al (2009) Oncologic results of laparoscopic versus open hepatectomy for colorectal liver metastases in two specialized centers. Ann Surg 250:849–855

Minimally Invasive Procedures for Liver Trauma

17

Graziano Ceccarelli, Raffaella Niola, Alberto Patriti, Fulvio Calise and Luciano Casciola

17.1 Introduction

The liver is one of the most frequently injured abdominal organs in trauma (about 35–45% of blunt abdominal trauma, 40% of stab wounds, and 30% of gunshot wounds); about 80% occur in men between 20 and 40 years of age [1]. Liver trauma is classified according to the Liver Injury Scale from the American Association for the Surgery of Trauma (AAST), which lists six grades with progressive severity depending above all on two parameters: depth of parenchyma laceration, and the area of hematoma with or without vascular involvement. The diagnostic and therapeutic approach to abdominal and liver trauma has evolved in recent decades, leading to a reduction in deaths [1, 2].

Conservative treatment in intensive care in the majority of liver injuries is supported by a high level of practitioner experience, with a successful outcome in approximately 70% of patients and a very low morbidity rate [3–5]. Treatment requires hemodynamic stability, accurate and close imaging evaluation [computed tomography (CT) scan, ultrasound (US)] and the absence of other serious abdominal injuries (bowel perforations, vascular or thoracic lesions). Considerations leading to a nonoperative approach are that many liver injuries stop bleeding spontaneously; too many nontherapeutic laparotomies in the past, and fewer complications with the nonoperative approach. The major drawbacks of the conservative approach are the need for close radiological and clinical monitoring and the possibility of missing lesions.

Minimally invasive techniques, such as image-guided percutaneous drainage, endoscopic retrograde cholangiopancreatogram (ERCP), operative

G. Ceccarelli (✉)
Department of Surgery, Division of General, Minimally Invasive and Robotic Surgery,
ASL3 Umbria, San Matteo degli Infermi Hospital, Spoleto, Italy
e-mail:g.cecca2003@libero.it

F. Calise and G. Casciola (eds.), *Minimally Invasive Surgery of the Liver,*
Updates in Surgery
DOI: 10.1007/978-88-470-2664-3_17, © Springer-Verlag Italia 2013

angiography, and – in particular – laparoscopy, may be useful in selected trauma cases and for treating complications correlated with the trauma.

17.2 Approach to Patients with Abdominal Trauma

A multimodality intervention strategy is required. Patients with liver trauma can be triaged into two subsets upon arrival in the emergency room: hemodynamically stable patients; and patients in critical condition from the outset, with hemodynamic instability and risk of consumption coagulopathy. Hemodynamic stability is defined as a systolic pressure >90 mmHg and a pulse rate <100 bpm. A tensely distended abdomen with acute anemia, persistent hypotension despite replacement with a liter of colloid infusion during transport, and the need for transfusion of ≥5 U blood require immediate operative intervention due to the of risk of coagulopathy and multiorgan failure. The presence of extraintestinal free air indicates perforation of a hollow viscus requiring surgery. There is no established consensus on how much blood loss or transfusion requirement mandates the decision to operate. Some authors consider a simple liver laceration with <250 ml of hemoperitoneum (limited bleeding) an indication for conservative management, whereas others accept as much as 500 ml of blood filling the cul-de-sac (moderate bleeding) before opting for surgery (Fig. 17.1a-c) [3, 6]. Multidetector-row CT (MDCT) is the main diagnostic tool in trauma evaluation to define liver injury severity, quantify hemoperitoneum, and identify other intraperitoneal or retroperitoneal injuries (duodenum, pancreas, kidneys) or the presence of pneumoperitoneum. It also offers the advantage of identifying ongoing active bleeding [4, 7, 8]. The value and clinical use of diagnostic peritoneal lavage (DPL), introduced in 1965 by Root [9], has been reduced with the diffusion of US and CT-scan procedures. US performed in casualty [focused assessment sonography in trauma (FAST) technique for four-quadrant evaluation and cardiac region], may be useful but has low efficacy in identifying bowel injury.

17.3 Imaging Findings

CT-based criteria guide the diagnostic management of blunt hepatic trauma. They include intrahepatic or subcapsular hematoma, evidence of arterial vascular injury such as pseudoaneurysms and traumatic arteriovenous fistulas, presence of hepatic venous involvement, devascularized tissue, hemoperitoneum grade, and associated injured abdominal organs (Fig. 17.1d-f). These findings indicate either conservative or operative management of hepatic injury and help in the selection of patients who should undergo hepatic angiography and possibly embolization.

Hepatic contusions appear on contrast-enhanced MDCT as poorly marginated, low-attenuation areas compared with the normal, enhancing, hepatic

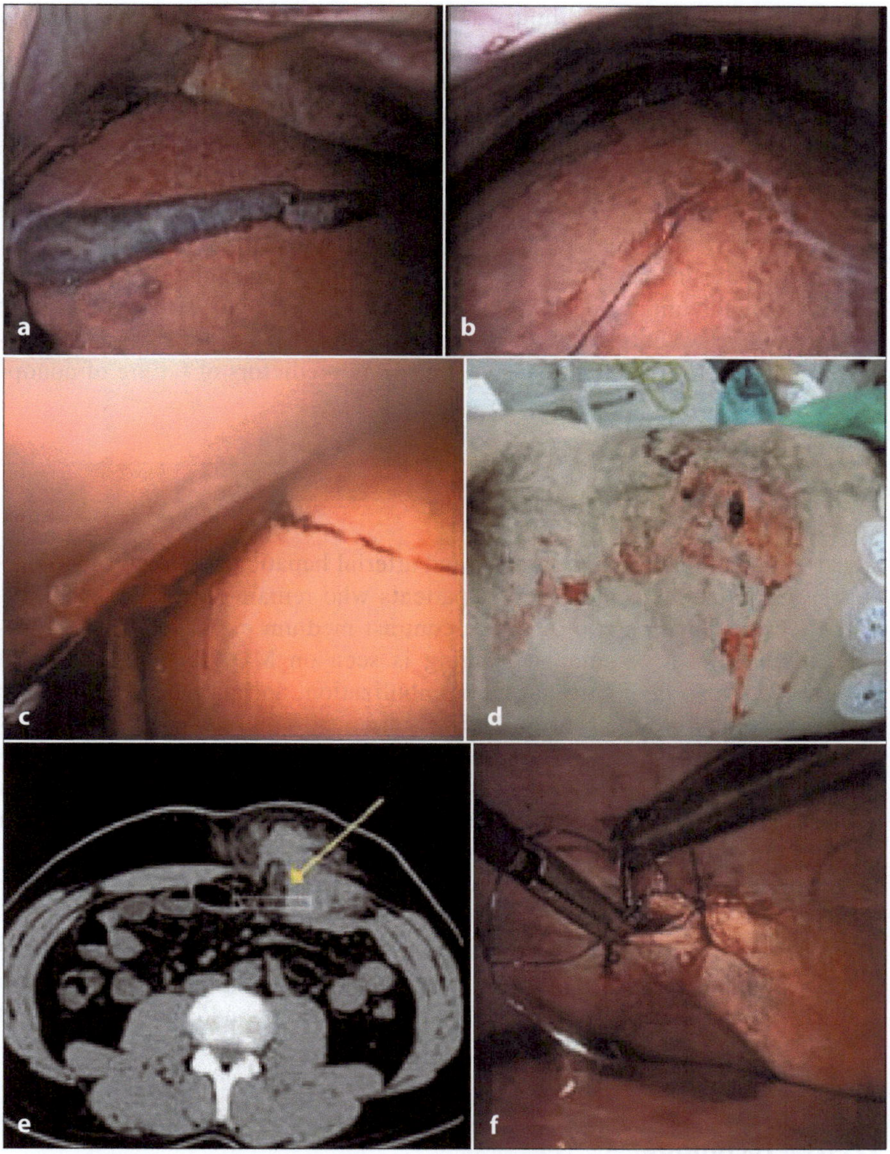

Fig 17.1 a Hepatic subcapsular hematoma. b Hematoma lacerations. c Hematoma laceration with hemoperitoneum. d Abdominal stab wound. e Abdominal stab wound TC imaging. f Laparoscopic repair of diaphragmatic laceration in a stab wound involving liver and diaphragm

parenchyma. Hepatic lacerations appear as well-defined, linear, or branching areas of low attenuation within the enhancing liver parenchyma. Perihepatic and intraperitoneal blood are common with liver lacerations and indicate tearing of the hepatic capsule [10]. A bare-area laceration is commonly associated with a right adrenal or renal injury and hematoma in the right retroperitoneum [8]. Hepatic subcapsular hematomas are seen as a low-density crescentic collection of blood, typically around the right lateral hepatic margin. A subcapsular hematoma often compresses the underlying hepatic parenchyma, a characteristic that is useful for distinguishing this lesion from perihepatic blood. Active bleeding within the liver appears as an irregular or linear high-attenuation focus of extravasated intravenously administered contrast material that remains persistently high in attenuation and typically increases in size on excretory-phase images. Active extravasation in the liver may be seen in the hepatic parenchyma, subcapsular space, or peritoneum. Hepatic vascular injuries and active extravasation resulting in bleeding into the peritoneal cavity through a tear in the liver capsule are strong predictors of failure of nonoperative management [11, 12].

17.4 Hepatic Embolization

General indications for angiography and arterial hepatic embolization are evidence of continuous hemorrhage or patients who remain in borderline condition after resuscitation and in whom contrast medium extravasation or intraparenchymal blush or contrast staining is seen on MDCT. Hepatic arterial bleeding can also be controlled with embolization, and patients with hepatic trauma generally are directed to the interventional radiology suite in two scenarios: The first is when a patient continues to hemorrhage after damage-control laparotomy. The second is when a hemodynamically stable trauma patient's imaging shows hepatic hemorrhage; in this case, an embolization procedure may be used to obviate surgery.

In their study of 230 consecutive patients with grade 3, 4, or 5 hepatic injuries treated with nonoperative management, Kozar et al. reported an 11 % overall complication rate. Complications included persistent bleeding, biliary-tract complications, and abdominal compartment syndrome. Twelve patients with persistent bleeding were treated with embolization, and the bleeding was successfully controlled in all cases [13].

17.4.1 Technique

An abdominal aortogram is the first step before selective hepatic angiography. Knowledge of the hepatic vascular variants is essential. Selective celiac and superior mesenteric artery (SMA) angiography should be performed. Images should be evaluated for the presence of extravasation, pseudoaneurysm, arteriovenous fistula, anteriobiliary fistula, or traumatic occlusion. Extravasation is recognized by a dense extravascular collection of contrast medium (Fig. 17.2a-g) Pseudoaneurysms are well circumscribed collections of contrast material that extend beyond the expected vascular wall and in the acute setting

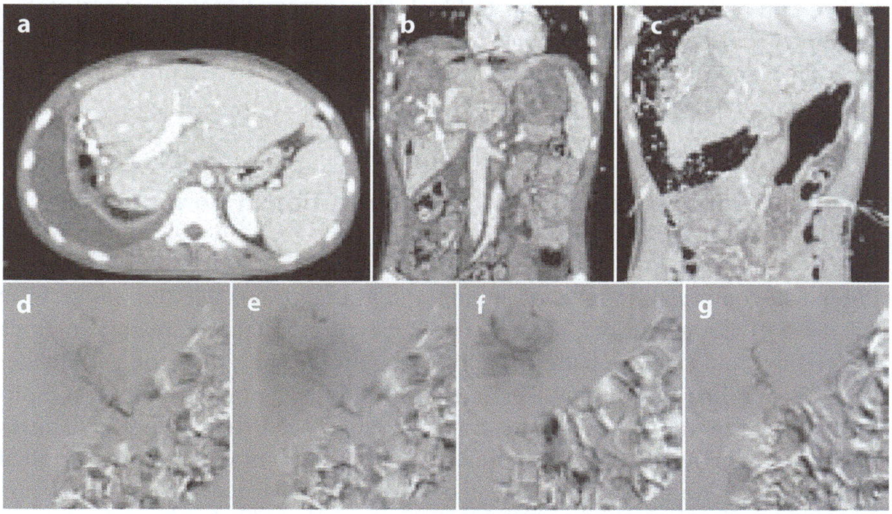

Fig 17.2 a Axial CT post-contrast image demonstrates perihepatic packing that surrounds a wide laceration of liver parenchyma. b Coronal CT post-contrast image shows a wide laceration of right lobe liver parenchyma (black arrows), in which an ongoing hemorrhage is found, demonstrated by the contrast medium extravasation (white arrow). c Coronal CT post-contrast image shows a pseudoaneurysm of the hepatic artery (black arrow). There is also a subcapsular hematoma (white arrow). d Selective hepatic arteriogram shows a contrast medium extravasation that looks like (but it is not) a small intraparenchimal pseudoaneurism, at the eight segment; arterial phase. e Late angiography phase. f coaxial technique by microcatheter allows a selective embolization of the lesion. g At the end of the procedure a devascularization of the lesion is evident

are nothing more than contained hematomas; these are unstable and should be treated. Traumatic occlusions may represent dissection without bleeding but cannot be differentiated angiographically from a transected vessel with intermittent bleeding, which may resume if the hemodynamic parameters of the patient change. Assuming normal anatomy, a diagnostic celiac angiogram though a 4- or 5-F catheter is performed and followed by subselection of small vessels as the active bleeder is approached. As with any embolization, a sheath at the access site is recommended to avoid losing access if the embolic agent occludes the catheter. The hepatic arteries are prone to spasm, and branches distal to the left or right main trunks may be too small to engage without a microcatheter. High-flow microcatheters, such as the Renegade HI-FLO (Boston Scientific, Watertown, MA, USA) and the Progreat (Terumo, Tokyo, Japan) in 100- and 110-cm lengths are preferred for use in visceral trauma. These catheters accept Gelfoam (Upjohn, Kalamazoo, MI, USA) slurry and large particles with less risk of catheter occlusion. Ideally, solitary bleeding sublobular branches are selected and embolized with 2- to 3-mm coils. If distal as well as proximal embolization across a bleeding site cannot be performed, Gelfoam or large-particle embolization should be performed before coil embolization in order to block flow reversal in the artery distal to the injury and continued backbleeding via these collateral pathways. Follow-up celiac angiography should always be performed to ensure pan-hepatic hemostasis before leaving the angiography suite.

17.4.2 Complications of Hepatic Embolization in Nonoperative Treatment

Published complication rates of angiography in the trauma setting range from 4% to 5% Complications of arterial hepatic embolization mainly include rebleeding, and later, hepatic infarction or abscess, bile duct, or gallbladder necrosis or biliary hematoma. Persistent bleeding is the most common complication of conservative nonoperative management, occurring up to 9 % of cases [14], and is the most common indication for surgical intervention. Abdominal hypertension (compartment syndrome), defined as vesical pressure >20 cm H_2O, with associated organ failure, may be the consequence of retroperitoneal bleeding, where intestinal edema and overgenerous repletion of vascular volume with crystalloid may play an important role [15]. Biliary complications typically occur later. The primary cause of these complications is biliary fistula, which often derives from the biliary tree or, more rarely, from the gallbladder or duodenum [16]. A bile leakage results in a direct chemical insult to the peritoneum. When the collection is well circumscribed, the treatment of choice is CT or echo-guided percutaneous drainage, which has a success rate of 70–100 % [14]. Collections that are difficult to access percutaneously can be managed by an open or laparoscopic surgical approach [17]. Hemobilia (caused by a communication between a bile duct and a blood vessel), biliopor-tal fistula, hepatic necrosis, abscess, and sepsis are other possible complications. Extensive hepatic necrosis exposes the patient to hepatocellular insufficiency and abscess formation and often requires surgical excision. Intrahepatic abscesses occur in 0.7–1.5 % of blunt hepatic trauma, and their time of onset ranges from 1 to 12 days [18]. Other complications associated with embolization are contrast reactions and vascular injury/dissection related to the procedure [19].

17.5 Laparoscopy in Abdominal Trauma Management

The role of laparoscopy in abdominal trauma is controversial and has a long history. During the early 1970s, US surgeons developed this approach, demonstrating its usefulness, efficacy, and safety in the emergency setting [20–23]. Berci et al., in 1991, evaluated a series of 150 cases of hemoperitoneum and demonstrated reduction in about 25 % of nontherapeutic laparotomies using diagnostic laparoscopy. Ivatury et al., in 1993 [33] and Brandt in 1994 [45] reached the same conclusions and moreover described abdominothoracic pen-

Table 17.1 Indications to laparoscopy in abdominal trauma (who and where)

Who	- Hemodynamically stable patients candidate for explorative laparotomy - Polytrauma patients candidates for general anesthesia, with hemoperitoneum and suspicion of hepatic injury - Abdominal compartment syndrome - Signs of abdominal complications after conservative treatment (fever, anemia with increase abdominal fluid, suspected bowel perforation) - Stab- or gunshot-wound trauma with liver or abdominal cavity involvement	Where	- Centers with expert teams in laparoscopy, available 24 h a day - Availability of necessary equipment and instruments

etrating injuries in about 40 % of cases associated with a diaphragmatic injury. Fabian et al., in 1993, concluded that laparoscopy "is a safe modality for trauma, it is most efficacious for evaluation of equivocal penetrating wounds. Significant cost savings would be gained by performance under local anesthesia..." [24]. Local anesthesia at bedside has now been completely abandoned, as operating theater and general anesthesia are considered safer, allowing more information to be gathered and better patient compliance.

The spread of laparoscopic and thoracoscopic approaches and the development of new endoscopic tools (5-mm high-definition cameras, ultrasonic scalpel, etc) have contributed to the diffusion of the minimally invasive surgical approach. Therapeutic laparoscopic options have increased in recent years to manage emergency situations. Nontherapeutic laparotomy has a morbidity rate ranging between 22% and 61%, with very high hospital stay [25–30].

To date, a laparoscopic approach is reserved for patients in hemodynamically stable conditions. Surgery is required only for persisting doubt or increasing hemoperitoneum after conservative management, the presence of shock signs, peritoneal irritation, or abdominal free fluid or free air in the follow-up period [31]. Indications to laparoscopy are reported in Table 17.1. General contraindications to a laparoscopic procedure are primarily hemodynamic instability and the patient's general condition (American Society of Anesthesiologists, ASA, grade IV).

Emergency laparoscopic treatment of hepatic injury consists primarily of hemostatic procedures (coagulation, or application of hemostatic devices), small atypical resections of devascularized tissue, positioning of stitches for hemostasis or biliary stasis. When the surgical treatment cannot be performed safety by laparoscopy or the surgeon has not enough experience, then a conversion to laparotomy is necessary.

17.6 Penetrating Abdominal and Liver Trauma

In penetrating abdominal trauma, especially those from gunshot wounds, the role of laparoscopy is still under debate; the bowel is the most frequently involved abdominal organ (about 50%). The type of penetrating agent and its trajectory can help the surgeon to evaluate the gravity of the wound. In 1999, Demetriades et al. reported successful nonoperative management in 21% of 52 patients with isolated gunshot wounds to the liver; that percentage increased to 28.4% in a study published in 2006 by the same authors [32].

Diagnostic laparoscopy for evaluating penetrating trauma is more defined in thoracic–abdominal stab wounds; laparoscopy may aid in the diagnosis of diaphragmatic and other intra-abdominal injuries, thus avoiding nontherapeutic laparotomies. Results of large case series are interesting, reporting that laparoscopy is associated with a low number of complications and reduces the number of unnecessary laparotomies. However, Ivatury and colleagues reported that laparoscopy identified only 20% of intestinal lesions caused by penetrating trauma [33]. A gunshot wound to the anterior abdomen with questionable tangential trajectory may be assessed in the same manner. The argument is that even if there are no clinical signs of intra-abdominal injuries, the disadvantages associated with an unnecessary laparotomy are minor compared with the danger of peritonitis in cases of delayed diagnosis of intestinal perforation. Laparoscopy allows inspection of the peritoneum for signs of perforation. In selected cases, several studies report treatment of intra-abdominal injuries [32, 34–36], describing an algorithm for treating hemodynamically stable patients with thoracoabdominal wounds but without peritoneal signs, suggesting videolaparoscopy is indicated if the wound is located in the left thoracoabdominal area [32] .

17.7 Laparoscopic Management of Complications

Nonoperative or conservative treatment has been associated with about 12% of in-hospital morbidity (abscess, biloma, etc). Management of these complications consists of a multimodal approach: radiological drainage, endoscopic stenting, and surgery. Laparoscopy recently gained a role as diagnostic and therapeutic techniques, with favorable results [33, 37, 38]. In some referral centers, delayed laparoscopy is even routinely proposed [39]. Thus, laparoscopy should not be considered a failure of nonoperative management but as a part of this therapeutic strategy.

17.8 Laparoscopic Procedure: Technical Notes

Laparoscopic exploration is performed under general anesthesia. The patient lies in the supine position; a nasogastric tube must always be introduced before anesthesia, and a bladder catheter should be already positioned. Pneumoperitoneum should be induced slowly and carefully, which is easier with an open technique. Special attention should be given to the possibility of a tension pneumothorax caused by an unsuspected diaphragmatic rupture. Generally, the first trocar, of 10–12 mm, is introduced at the umbilicus. A 30° camera is used. The peritoneal cavity should be examined systematically, beginning with the right upper quadrant and proceeding clockwise. Suction/irrigation may be needed for optimal visualization, and methylene blue can be administered to help identify gastrointestinal injuries. Trocar position is related to the area of injury, so the number and the seat of trocar positioning vary from patient to patient and must be decided upon as necessary. The presence and localization of blood clots inside the abdominal cavity is an important landmark of where the bleeding has started. The surgeon should not hesitate to convert to laparotomy if not confident that there are no missed injuries (especially retroperitoneal colon, pancreas, duodenum, and kidneys). Bowel distension increases the risk of iatrogenic injury or lesion underestimation. Once the bleeding source is identified, the surgeon can decide upon either conservative treatment or proceed to surgical repair, which can be done completely by laparoscopy or by an open approach. Laparoscopic sutures, intestinal resection, and intracorporeal anastomosis often require additional trocars and extensive surgeon experience. Laparoscopic liver resections and total or partial splenectomies require the use of devices such as ultrasonic or radiofrequency dissectors; anywhere else, simple hemostasis bipolar or monopolar dissectors are enough. To manage parenchymal bleeding, a wide variety of hemostatic agents are now available that can be applied by laparoscopy (see video).

Evidence in Minimally Invasive Surgery for Abdominal Trauma

Consensus Developement Conference of the Società Italiana Chirurgia Endoscopica e nuove tecnologie (SICE); Associazione Chirurghi Ospedalieri Italiani (ACOI); Società Italiana di Chirurgia (SIC); Società Italiana Chirurgia d'Urgenza e Trauma (SICUT), Società Italiana Chirurghi dell'Ospedalità Privata (SICOP), and the European Association for Endoscopic Surgery (EAES), 2010 [40]:

In stable penetrating trauma of the abdomen, laparoscopy may be useful in patients with documented or equivocal penetration of the anterior fascia – grade of reference (GoR) B – and in stable blunt trauma patients with suspected intra-abdominal injury and equivocal findings on imaging studies, or even in patients with negative studies but with a high clinical likelihood for intra-abdominal injury ('unclear abdomen') to exclude relevant injury (GoR C). To optimize results, the procedure should be incorporated in institutional diagnostic and treatment algorithms for trauma patients (Gor D).

17.9 Conclusions

Hemodynamic stability is a prerequisite for an accurate work up in emergency situations. Contrast-enhanced computed tomography (CE-CT) can be used to assess injury severity and to determine subsequent strategies [26]. Approximately 70 % of liver injuries do not require a surgical procedure; including select patients with most complex hepatic injuries (grades 4 and 5). Angiography is indicated to identify and treat active bleeding detected by CT. Diagnostic accuracy of laparoscopy has been reported to be as high as 75 % (Level of evidence, LE IIb) [41] and is indicated in patients with suspected intra-abdominal lesions, equivocal findings on imaging studies, and when non-operative management has failed or there are suspected hollow viscus injuries with peritonitis and a potential diaphragmatic lesion. The procedure effectively decreases the rate of negative laparotomies and minimizes patient morbidity [29, 40].

To approach abdominal and liver trauma by laparoscopy, an expert laparoscopic surgical and anesthesiological team is as necessary as adequate and appropriate equipment. Laparoscopy must be considered a therapeutic procedure in selected cases as well as being a diagnostic tool. Laparoscopy limitations are the impossibility of exploring the abdominal cavity, particularly the retroperitoneal space, so conversion to open surgery is required in many cases. Complications related to carbon dioxide (CO_2) diffusion to the chest through diaphragmatic injuries can also occur. Very rare cases of gas embolism, found mainly after hepatic lesions, have also been reported [42, 43]. The presence of diaphragmatic rupture is not a contraindication to a laparoscopic approach, but a thoracic tube must be placed before laparoscopy to avoid hypertensive pneumothorax. Among the absolute contraindications are: hemodynamic instability, poor or critical general conditions (ASA III–IV), gunshot wounds with multivisceral involvement and evisceration, severe head injury (where the high intra-abdominal pressure may increase intracranial pressure), and the presence of severe thoracic trauma.

Others benefits of minimally invasive surgery are represented by fewer wound complications, early patient recovery and resumption of respiratory function, and significant reduction in hospitalization. Retrospective cost-analysis studies compared total hospital costs of exploratory laparotomy and diagnostic laparoscopy with penetrating abdominal trauma, showing that laparoscopy is cheaper than exploratory laparotomy (level of evidence 4) [44].

References

1. Richardson DJ, Franklin GA, Lukan JK et al (2000) Evolution in the management of hepatic trauma: a 25-years perspective. Ann Surg 232(3):324–330
2. Awad AT, El Hammadi HA (1999) Hepatic injuries: management and outcome. Int Surg 84(3):266–270

3. Croce MA, Fabian TC, Menke PG et al (1995) Non-operative management of blunt hepatic trauma is the treatment of choice for hemodynamically stable patients: results of a prospective trial. Ann Surg 221:744–753
4. Cuff RF, Cogbill TH, Lambert PJ (2000) Nonoperative management of blunt liver trauma: the value of follow-up abdominal computed tomography scans. Am Surg 66:332–336
5. Franklin GA, Casós SR (2006) Current advances in the surgical approach to abdominal trauma. Injury 37(12):1143–1156
6. Chen RJ, Fang JF, Chen MF (2001) Intra-abdominal pressure monitoring as a guideline in the non-operative management of blunt hepatic trauma. J Trauma 51(1):44–50
7. Woong YoonW, Jeong YY, Kim JK et al (2005) CT in blunt liver trauma. Radiographics 25(1):87–104
8. Shanmuganathan K (2004) Multi-detector row CT imaging of blunt abdominal trauma. Semin Ultrasound CT MR 25:180–204
9. Root HD, Hauser CW, McKinley CR et al (1965) Diagnostic peritoneal lavage. Surgery 57:633–637
10. Patten RM, Spear RP, Vincent LM et al (1993) Traumatic lacerations of the liver limited to the bare area: CT findings in 25 patients. AJR Am J Roentgenol 160:1019–1022
11. Fang JF, Chen RJ, Wong YC et al (2000) Classification and treatment of pooling of contrast material on computed tomographic scan of blunt hepatic trauma. J Trauma 49:1083–1088
12. Wong YC, Wang LJ, See LC et al (2003) Contrast material extravasation on contrast-enhanced helical computed tomographic scan of blunt abdominal trauma: its significance on the choice, time, and outcome of treatment. J Trauma 54:164–170
13. Kozar RA, Moore JB, Niles SE et al (2005) Complications of nonoperative management of high-grade blunt hepatic injuries. J Trauma 59(5):1066–1071
14. Carillo EH, Spain DA, Wohltmann CD et al (1999) Interventional techniques are useful adjuncts in nonoperative management of hepatic injuries. J Trauma 46:619–622
15. Malbrain ML, Cheatham ML, Kirkpatrick A et al (2006) Results from the International Conference of Experts on Intra-abdominal Hypertension and Abdominal Compartment Syndrome. I. Definitions. Intensive Care Med 32(11):1722–1732
16. Griffen M, Ochoa J, Boulanger BR (2000) A minimally invasive approach to bile peritonitis after blunt liver injury. Am Surg 66:309–312
17. Franklin GA, Richardson JD, Brown AL et al (2007) Prevention of bile peritonitis by laparoscopic evacuation and lavage after nonoperative treatment of liver injuries. Am Surg 73(6):611–616, discussion 616–617
18. Hsieh CH, Chen RJ, Fang JF et al (2003) Liver abscess after nonoperative management of blunt liver injury. Langenbecks Arch Surg 387(9-10):343–347
19. Monnin V, Sengel C, Thony F et al (2008) Place of arterial embolization in severe blunt hepatic trauma: a multidisciplinary approach. Cardiovasc Intervent Radiol 31(5):875–882
20. Berci G, Dunkelman D, Michel S et al (1983) Emergency minilaparoscopy in abdominal trauma. Am J Surg 146:261–265
21. Gazzaniga AB, Stanton WW, Bartlett RH (1976) Laparoscopy in the diagnosis of blunt and penetrating injuries to the abdomen. Am J Surg 131(3):315–318
22. Carnevale N, Baron N, Delany HM (1977) Peritoneoscopy as an aid in diagnosis of abdominal trauma: a preliminary report. J Trauma 17(8):634–641
23. Sherwood R, Berci G, Austin E, Morgenstern L (1980) Minilaparoscopy for blunt abdominal trauma. Arch Surg 115(5):672–673
24. Fabian TC, Croce MA, Stewart RM et al (1993) A prospective analysis of diagnostic laparoscopy in trauma. Ann Surg 217:557–565
25. Taner AS, Topgul K, Kucukel F, Demir A, Sari S (2001) Diagnostic laparoscopy decreases the rate of unnecessary laparotomies and reduces hospital costs in trauma patients. J Laparoendosc Adv Surg Tech 11:207–211
26. Uranüs S, Dorr K (2010) Laparoscopy in abdominal trauma. Eur J Trauma Emerg Surg 36:19–24
27. Smith RS, Fry WR, Morabito DJ, Koehler RH, Organ CH Jr (1995) Therapeutic laparoscopy in trauma. Am J Surg 170:632–637

28. Smith CH, Novick TL, Jacobs DG, Thomason MH (2000) Laparoscopic repair of a ruptured diaphragm secondary to blunt trauma. Surg Endosc 14:501–502
29. Choi YB, Lim KS (2003) Therapeutic laparoscopy for abdominal trauma. Surg Endosc 17:421–427
30. Mathonnet M, Peyrou P, Gainant A, Bouvier S, Cubertafond P (2003) Role of laparoscopy in blunt perforations of the small bowel. Surg Endosc 17:641–645
31. Kozar RA, Moore FA, Cothren CC et al (2006) Risk factors for hepatic morbidity following nonoperative management: multicenter study. Arch Surg 141(5):451–458
32. Demetriades D, Hadjizacharia P, Constantinou C et al (2006) Selective nonoperative management of penetrating abdominal solid organ injuries. Ann Surg 244(4):620–628
33. Ivatury RR, Simon RJ, Stahl WM (1993) A critical evaluation of laparoscopy in penetrating abdominal trauma. J Trauma 34(6):822–827
34. Ferrada R, Birolini D (1999) New concepts in the management of patients with penetrating abdominal wounds. Surg Clin North Am 79(6):1331–1356
35. Sosa JL, Arrillaga A, Puente I, Sleeman D, Ginzburg E, Martin L (1995) Laparoscopy in 121 consecutive patients with abdominal gunshot wounds. J Trauma 39:501–506
36. Mazuski JE, Shapiro MJ, Kaminski DL (1997) Diagnostic laparoscopy for evaluation of penetrating abdominal trauma. J Trauma 42:163
37. Carrillo EH, Reed DN Jr, Gordon L, Spain DA, Richardson JD (2001) Delayed laparoscopy facilitates the management of biliary peritonitis in patients with complex liver injuries. Surg Endosc 15(3):319–322
38. Marzano E, Rosso E, Oussoultzoglou E, Collange O, Bachellier P, Pessaux P (2010) Laparoscopic treatment of biliary peritonitis following nonoperative management of blunt liver trauma. World J Emerg Surg 5:26
39. Feliciano DV, Rozycki GS (2002) Hepatic trauma. Scand J Surg 91(1):72–79
40. Agresta F, Ansaloni L, Baiocchi L et al (2011) Consensus Development Conference of the Società Italiana Chirurgia Endoscopica e nuove tecnologie (SICE); Associazione Chirurghi Ospedalieri Italiani (ACOI); Società Italiana di Chirurgia (SIC); Società Italiana Chirurgia d'Urgenza e Trauma (SICUT), Società Italiana Chirurghi dell'Ospedalità Privata (SICOP) and the European Association for Endoscopic Surgery (EAES)
41. Mallat AF, Mancini ML, Daley BJ, Enderson BL (2008) The role of laparoscopy in trauma: a ten-year review of diagnosis and therapeutics. Am Surg 74(12):1166–1170
42. Jayaraman S, Khakhar A, Yang H, Bainbridge D, Quan D (2009) The association between central venous pressure, pneumoperitoneum, and venous carbon dioxide embolism in laparoscopic hepatectomy. Surg Endosc 23(10):2369–2373
43. Min SK, Kim JH, Lee SY (2007) Carbon dioxide and argon gas embolism during laparoscopic hepatic resection. Acta Anaesthesiol Scand 51(7):949–953
44. Marks JM, Youngelman DF, Berk T (1997) Cost analysis of diagnostic laparoscopy vs laparotomy in the evaluation of penetrating abdominal trauma. Surg Endosc 11:272–276
45. Brandt CP, Priebe PP, Jacobs DG (1994) Potential of laparoscopy to reduce non-therapeutic trauma laparotomies. Am Surg 60:416-420

Minimally Invasive Surgery in Cirrhotic Patients

18

Giulio Belli, Paolo Limongelli, Andrea Belli, Gianluca Russo and Alberto D'Agostino

In Western countries, hepatocellular carcinoma (HCC) is the most common primary liver cancer. Its incidence is increased in patients with chronic liver disease mostly due to hepatitis B or C infections [1]. In patients with cirrhosis, screening can improve survival by diagnosing the tumor while it is still small and asymptomatic [2]. Liver transplantation is a curative option for HCC in patients with underlying chronic liver disease but cannot be applied on a large scale due to patient age and alcohol abuse, associated diseases, and shortage of donors [3, 4]. Therefore, other treatments, such as hepatic resection and percutaneous techniques such as ethanol injection and radiofrequency ablation (RFA), are needed either as bridging treatments or alternative management for those unsuitable for a liver transplant. Although hepatic resection performed in cirrhotic patients remains a surgical challenge for both surgeons and patients, it can attain a high curative rate, a morbidity rate of 10–40%, and a mortality rate <10% [5–7]. Open liver resection plays a paramount role in the curative treatment of HCC in patients with adequate liver function [8]. With improvements in technology and equipment, laparoscopic liver resection (LLR) is now considered a safe procedure, even for managing liver tumors, if performed by experienced surgeons [6].

Since the early 2000s, the minimally invasive approach has been used increasingly to manage hepatic diseases, even in presence of cirrhosis [9]. Despite its technical challenges, reduced operative blood loss, fewer early postoperative complications, lower analgesic drug requirements, and shorter hospital stay are attainable. Besides, oncologic clearance and survival rate are similar to those of open surgery. The role of the minimally invasive approach in liver surgery continues to increase, and many types of liver resections,

G. Belli (✉)
General and HPB Surgery, Loreto Nuovo Hospital, Naples, Italy
e-mail: chirurgia.loretonuovo@tin.it

F. Calise and G. Casciola (eds.), *Minimally Invasive Surgery of the Liver*,
Updates in Surgery
DOI: 10.1007/978-88-470-2664-3_18,

including major hepatectomies, are now performed laparoscopically in specialized centers [10–12]. In addition, laparoscopy seems to improve the patient's postoperative course and avoid abdominal-wall, and probably parenchymal, injury. In fact, the benefit of laparoscopic over open procedures seems to be greater, especially in cirrhotic patients, in whom postoperative ascites and abdominal-wall complications can be decreased. Nevertheless, close monitoring to detect and manage recurrence by bridge treatments, redo liver resection and salvage transplantation in eligible patients is required to improve survival [13, 14]. Recurrence and the need for repeat treatments are the Achilles' heel of hepatic resection for HCC in cirrhosis. Therefore, as with the first operation, selection criteria using the laparoscopic approach in patients with recurrent tumor include generally well-compensated chronic liver disease (Child–Pugh class A) without signs of severe portal hypertension (esophageal varices F2); exophytic or subcapsular tumors located in the left (2, 3, 4b) or peripheral right (5–6) segments; a maximum lesion size of 4–5 cm; and limited resection (less than three segments). Patients with complicated cirrhosis (Child–Pugh class B–C) or an American Society of Anesthesiology (ASA) classification >2 are generally excluded from this procedure. In recent years, thanks to improvements in technology and equipment and accumulated surgical-team experience, any type of liver resection, including posterosuperior segments and left and right hemihepatectomy in well-selected patients can be performed successfully.

The most frequently required type of liver resection for small, single HCC is limited anatomic resection. Such resections are particularly suitable for laparoscopy. In our experience, laparoscopic resection has been successfully performed in >90 % of our patients, with only few procedures requiring conversion to open surgery. Our institution started a specific program of LLR and is now a referral center for managing HCC patients eligible for the approach, accounting for the fact that more than one third of liver resections are now performed laparoscopically. Good perioperative results were attained even at the beginning of our experience when highly selected, small subcapsular lesions were the main indication for HCC resection in cirrhosis. These results were confirmed in the second half of our experience, when we included larger tumors, major resections, and a significant increase in bisegmentectomies. These findings reflect the impact of the surgical team's learning curve on patients previously deemed unsuitable for a laparoscopic approach [6].

However, liver resections in cirrhotic patients are technically more difficult than in patients with normal liver, presenting added complications such as profuse bleeding during liver mobilization and parenchymal transection. Intraoperative bleeding is of great concern in LLR, with a reported incidence of 7%, and controlling it remains of key importance [15]. The development of new surgical devices and the hemostatic effect of pneumoperitoneum on hepatic vein branches partly explain comparable blood loss between open and LLRs [3, 16]. Operative durations of LLRs are significantly longer than those of matched open counterparts [17, 18], but the procedure's safety is attested to by the absence of mortality and the low specific morbidity rates (15%), with low rates of transient postoperative liver failure. These findings are important, as ascites and jaundice are the main complications of liver resection – even of minor ones – in cirrhotic patients.

Laparoscopic resection of HCC in cirrhotic liver is feasible and safe in selected patients and achieves adequate long-term survival and recurrence rates compared with open surgery when stratified for tumor characteristics known to be related to survival outcome [6].

References

1. Cherqui D, Laurent A, Tayar C, Chang S, Van Nhieu JT, Loriau J et al (2006) Laparoscopic liver resection for peripheral hepatocellular carcinoma in patients with chronic liver disease. Ann Surg 243:499–506
2. Chan AC, Poon RT, Ng KK, Lo CM, Fan ST, Wong J (2008) Changing paradigm in the management of hepatocellular carcinoma improves the survival benefit of early detection by screening. Ann Surg 247:666–673
3. Llovet JM, Fuster J, Bruix J (1999) Intention-to-treat analysis of surgical treatment for early hepatocellular carcinoma: resection versus transplantation. Hepatology 30:1434–1440
4. Yao FY, Bass NM, Nikolai B, Davern TJ, Kerlan R, Wu V et al (2002) Liver transplantation for hepatocellular carcinoma: analysis of survival according to the intention-to-treatprinciple and dropout from the waiting list. Liver Transpl 8:873–883
5. Belli G, Fantini C, D'Agostino A, Cioffi L, Langella S, Russolillo N et al (2007) Laparoscopic versus open liver resection for hepatocellular carcinoma in patients with histologically proven cirrhosis: short- and middle-term results. Surg Endosc 21(11):2004–2011
6. Belli G, Limongelli P, Fantini C, D'Agostino A, Cioffi L, Belli A et al (2009) Laparoscopic and open treatment of hepatocellular carcinoma in patients with cirrhosis. Br J Surg 96(9):1041–1048
7. Dagher I, Belli G, Fantini C, Laurent A, Tayar C, Lainas P et al (2010) Laparoscopic hepatectomy for hepatocellular carcinoma: a European experience. J Am Coll Surg 211(1):16–23
8. Cha CH, Ruo L, Fong Y, Jarnagin WR, Shia J, Blumgart LH et al (2003) Resection of hepatocellular carcinoma in patients otherwise eligible for transplantation. Ann Surg 238:315–321
9. Belli G, D'Agostino A, Fantini C, Cioffi L, Belli A, Russolillo N et al (2006) Laparoscopic incisional and umbilical hernia repair in cirrhotic patients. Surg Laparosc Endosc Percutan Tech 16(5):330–333
10. Buell JF, Cherqui D, Geller DA, O'Rourke N, Iannitti D, Dagher I et al (2009) The international position on laparoscopic liver surgery: The Louisville Statement, 2008. Ann Surg 250(5):825–830
11. Belli G, Fantini C, D'Agostino A, Cioffi L, Russo G, Belli A, Limongelli P (2011) Laparoendoscopic single site liver resection for recurrent hepatocellular carcinoma in cirrhosis: first technical note. Surg Laparosc Endosc Percutan Tech 21(4):e166–168
12. Belli G, Gayet B, Han HS et al (2012) International Consensus Group for Laparoscopic Liver Surgery Laparoscopic Hemi-Hepatectomy: a consideration for acceptance as standard of care. New Orleans, LA, 10th World Congress of IHPBA, Paris, July 1-5, 2012
13. Belli G, Fantini C, Belli A, Limongelli P (2011) Laparoscopic liver resection for hepatocellular carcinoma in cirrhosis: long-term outcomes. Dig Surg 28(2):134–140
14. Belli G, D'Agostino A, Fantini C, Cioffi L, Belli A, Russolillo N et al (2007) Laparoscopic radiofrequency ablation combined with laparoscopic liver resection for more than one HCC on cirrhosis. Surg Laparosc Endosc Percutan Tech 17(4):331–334
15. Chen HY, Juan CC, Ker CG (2008) Laparoscopic liver surgery for patients with hepatocellular carcinoma. Ann Surg Oncol 15:800–806

16. Dagher I, Lainas P, Carloni A, Caillard C, Champault A, Smadja C et al (2008) Laparoscopic liver resection for hepatocellular carcinoma. Surg Endosc 22:372–378
17. Laurent A, Cherqui D, Lesurtel M, Brunetti F, Tayar C, Fagniez PL (2003) Laparoscopic liver resection for subcapsular hepatocellular carcinoma complicating chronic liver disease. Arch Surg 138:763–769
18. Lesurtel M, Cherqui D, Laurent A, Tayar C, Fagniez PL (2003) Laparoscopic versus open left lateral hepatic lobectomy: a case-control study. J Am Coll Surg 196:236–242

Reopercative Laparoscopic Hepatectomy

19

Joseph F. Buell, Alan Koffron, Bjørn Edwin,
Robert Cannon and Brice Gayet

Since the inception of laparoscopic liver resection (LLR), the indication for this procedure has evolved from peripheral wedge resections to formal lobectomies and now even to reoperative hepatectomy. Multiple centers worldwide have reported their experiences using various laparoscopic surgical techniques for hepatectomy, including pure laparoscopy, hand assist, and the hybrid technique [1–10]. Concurrent data clearly report the efficacy of repeat hepatectomy, particularly for colorectal metastases resection, supporting the role for secondary and even tertiary redo hepatectomy. This increase in technical field strength in conjunction with efficacy data led to the evolution of reoperative LLR strategies [11–14].

Reoperative LLR is inherently complex due to the presence of adhesions, atypical anatomy, and – in the case of redo liver surgery – a potentially compromised liver remnant. These technical considerations often prohibited reoperative laparoscopic hepatectomy to all but a few highly experienced surgical teams and large-volume centers. Prior to embarking on a laparoscopic approach to reoperative hepatectomy, all three institutions participating in our collaborative group had extensive experience in both open redo hepatectomies and LLRs.

This chapter is dedicated to examining the evaluation process of potential candidates for the technical approaches used to execute the reoperative laparoscopic hepatectomy. To achieve this, we examined our clinical experience with reoperative LLRs from a collaborative group comprising: (1) Institute Mutualiste Montsouris, Paris, France; (2) Rikshospitalet, Oslo University Hospital, Oslo, Norway, and (3) Tulane University, New Orleans, Louisiana, USA. Each group previously published their separate and extensive technical approaches to primary LLRs [5–10].

J. F. Buell (✉)
Professor of Surgery and Pediatrics, Tulane Transplant Institute,
New Orleans, Louisiana, USA
e-mail: Jbuell1@tulane.edu

F. Calise and G. Casciola (eds.), *Minimally Invasive Surgery of the Liver,*
Updates in Surgery
DOI: 10.1007/978-88-470-2664-3_19, © Springer-Verlag Italia 2013

In a 14-year period, 74 patients underwent 76 repeat LLRs [10]. They were most commonly performed for resection of a liver lesion in the same or adjacent segment resected during a prior resection (75 %). Notably, few selected patients had steatosis or fibrosis, and none had cirrhosis. A significantly higher incidence of operative conversions was observed than with traditional laparoscopic hepatectomy (10 % vs. 2 %). The two principal causes for operative conversions were uncontrolled hemorrhage or failure to progress. Failure to progress arose from dense adhesions, posterior-segment tumors, and lesion multiplicity. Despite these complexities, an R0 resection was achieved in 92 % of patients. However, the resection margin was <1 mm in nearly 10 % of cases. Despite this limited margin, no local tumor recurrences were observed. Tumor recurrence was observed in 13 % of patients with liver disease as the only site of cancer recurrence. There were no cancer recurrences observed at the port sites. The overall incidence of postoperative complications was significantly higher in the laparoscopic reoperative hepatectomy group (30 %) compared with a collaborative incidence for primary laparoscopic hepatectomy (30 % vs. 15 %). Interestingly, postoperative bile leaks also occurred at a higher rate (6.5 % vs. 4 %).

The most common indication in the literature for recurrent open hepatic resection is regional recurrence of colorectal liver metastasis [11–14]. Repeat hepatectomy in itself is among one of the most challenging technical procedures in the field of hepatobiliary surgery. Concerns with repeat liver resections include the difficulty of the procedure itself and the potential for hemorrhagic complications in the setting of uncertain remnant anatomy, dense adhesions, or even compromised intrinsic hepatic disease [15]. Reported morbidity and mortality rates for open repeat liver resections varies from 22 % to 28 % and 0 % to 5 %, respectively [11–14, 16] (Table 19.1). Laparoscopic hepatectomy results in decreased blood loss and postoperative pain, shortened hospital stay, and improved recovery times. Despite initial concerns over oncological integrity, the laparoscopic approach to cancer is not inferior to that of open resection [16–19]. This is consistent with findings from primary laparoscopic colon resection data that demonstrates survival is improved with a laparoscopic approach to colon resection in advanced-stage colorectal cancer (Table 19.2).

Belli et al. reported similar findings in their experience with 12 reoperative laparoscopic resections for hepatocellular carcinoma (HCC) in the setting of cirrhosis [20]. Both Belli's and our group concluded that re-resection after primary open liver resection requires more extensive intra-abdominal adhesiolysis than prior laparoscopic resection, leading to longer operative times and higher bleeding risk. This is consistent with findings of several open repeat hepatectomy series for colorectal metastasis, which showed higher blood loss and longer hospital stays compared with primary open resections [21-22].

Critical to reoperative liver resection is extrahepatic disease detection. Excluding peritoneal carcinomatosis in dense intra-abdominal adhesions requires extensive adhesion lysis and suspicious lesion biopsy. Intraoperative ultrasound is also critical to define remnant vascular anatomy affected by prior

Table 19.1 Comparison of reoperative liver resection for colorectal metastases

Author	Cases	Re-resection	Formal lobe (%)	EBL (ml)	Morbidity (%)	Mortality (%)	5-year survival (%)
Tuttle et al. [12]	23	Open	30	800	22	0	32
Petrowsky et al. [13]	126	Open	28	900	28	1.5	34
Ishiguro et al. [14]	111	Open	11	913	14	0	41
de Jong et al. [16]	246	Open	21	NR	21	0.4	47
Shafaee et al. [10]	55	Laparoscopic	29	300	27	0	55

EBL, estimated blood loss; *Formal lobe*, right lobe segments 5-7, left lobe segments 4, 2, 3, ± 1; *NR* , not reported.

Table 19.2 Comparison of high-volume laparoscopic liver resection with our reoperative laparoscopic liver resection experience

Author	Cases	Formal lobe (%)	OR time (min)	EBL (ml)	Conversion (%)	Morbidity (%)	Mortality (%)	Cirrhosis (%)
Buell et al. [24]	342	23	150	221	1.4	19.1	1.1	16
Kazaryan et al. [9]	149	32	164	350	3.4	12.6	0.7	NR
Gayet et al. [7]	313	31	160	150	5.4	15.3	0	9.2
Shafaee et al. [10]	76	25	195	300	10	32	0	1.3

EBL, estimated blood loss; OR time, operating room time; *Formal lobe*, right lobe segments 5-7, left lobe segments 4, 3, 2, ± 1; *NR* , not reported.

resection. Alterations in major vascular distributions must be defined to develop an operative strategy and to prevent inadvertent vascular injury or remnant compromise.

The observation of smaller surgical margins in reoperative laparoscopic hepatectomy is concerning but may be dictated by the altered vascular anatomy present in the remnant liver. Historically, 10 mm is advocated as an appropriate surgical margin for hepatic resection. In the setting of redo liver resections, this may not be achievable due to the risk of remnant liver compromise. Additionally, stapler hepatectomy has been postulated, with the devices compression requirements to artificially reduce the pathologic margin. Multiple publications report the oncologic benefit of 10-mm margins, with several arguing the adequacy of a negative margin. de Haas et al. reported that R1 resections did not adversely affect survival in patients with hepatic colorectal metastasis where an R0 resection was not feasible [22].

Technically, the use of a hand-assist device appears to be beneficial in the setting of reoperative laparoscopic hepatectomy [23]. In patients with prior open hepatectomy, the use of the prior right subcostal portion of the chevron is an ideal location for placing the hand-assist device (Figs. 19.1 and 19.2). This incision also allows significant adhesion lysis using a hybrid approach and minimizes the amount of dissection performed through a laparoscopic

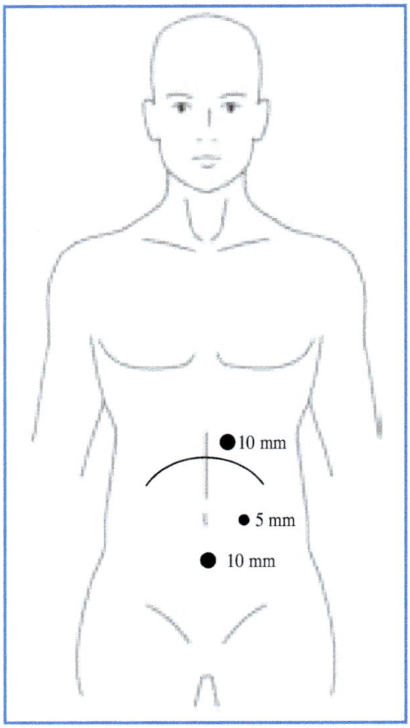

Fig. 19.1 Port placement for reoperative laparoscopic liver resection with prior open hepatic resection

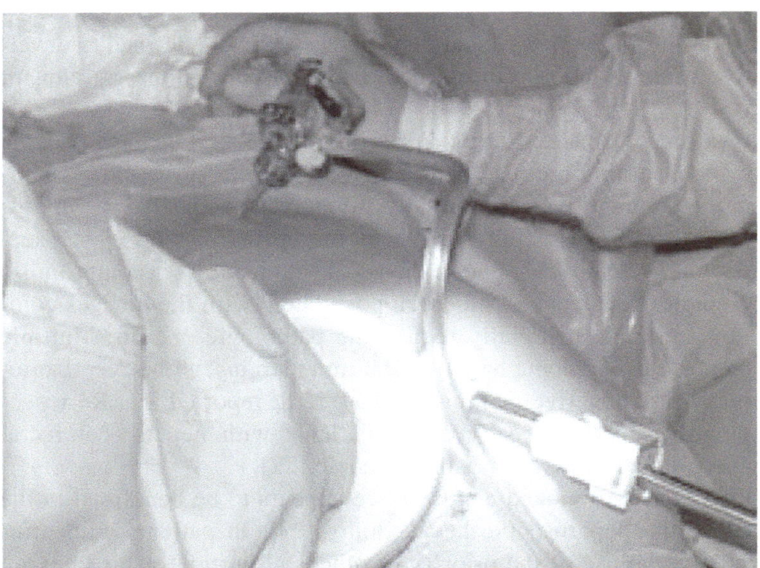

Fig. 19.2 Intraoperative and hand-assist device port placement following prior chevron incision; laparoscopic approach for redo hepatic resection to segment 4 after prior open hepatic resection of segments 6–7 for colorectal metastases

approach. Data support the feasibility of reoperative laparoscopic hepatectomy whether the primary resection was laparoscopic or open.

Reoperative laparoscopic hepatectomy is feasible, safe, and efficacious in high-volume centers in which the surgical team is experienced and comfortable with complex open and laparoscopic resections. Patients with prior laparoscopic resections are felt to be better candidates than patients with prior open hepatic resections. Critical to the success of reoperative laparoscopic hepatectomy is detailed intraoperative laparoscopic ultrasound and meticulous and extensive adhesion lysis to exclude the presence of extrahepatic disease. Attention to detail results in precise resection with minimized risk to remnant liver function.

Acknowledgements A contribution to this chapter is gratefully acknowledged to Dr. Bob Saggi.

References

1. Lefor AT, Flowers JL (1994) Laparoscopic wedge biopsy of the liver. J Am Coll Surg 178(3):307–308
2. Cherqui D, Husson E, Hammoud R et al (2000) Laparoscopic liver resections: a feasibility study in 30 patients. Ann Surg 232:753–762
3. Descottes B, Lachachi F, Sodji M et al (2000) Early experience with laparoscopic approach for solid liver tumors: initial 16 cases. Ann Surg 232:641–645
4. Gigot JF, Glineur D, Santiago Azagra J et al (2002) Laparoscopic liver resection for malignant liver tumors: preliminary results of a multicenter European study. Ann Surg 236:90–97
5. Vibert E, Kouider A, Gayet B et al (2004) Laparoscopic anatomic liver resection. HPB (Oxford) 6:222–229
6. Buell JF, Thomas MJ, Doty TC et al (2004) An initial experience and evolution of laparoscopic hepatic resectional surgery. Surgery 136:804–811
7. Gayet B, Cavaliere D, Vibert E et al (2007) Totally laparoscopic right hepatectomy. Am J Surg 194:685–689
8. Buell JF, Thomas MT, Rudich S et al (2008) Experience with more than 500 minimally invasive hepatic procedures. Ann Surg 248:475–485
9. Kazaryan AM, Marangos IP, Rosseland AR et al (2010) Laparoscopic liver resection for malignant and benign lesions: 10 year Norwegian single centre experience. Arch Surg 145(1):34–40
10. Shafaee Z, Kazaryan AM, Marvin M et al (2011) An initial assessment of laparoscopic repeat hepatectomy. A tri-institutional analysis. J Am Coll Surg 212(2):171–179
11. Fong Y, Blumgart LH, Cohen A et al (1994) Repeat hepatic resections for metastatic colorectal cancer. Ann Surg 200:657–662
12. Tuttle TM, Curley SA, Roh MS et al (1997) Repeat hepatic resection as effective treatment for recurrent colorectal liver metastases. Ann Surg Oncol 4(2):125–130
13. Petrowsky H, Gonen M, Jarnagin W et al (2002) Second liver resections are safe and effective treatment for recurrent hepatic metastases from colorectal cancer, a bi-institutional analysis. Ann Surg 235:863–871
14. Ishiguro S, Akasu T, Fujimoto Y et al (2006) Second hepatectomy for recurrent colorectal liver metastasis: analysis of preoperative prognostic factors. Ann Surg Oncol 13(12):1579–1587
15. Elias D, Lasser P, Hoang JM et al (1993) Repeat hepatectomy for cancer. Br J Surg 80:1557–1562

16. de Jong MC, Mayo SC, Pulitano C et al (2009) Repeat curative intent liver surgery is safe and effective for recurrent colorectal liver metastasis: Results from an International multi-institutional analysis. J Gastrointest Surg 13:2141–2151
17. Simillis C, Constantinides VA, Tekkis PP et al (2007) Laparoscopic versus open hepatic resections for benign and malignant neoplasms—a meta-analysis. Surgery 141: 203–211
18. Mala T, Edwin B, Gladhaug I et al (2002) A comparative study of the short-term outcome following open and laparoscopic liver resection of colorectal metastases. Surg Endosc 16(7):1059–1063
19. Viganò L, Tayar C, Laurent A et al (2009) Laparoscopic liver resection: a systematic review. J Hepatobiliary Pancreat Surg 16:410–421
20. Belli G, Cioffi L, Fantini C et al (2009) Laparoscopic redo surgery for recurrent hepatocellular carcinoma in cirrhotic patients: feasibility, safety, and results. Surg Endosc 23:1807–1811
21. Castaing D, Vibert E, Ricca L et al (2009) Oncologic results of laparoscopic versus open hepatectomy for colorectal liver metastases in two specialized centers. Ann Surg 250(5):849–855
22. de Haas RJ, Wicherts DA, Flores E et al (2008) R1 resection by necessity for colorectal liver metastases. Is it still a contraindication to surgery? Ann Surg 248(4):626–636
23. Cannon RM, Brock GN, Marvin MR et al (2011) Laparoscopic liver resection: an examination of our first 300 patients. J Am Coll Surg 213(4):501–507
24. Buell JF, Cherqui D, Geller DA et al (2009) World Consensus Conference on Laparoscopic Surgery. The international position on laparoscopic liver surgery: The Louisville Statement. Ann Surg 250(5):825–830

Treating the Resected Surface 20

Vincenzo Scuderi, Antonio Ceriello, Giuseppe Surfaro,
Gianpaolo Marte and Fulvio Calise

20.1 Introduction

Postoperative complications directly due to liver resection are hepatic failure and abdominal fluid collections due to bleeding or biliary leaks from the resection plane. The decrease in complication rates is due to technological advances and the widespread use of anatomically oriented resection techniques that significantly minimize liver-tissue necrosis. Hepatic parenchyma has in fact a complex, high-density framework of vascular and biliary structures, and even the most meticulous and appropriate approach to resection does not exclude the risk of postoperative bleeding and/or biliary leaks.

The intriguing concept of "dressing" the resection area with topical (hemostatic/sealant) agents capable of stopping biliary leaks and bleeding has gradually become widespread. Success in this area perhaps lies in the desire of many surgeons for a safe and effective biocompatible "ally" that once applied would remain in the patient's abdomen to assist the healing process. The list of characteristics necessary to create an ideal ally is long: has no adverse effects, is completely absorbable, effectively stops all fluid flow, is inexpensive, is easily preserved without particular temperature requirements, is ready to use or requires quick preparation only, and requires minimal surgeon training.

The use of hemostatic and sealant agents on the cross-section of the resected liver surface can prevent not only postoperative complications but improve resection time by obtaining a dry liver surface during the parenchyma transection; this last consideration is particularly important during minimally invasive liver surgery (MILS) when treating both liver resection surfaces provides a clear surgical field and thus eliminates light absorption due to bleeding.

V. Scuderi (✉)
Hepatobiliary Surgery and Liver Transplant Unit, A. Cardarelli Hospital,
Naples, Italy
e-mail: scuderi.enzo@gmail.com

F. Calise and G. Casciola (eds.), *Minimally Invasive Surgery of the Liver*,
Updates in Surgery
DOI: 10.1007/978-88-470-2664-3_20, © Springer-Verlag Italia 2013

In the literature are many articles reporting the use of such agents in different kinds of surgical procedures. Kraus et al. [1] report on >2,000 articles related to using hemostatics, glues, and sealants in various fields of surgery. The surgeons' knowledge concerning pharmacodynamic characteristics and strengths of products often appears to be rather poor. Current indications for agent choice and usage are heterogeneous and mostly based on the individual surgeon's preference, particularly in relation to the type of surgical procedure (open or MILS), bleeding severity, surgeon's personal experience, cost considerations, and availability in the operating room. On the basis of recent evidence using the traditional resection approach, there is potential for widespread use of such strategies also in MILS.

All such products are virtually suitable for use during laparoscopic liver resection (LLR); in fact, there are several anecdotal or case reports of their effective use during MILS. The aim of this chapter is to report briefly the rational use of these agents and their characteristics considering the feasibility of their use during MILS.

20.2 Rationale for Use of Hemostats and Sealant in Liver Surgery

Bleeding and bile leaks from the parenchymal surface are often particularly troublesome resection-related abdominal complications and are especially difficult to manage in patients with liver cancer or cirrhosis due to reduced platelet counts and blood coagulation activity.

Several agents are available that, depending on composition and structure and, through different mechanisms, obtain and/or improve hemostasis and reduce the postoperative risk of bile leaks. They can be broadly categorized into the following: fibrin sealants, collagen hemostatic absorbable sponge, gelatin matrix, and oxidized regenerated cellulose fleece.

Some agents improve endogenous hemostasis by stimulating clot formation and inhibiting fibrinolysis; others are a combination of a procoagulant substance with a vehicle such as collagen matrix; yet others provide a mechanical barrier to bleeding and a matrix to act as a template for the endogenous coagulation cascade.

Sealants can be used to treat and prevent blood and/or bile leakage. Applied to a relatively bloodless field, they create a barrier to fluid flows, reducing the need for red cell transfusion and preventing postoperative bleeding or bile leakage.

Fibrin sealants consist of a source of human fibrinogen combined with human or bovine thrombin, which then forms a cross-linked, insoluble fibrin matrix. Some agents contain an antifibrinolytic agent, such as tranexamic acid or bovine aprotinin, to stabilize the clot.

These products do not require active bleeding or blood-derived fibrinogen to achieve polymerization. The quantitative composition of the fibrin glue affects the properties of the resulting clot: high fibrinogen concentrations produce strong, slowly forming clots; high thrombin concentrations produce weaker but rapidly forming clots [2]. There are also systems available that obtain the fibrinogen component of fibrin sealant from the patient's own plasma.

Hemostatics can be used to arrest bleeding if directly applied to a bleeding site; they work in the presence of blood flow and may even require the patient's blood factors before they can begin having an effect.

20.2.1 Bleeding

Blood loss is a prognostic factor of postoperative morbidity and mortality after liver resection. The liver has a vascular structure with a high predisposition to bleeding, particularly during major resection. Its hepatic sinusoidal structure is not capable of smooth-muscle contraction to induce vasoconstriction.

Hepatocellular carcinoma (HCC), an indication for hepatic resection, usually occurs as a result of a *cirrhotic liver*, a condition that often leads to clotting alterations. The other frequent indications for hepatic resection are metastases from colorectal cancer in patients treated with chemotherapy, which can damage the non-neoplastic liver parenchyma by causing chemotherapy-associated liver injury (*CALI*; staged in steatosis, steatohepatitis, sinusoidal obstruction syndrome) that may enhance the risk of bleeding.

A series of 19 MILS left lateral sectionectomy for malignant lesion and 18 left lateral sectionectomies for benign conditions (17 solid lesion and one living donor) recorded significantly more blood loss with benign than with malignant tumors. This could be explained not only as a learning-curve effect (more frequent resection of benign lesions at the beginning of the learning curve) but by the fact that severe *steatosis* was mostly predominant in benign solid tumors [3]. This pathological finding may increase bleeding predisposition of the parenchyma transection and have consequences on postoperative morbidity, mortality, and cancer recurrence rates [4].

Applying direct pressure at a bleeding site frequently provides either complete arrest or adequate control to enable more definitive measures to be taken. However, this method may not be sufficient in case of diffuse venous bleeding or when severe coagulopathy is present. Topical hemostatic agents can be therefore helpful when conventional surgical methods, such as compression, ligation, clipping, and electrocautery, are not sufficient. A wide variety of products has been developed and is now available to assist surgeons in bleeding management.

The often large variation in efficiency detected when comparing studies can be attributed to the diversity of agents applied, the rather heterogeneous compositions of classic fibrin glues tested, poor standardization of application techniques, and overall differences in clinical settings [1].

During laparoscopic surgery for parenchymal organs (liver, pancreas, kidney), the major problem is successfully performing hemostasis from the resected stump [5–8]. In particular, MILS remains a highly specialized field due to potential bleeding from the transection surface, which may be more difficult to control laparoscopically than during the open procedure.

20.2.2 Biliary Leaks

One of the most feared complications involved with hepatic resection is biliary fistula through small lesions in the biliary ductules during parenchymal transection. Despite the decrease in overall postoperative complications the incidence of bile leakage remains from 3.6 % to 12 % [9, 10]. Bile leaks and its outputs are often difficult to detect and quantify. Several different criteria have been used so far to establish the presence of bile leaks: drain volume; drain output bilirubin concentration in comparison with serum bilirubin concentration; cholangiography.

The biliary system is a very low-pressure system, and the watertight sealing of small, undetected duct lesions in this system would be of great benefit in hepatobiliary surgery [11]. The potential for mechanical sealant occlusion of small lesions of tubular structures is well demonstrated in most experiences in neurosurgery and pulmonary surgery. Nevertheless, it must be assumed that these results cannot be directly applied to the biliary system. The variable types and aggressive nature of different bodily fluids probably must be taken into account. Bile leaks are most likely far more difficult to control than leakage of pulmonary air or cerebral fluids.

In a review of 2,804 laparoscopic liver surgeries (LLSs) [12], postoperative complications were primarily related to biliary leaks (42; 1.5 %). Although LLS has a lower incidence of biliary leaks with respect to the open approach, it is correct to consider that the former comprises a minor number of major procedures. Nevertheless, in some series comparing LLS with the open approach, it seems that the former approach would result in an increased incidence in bile leaks, particularly those due to the use of staplers to transect the parenchyma.

20.3 Hemostats and Sealants in MILS

Limiting intraoperative blood loss is an important aspect of MILS because improving view of the operative field reduces the rate of open conversion.

A peculiar aspect of MILS is the extreme attention required to stop any bleeding, not only from the remaining parenchymal surface (as in the open approach), but also from the resected parenchymal surface.

During laparoscopic hepatic transection, a variety of techniques are available to stop oozing and/or major bleeding. The first, and easiest, maneuver is pressure application to the transection surface. This maneuver stops small hemorrhages and gains surgical time in case a greater and more serious hemorrhage requires a suitable hemostatic approach (intracorporeal sutures, bipolar diathermy, an appropriate clip). Applying pressure on the transection surface is done by gently pressing a small gauze swab against the bleeding point or by re-closing the resection line; the latter procedure, in addition to the effect of pressure, allows the bleeding surface to come into contact with the opposite procoagulant- rich surface of the transected liver [13].

At the end of parenchymal transection, central venous pressure (CVP) and blood pressure are restored to normal physiological parameters, the pneumoperitoneum can be eliminated and, after 10 min, resection margins can be carefully inspected for any bleeding and bile leaks; a Valsalva maneuver has been proposed to confirm hemostasis [13].

At the end of traditional hepatic resection, many authors report managing hemorrhagic sites by applying an argon beam coagulator (ABC) to the final raw liver surface. Its use during laparoscopic procedures is, however, associated with reports of serious complications, including gas embolism and tension pneumothorax [14, 15]. The ABC should only be used during laparoscopic surgery once all other possible methods of achieving hemostasis have been attempted unsuccessfully. However, ABC is safe as long as pressures are closely monitored and evacuation of flume and pressure is appropriately controlled.

Few papers in the literature focus solely on concern with bleeding, bile leaks, and the use of glues and sealants during LLS [13, 16–18]. Several papers on MILS, however, report that topical hemostatic agents are being used on a routine basis. A recent literature review by Saif et al. [19] regards the use of sealants and hemostats during MILS. Below we address the characteristics and methods of hemostatic and sealant applications that are most often used in MILS:

- Liquid fibrin sealant glues (Tisseel®, Evicel®)
- Gelatin matrix (FloSeal®)
- Collagen sponges (TachoSil®)
- Oxidized regenerated cellulose fleece (Surgicel®).

20.3.1 Liquid Fibrin Sealants and Glues

Fibrin sealants consist of a source of human fibrinogen combined with human or bovine thrombin to form a cross-linked, insoluble fibrin matrix. Some agents also contain an antifibrinolytic agent such as tranexamic acid or bovine aprotinin to stabilize the clot. These products do not require active bleeding or blood-derived fibrinogen to achieve polymerization.

Tisseel® (Baxter, USA) (2 ml, 4 ml, 10 ml) is a fibrin glue with two main human components: fibrinogen and thrombin. It is provided as two prefilled syringes, the contents of which are mixed on the bleeding site. Mixing the two components forms a clot that reproduces and enhances the last phase of the coagulation cascade independently from the patient's own coagulation process.

- First syringe: fibrinogen 70–110 mg/ml; fibronectin 2–9 mg/ml; Factor XIII 10–15 U/ml; plasminogen 40–80 μg/ml; aprotinin 3,000 kiu/ml; growth factors (epithelial growth factor or EGF, basic fibroblast growth factor or bFGF, transforming growth factor beta-1 or TGF-β1, vascular endothelial growth factor or VEGF)
- Second syringe: human thrombin (500 UI/ml, 4 UI/ml); calcium.

Due to the presence of growth factor, Tisseel® is not only a hemostatic and adhesive agent but also promotes tissue healing. The manufacturer has developed laparoscopic applicators for this fibrin glue: 1) the Duplocath 35 applicator, which has a dual-lumen; 2) 35-mm-long catheter designed for use in a 5-mm trocar (Fig. 20.1); the Duplospray system is designed for aerosolized spray application of fibrin glue to the raw surface of the liver during MILS; it has a feedback system to maintain a steady pneumo pressure (Fig. 20.2).

Evicel® (Ethicon, USA) is a fibrin sealant consisting of two clottable human proteins: fibrinogen and thrombin. In this fibrin glue, chromatographic filtering techniques have reduced the concentration of plasminogen and therefore abolished the need for an antifibrinolytic.

It is the new formulation of Quixil® (in Europe) or Crosseal® (in the USA), differing from them by not contain the potentially neurotoxic antifibrinolytic agent, tranexamic acid; this change does not affect hemostatic efficacy or longevity of its fibrin clots. It is also available a laparoscopic spray system for this glue.

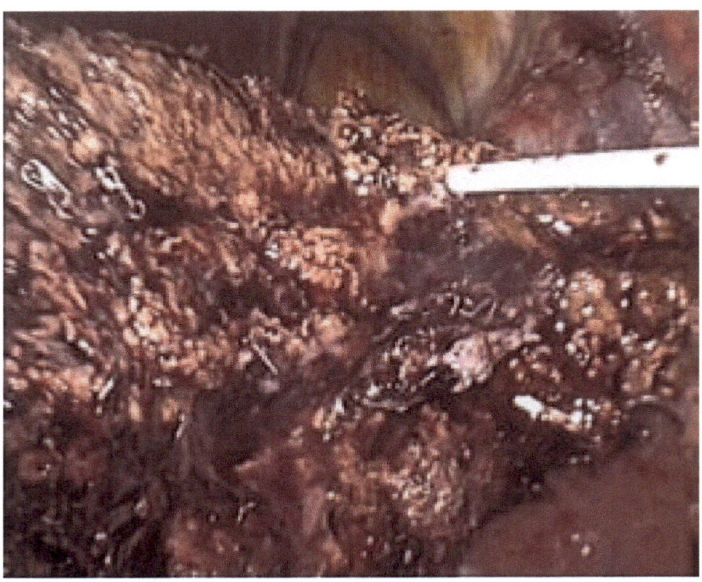

Fig. 20.1 Laparoscopic application of Tisseel with Duplocath 35 applicator

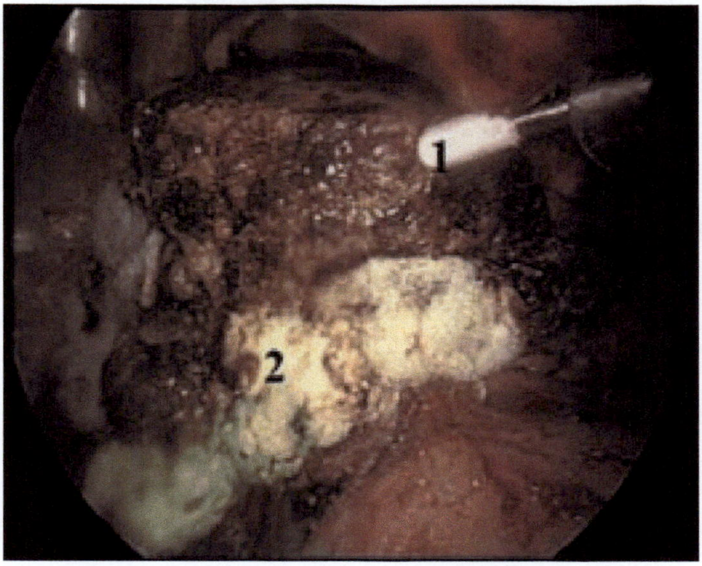

Fig. 20.2 Aereosolized spray application of Tisseel. *1*, Duplospray system; *2*, fibrillar regenerated cellulose

20.3.2 Gelatin Matrix

The gelatin matrix FloSeal® (Baxter, USA) consist of granules with a mean size of 500–600 μm crosslinked with glutaraldehyde and packaged in a syringe in a hydrated state below the equilibrium swell of the gelatin. Before use, the gelatin matrix is prepared by mixing together human-derived thrombin solution in a process that can be completed by the nurse in <2 min. It provides adequate hemostasis via a unique mechanism in which both components act independently and synergistically to promote clot formation at the bleeding site, thus leading to hemostasis (mechanical and biological effect). FloSeal® is delivered to the bleeding site through a single-barrel syringe and held in place with a gauze sponge for 10 min or until the bleeding stops.

The granular nature of the thrombin-coated gel allows conformation to irregular liver surfaces, resulting in intimate gelatin contact with the tissue surface at the bleeding site. Upon contact with blood, the gelatin granules swell (approximately 20 % within 10 min), creating a tamponade effect, physically restricting blood flow, and thus providing a mechanically stable matrix. The blood, percolating between the granules, is exposed to high concentrations of human thrombin that rapidly catalyze fibrinogen conversion to fibrin monomers, accelerating formation of a clot that is reinforced by the incorporation of granules within the fibrin mesh of the clot. Reinforcement of the clot makes it less susceptible to coagulopathies due to clotting factor deficiencies

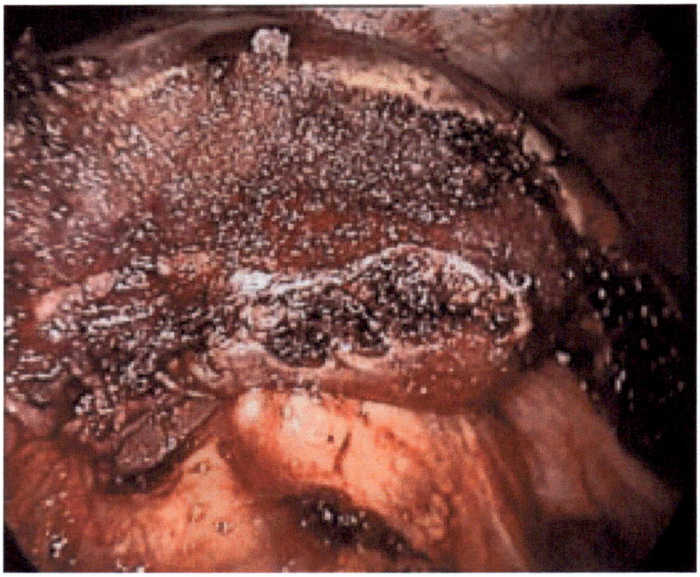

Fig. 20.3 Resected surface of S3 segmentectomy treated with FLOSEAL®

or platelet malfunction, as observed in cirrhotic patients or those treated with chemotherapy.

During MILS, the bleeding site is covered by FloSeal® using the specifically developed laparoscopic applicator, and a moist gauze pad introduced into the abdomen is used to place pressure on the bleeding site (Fig. 20.3). Structure of the fibrin matrix keeps it in place on the liver surface. Granules not incorporated into the clot can and should be removed with gentle irrigation without disrupting the clot to allow verification of hemostasis, avoid artifacts in postresection imaging, reduce the risk of adhesions, and prevent the risk of infections and foreign-body reactions. Preclinical studies show the material to be fully resorbed within 6–8 weeks, consistent with the normal healing process.

Care must be taken by the surgeon to deliver an adequate volume to the bleeding site and ensure that the gel makes intimate contact with bleeding tissue. Because of the unique mechanism by which it provides hemostasis, FloSeal® does not work in the absence of bleeding. This differentiates it from other fibrin sealants, which initially require a relatively dry surface for the clot to adhere to the underlying tissue. Furthermore, as FloSeal® does not rely on the presence of functional platelets or other coagulation factors, except fibrinogen, to produce its hemostatic effect, it is effective when used in patients with coagulopathy.

20.3.3 Collagen Sponges

TachoSil® (Nycomed: a Takeda Company) is a hemostatic and tissue sealant consisting of a ready-to-use equine collagen sponge carrying the human coagulation factors, is sterile, and is absorbable within 12 weeks. It is a development of TachoComb® and TachoComb H® but differs from them by containing only human coagulation factor without bovine aprotinin; therefore, the theoretical risk of transmitting bovine diseases (including bovine spongiform encephalopathy) is eliminated. The active (hemostatic-sealant) side of the patch is yellow (due to use of riboflavin); the other side is white (Fig. 20.4a, b). The active side of the patch carries a fixed combination of human fibrinogen (5.5 mg) and thrombin (2.0 IU). TachoSil® is applied with the yellow side onto the transected liver surface. Because moisture activates the adhesive, the patch must be applied dry, and therefore, forceps and gloves used to manipulate it until in place must also be dry. After application, TachoSil® can be touched on the white side, and gentle compression using a wet gauze swab is required to

Fig. 20.4 Collagen sponge TachoSil (**a** yellow side and **b** white side)

attain complete contact with the parenchyma surface. Application can be difficult during MILS; therefore, various tricks and techniques have been proposed to introduce TachoSil® into the abdomen (see video related to this chapter). This pliable patch is easily modeled using gentle compression with a swab moistened with saline solution and other usual instrumentation, such as forceps, to improve close adhesion to the wet, irregular cut surface. After 3–5 min of compression, the gauze is removed, and close adhesion of the patch to the parenchyma surface is verified. Patches are available in three sizes: 9.5 cm × 4.8 cm; 4.8 cm × 4.8 cm; 3.0 cm × 2.5 cm.

20.3.4 Oxidized, Regenerated Cellulose Fleece

Oxidized cellulose and oxidized, regenerated cellulose fleece are ready-to-use hemostatic devices that use mechanical effects to absorb liquids and compress tissue; these effects are activated by swelling of the resulting gelatinous mass. Denaturation of blood proteins by cellulosic acid within the product probably contributes to the hemostatic effect. Their acidic nature also explains why they may have a vasoconstrictive and bactericidal effect. Surgicel® (Ethicon, USA) is the most commonly used product of this nature and is available as a tightly woven, knitted patch or in a fibrillar form (Fig. 20.2). These agents can be easily applied during LLS because they are readily inserted through the trocars.

20.4 Conclusion

Thanks to technical progress in liver surgery, uncontrollable intraoperative bleedings are exceptional. On the other hand, postoperative mortality and morbidity are still related to postoperative bleeding, bile leakage, and fluid accumulation, potentially leading to repeated surgical or interventional treatment. Despite all available pre- and intraoperative information, the risk for bleeding and bile leakage is difficult to define.

Thanks to advances in surgical techniques and overall management of the resection area, the operative risks of liver surgery are now minimized. Applying an absorbable "dressing" to the raw resected liver surface and creating a watertight occlusive effect on blood, bile, and lymphatic vessels and triggering clot formation is an intriguing concept.

Difficulty obtaining hemostasis continues to be the Achilles' heel of LLS. MILS is a potentially high-risk approach because the surgeon's capacity to control troublesome, persistent bleeding sites are reduced. During LLS blood obscures the surgical field, making surgical maneuvers difficult to perform, and the absence of direct manual compression can necessitate conversion to the open procedure.

The availability and development of agents to control bleeding will potentially enhance the safety of MILS. To prevent stump-related resection complications, topical agents, routinely used during open liver resection, are now used during MILS. Though this application can never replace an adequate surgical technique, familiarity with tools, methods, and techniques for hemostasis and sealant application is essential for laparoscopic surgeons.

The widespread use of LLS will depend also on the ability and skill of expert surgeons and technological advances to address managing risks related to bleeding and bile leaks using any new approach. The development of new agents, technique, specialized equipment, and applicators for use during laparoscopic procedures means that applications of this sort will increase and their use become more extensive. Successful hemostasis in MILS extends and confirms laparoscopic indications. Limitations due to insufficient hemostasis and the need to convert to the open approach due to uncontrollable bleeding are becoming rare thanks to progressing technological developments and increased surgical-team expertise.

References

1. Kraus TW, Mehrabi A, Schemmer P, Kashfi A et al (2005) Scientific evidence for application of topical hemostats, tissue glues and sealants in hepatobiliary surgery. J Am Coll Surg 200(3):418–427
2. Yoshida H, Hirozane K, Kamiya A (2000) Adhesive strength of autologous fibrin glue. Biol Pharm Bul 23:313–317
3. Troisi RI, Van Huysse J, Berrevoet F et al (2011) Evolution of laparoscopic left lateral sectionectomy without the Pringle maneuver: through resection of benign and malignant tumors to living liver donation. Surg Endosc 25:79–87
4. Vauthey JN, Pawlik TM, Ribero D et al (2006) Chemotherapy regimen predicts steatohepatitis and an increase in 90-day mortality after surgery for hepatic colorectal metastases. J Clin Oncol 24:2065–2072
5. Lattouf JB, Beri A, Klinger CH et al (2007) Practical hints for hemostasis in laparoscopic surgery. Minim Invasive Ther Allied Technol 16:45–51
6. Msezane LP, Katz MH, Gofrit ON et al (2008) Hemostatic agents and instruments in laparoscopic renal surgery. J Endourol 22:403–408
7. Katkhouda N (2004) New hemostatic agents in general open and laparoscopic surgery. Surg Technol Int 13:65–70
8. Melman L, Matthews BD (2008) Current trends in laparoscopic solid organ surgery: spleen, adrenal, pancreas and liver. Surg Clin North Am 88:1033–1046
9. Imamura H, Seyama Y, Kokudo N et al (2003) One thousand fifty-six hepatectomies without mortality in 8 years. Arch Surg 138:1198–1206
10. Capussotti L, Ferrero A, Viganò L et al (2006) Bile leakage and liver resection: where is the risk? Arch Surg 141:690–694
11. Toti L, Attia M, Manzia M et al (2010) Reduction in bile leaks following adult split liver transplant using a fibrin-collagen sponge: a pilot study. Dig Liver Dis 42:205–209
12. Nguyen KT, Gamblin TC, Geller DA (2009) World review of laparoscopic liver resection-2,804 patients. Ann Surg 250(5):831–841
13. Hilal MA, Underwood T, Taylor MG et al (2010) Bleeding and hemostasis in laparoscopic liver surgery. Surg Endosc 24:572–577

14. Shanberg AM, Zagnoev M, Clougherty TP (2002) Tension pneumothorax caused by the argon beam coagulator during laparoscopic partial nephrectomy. J Urol 168:2162
15. Ikegami T, Shimada M, Imura S et al (2009) Argon gas embolism in the application of laparoscopic microwave coagulation therapy. J Hepatobiliary Pancreat Surg 16:394–398
16. Saxton ML (2007) Hemostasis in minimally invasive liver surgery. Surgery 142(suppl 4):S46–S49
17. Saxena AK, van Tuil C (2007) Advantages of fibrin glue spray in laparoscopic liver biopsies. Surg Laparosc Endosc Percutan Tech 17:545–547
18. Robinson S, Manas D, Noormohamed S et al (2009) The use of an impregnated collagen sponge (TACHOSIL) for laparoscopic radiofrequency ablation of liver tumours. HPB 11(suppl 2):150
19. Saif R, Jacob M, Robinson S et al (2011) Use of fibrin-based sealants and gelatin-matrix hemostats in laparoscopic liver surgery. Surg Laparosc Endosc Percutan Tech 21:131–141

Specimen Extraction

21

Alberto Patriti, Graziano Ceccarelli, Raffaele Bellochi
and Luciano Casciola

21.1 Introduction

Specimen extraction is the final step of every laparoscopic liver resection (LLR). An ideal method of extraction should minimize contamination of the peritoneum and port sites in the abdominal wall, have a good cosmetic result, avoid early (wound infection) and late (incisional hernia) complications, and allow specimen integrity for pathological examination. There is at present no consensus on the best site for specimen extraction. However, three different sites are adequate for using 10-mm endoscopic bags: an enlarged port site (umbilical area), a Pfannenstiel incision, and a pre-existing abdominal scar. Most of the time, the specimen dimension is the main limiting factor, followed by esthetic concerns, when choosing the extraction site.

21.2 Periumbilical Extraction

This site can be used when a trocar is inserted in the periumbilical area and when a minor hepatectomy is performed [1]. A 10-mm endoscopic bag is inserted through the umbilical trocar. Once the specimen is accommodated into the bag, the abdomen is deflated. The umbilical incision is extended along the left circumference of the umbilicus. When the subcutaneous connective tissue is stretched apart, the linea alba becomes evident. A 3- to 5-cm incision is made in the craniocaudal direction and in some cases can be extended transversely, entering the left rectus sheath. With dilation and slight traction

A. Patriti (✉)
Department of Surgery, Division of General, Minimally Invasive and Robotic Surgery,
ASL3 Umbria, San Matteo degli Infermi Hospital, Spoleto, Italy
e-mail: albertopatriti@gmail.com

F. Calise and G. Casciola (eds.), *Minimally Invasive Surgery of the Liver*,
Updates in Surgery
DOI: 10.1007/978-88-470-2664-3_21,

on the bag, specimen removal is carried out aseptically and without the risk of neoplastic seeding. The closure is performed in layers with an interrupted suture. Generally, the skin incision is hidden by the umbilical scar and achieves good long-term esthetic results. This technique avoids the disadvantage of an additional incision in the abdominal wall, but in patients with risk factors – diabetes, smoking, obesity, wound infection – may give rise to an incisional hernia with a higher incidence than the Pfannenstiel incision [2].

21.3 Pfannenstiel Incision

The Pfannenstiel site can be used when a major hepatectomy is performed. When liver resection is completed and the patient is tilted on the left flank, the procedure can be demanding and time consuming. The specimen is inserted into the endoscopic bag. The wire of the bag is cut and the bag is left in the pelvis. A curved, transverse abdominal incision with downward convexity is performed above the symphysis pubis. The incision passes through the skin, superficial fascia, and aponeurosis of the rectus sheath, exposing the pyramidalis and rectus muscles, which are separated at the midline, after which the peritoneum is opened vertically. The endoscopic bag is grasped and removed. The incision is sutured in layers. This technique has a very good cosmetic outcome and is associated with a lower risk of incisional hernia than with the other methods. However, it is an additional laparotomy in a site not used for trocar insertion during minimally invasive liver surgery and could expose the patient to an unnecessary hypothetical risk of infection and incisional hernia.

21.4 Pre-existing Abdominal Scar

Regardless of the site of a previous scar (appendectomy, oophorectomy...), the abdominal wall is incised to the fascia, and the endoscopic bag with the specimen is retained, as previously described. The specimen is then retracted against the fascia, and the fascia and peritoneum are opened only as much as necessary for retrieval. The incision is finally sutured in layers. This technique could be associated with a high risk of incisional hernia because it produces a new incision in scar tissue.

References

1. Casciola L, Codacci-Pisanelli M, Ceccarelli G, Bartoli A, Di Zitti L, Patriti A (2008) A modified umbilical incision for specimen extraction after laparoscopic abdominal surgery. Surg Endosc 22(3):784–786. doi:10.1007/s00464-007-9584-2
2. Bartone G, de Bellis M, Crovella F (2008) Trocar-Site Hernia. In: Crovella F, Bartone G, Fei L (eds) Incisional Hernia. Springer Milan, pp 175-185

22 Short- and Long-Term Follow-Up

Antonio Giuliani, Carla Migliaccio, Giuseppe Surfaro,
Antonio Ceriello and Maurizio Defez

22.1 Introduction

Laparoscopic liver surgery (LLS) is now a widely accepted procedure. It is no longer questionable whether short-terms results after liver resection are better for LLS than for open surgery. Postoperative morbidity and mortality rates, hospital stay, abdominal-wall complications, and intraoperative bleeding are reduced with LLS in both benign and malignant conditions [1–6]. In the absence of general contraindications, a laparoscopic approach should be indicated for the majority of benign liver nodules located in the anterior segments [7–9]. Many investigators in the early 2000s felt there were no benefits in long-term oncological outcomes for patients with hepatocellular carcinoma (HCC) and colorectal metastasis (CRM) undergoing LLS. In very early experiences with general laparoscopic surgery, there was evidence of an unusually high tumor recurrence rate, especially at port sites. This condition was not found following LLS, probably because this surgery became widely diffused among surgeons following technical refinements to reduce tumor manipulation and avoid tumor-cell seeding. Another early issue was the adequacy of surgical margins [10]. More recent series show that this condition is not sustained by evidence [2, 3, 7, 11], even if there are no controlled trials comparing open and LLS approaches.

However, in 2010, Hilal et al. showed that in LLS for CRM, they obtained a mean tumour-free margin of 17 mm, with >1 cm in 76% of patients [11]. Regardless, although it is clear that LLS is feasible and reproducible, some doubts remain concerning short- and long-term results compared with open surgery [1].

A. Giuliani (✉)
Unit of Hepatobiliary Surgery and Liver Transplant Center, A. Cardarelli Hospital, Naples, Italy
e-mail: giuldoc@hotmail.com

F. Calise and G. Casciola (eds.), *Minimally Invasive Surgery of the Liver*, Updates in Surgery
DOI: 10.1007/978-88-470-2664-3_22, © Springer-Verlag Italia 2013

22.2 General Considerations

LLS is a safe and reproducible procedure, with an overall mortality rate of 0.3%–0.6%, a morbidity rate of 10.5%, with no intraoperative deaths [5–6]. Intraoperative blood loss, hospital length of stay, postoperative complications, amount of intravenously administered narcotic agents, and days until oral intake are significantly reduced in patients undergoing LLS [5–7, 12]. Moreover, if we consider hospital readmission rates due to surgical complications, we observe a better trend in patients who underwent LLS.

Gustafson et al. showed that 30-day hospital readmission rates for surgery-related complications were higher, even though not significantly, in patients with open liver surgery compared with those who underwent LLS (9 of 49 vs. 2 of 27, $p = 0.20$) [13]. Furthermore, 1-year hospital readmission rates due to surgical complications were significantly lower in the LLS group (4 of 27 vs. 19 of 49, $p = 0.002$) [13]. Similarly, Vanounou et al., comparing 44 patients with LLS and 29 with open surgery, showed significant reduction in postoperative morbidity ($p = 0.001$) [14]. This observation was confirmed in a subgroup of patients with malignant diseases ($p = 0.003$) [14]. Interestingly, however, analysis of cost–benefit effect and economic impact of LLS vs open surgery showed that LLS was US$1,527–2,939 more cost efficient for the patient compared with the open technique [14]. The cost analysis of a surgical procedure should not be limited to the cost of instruments and technology but extended to other parameters, such as hospital stay or blood transfusions. Most likely, the observation of Vanounou et al. is related to the fact that with LLS there is a reduction in costs related to many indirect parameters, such as patients' loss of work days, the use of drugs for pain control at home, the need for new outpatient access, and the possibility of being autonomous in daily activities without the help of others. These parameters and the reduced rate of rehospitalization compared with open surgery (the latter due to complications such as incisional hernia) may explain the cost advantages for LLS.

22.3 Benign Tumors

There are no series reporting on long-term follow-up for benign liver tumors. Specific data may only be extrapolated from multicentric or single-center experiences. The most common benign indications are focal nodular hyperplasia (FNH), adenomas, and hemangiomas.

A meta-analysis from Croome and Yamashita [12] analyzed 26 randomized studies comparing LLS and open liver surgery, with a total number of 354 benign liver lesions of 851 liver resections (41.6%). Considering LLS only, Mirnezami et al. found that 27.3% of all resections were for benign diseases (196 of 717 LLS) [5]. Both studies showed good postoperative and short-term results for LLS, with no mention about long-term results. A series of 61 patients with hepatolithiasis undergoing a total of 28 LLS showed at a mean follow-up of 17 months (*r* 2–40), only one patient with stone recurrence (3.4%) [15]. A similar observation was made by Lai et al. in 55 patients who underwent LLS, with stone recurrence in three patients (5.4%) at a mean follow-up of 59 ± 30 months [16]. Furthermore, LLS for benign disease shows better health-related quality of life (HR-QoL) in the first year after surgery compared with the open approach. We retrospectively compared surgical results of two groups of patients with benign liver lesions (29 with LLS and 46 with open surgery) and evaluated HR-QoL using the 36-question Short Form (SF-36) Health Survey. Our results show a significantly better HR-QoL in the first year after LLS than with open surgery (unpublished data).

22.4 Malignancies

Most criticism of LLS for malignancy in the earlier experience seems to have abated. There are no reported cases of port-site recurrence after LLS for malignant disease [5]. Mirnezami et al., in their meta-analysis, showed that in eight studies with a total of 392 patients and available data on resection margins, there were no significant differences between LLS and open surgery. Furthermore, there was an increased incidence of margin resections <1 cm in patients with open resection, with positive resection margins being greater in open surgery (neither statistically significant) [5]. This is probably due to the larger number of anatomical resections observed in LLS series than in open surgery, in which nonanatomical and wedge resections are more frequent. Croome and Yamashita confirm that there are no significant differences in the risk of positive margins after LLS, although they found the risk of resection margins <1 cm to be approximately twice as high in the LLS group – relative risk (RR) 1.99; 95% confidence interval (CI) 1.31–3.02; $p = 0.001$) [12].

Many studies in the literature compare results of LLS and open surgery for HCC, whereas there are fewer studies about CRM. This is probably because patients with CRM, except those who undergo a one-stage procedure for synchronous metastases, have always had previous surgery for the primary malignancy. Previous surgery, in fact, in most surgical units, is still considered a contraindication to LLS.

22.4.1 Colorectal Metastases

Nguyen et al., in a review of world literature on LLS accounted of 2804 patients, found a 3 years overall survival from 80 % to 87 % and a 3 year disease survival of 51% respectively for patients with CRM [6].

So, are there differences between long-term results of LLS and open surgery? In 2009, Welsh et al. compared two groups of patients who underwent both LLS (group 1, $n = 266$) and open (group 2, $n = 886$) surgery for CRM. They made a careful selection, excluding from LLS patients with lesions ≥ 6 cm, lesions near or involving the hilum/internal vena cava (IVC), involving surrounding tissue, and patients with previous abdominal surgery [17]. Furthermore, in group 1, there were more patients with fewer than three metastases ($p < 0.001$), bilobar diffusion ($p < 0.001$), and fewer mean tumor diameter (3.3 cm vs. 5.3 cm; $p < 0.001$). The overall 5-year survival rate was better for group 1 (44.2% vs. 37.8% ; $p = 0.005$), with a 7-year survival rate of 36.9% vs. 32.1% ($p = 0.004$) [17]. In 2010, Nguyen et al. [18] evaluated four studies of LLS for CRM, finding a 5-year survival rate ranging from 46% to 64%, which is absolutely comparable with open surgery but with the advantages of LLS in the short term, such as small skin incisions, less pain, less narcotic requirement, and shorter length of hospital stay. In patients with CRM, part of the previous incision can be used to extract the specimen, which is an obvious advantage for the patient. Moreover, in this study, they found that LLS gave an actuarial hospital cost saving of US$2,939.00 per case compared with open surgery due entirely to shorter hospital stay [18].

It is obvious, therefore, that a controlled trial with similar groups for lesion number and size, resection type (anatomical/nonanatomical), and type of chemotherapy (neoadjuvant/adjuvant/perioperative) with a mean follow-up of 3 years should answer the question. Table 22.1 shows medium and long-term follow-up results after LLS for CRM in major series in the literature.

22.4.2 Hepatocellular Carcinoma

In patients undergoing surgery for HCC, we observed a cumulative survival rate at 1, 3, and 5 years of 60%, 38%, and 25%, respectively, in the paper by Poon et al. [19] who evaluated 377 patients. However, as is widely known, survival may consistently change according to HCC size, number, and grading [9]. The benefits of LLS for HCC seem to be greater than for other indications: advantages mainly are reduction of postoperative ascites and abdominal-wall complications without compromising porto-caval shunts [19, 20]. A large European series evaluated 163 LLS for HCC compared with 378 patients who underwent open surgery [20]. OS (overall survival) was good, with an incidence of 1-, 3-, and 5-year disease-free survival (DFS) compared with open surgery of 77.5%, 47.1%, and 32.2%, respectively. Aldrighetti et al. showed a mean DFS of 23.3 months, but more than half of their patients had a Child–Pugh score of B or C [21].

Truant et al. compared 36 LLS to 53 open resections. Mean age in the LLS group was 60.5 (±10.2) years, with 34 patients (94.4%) with a solitary tumor with a mean size of 2.9 (±1.2) cm. Two of these patients underwent liver transplantation for recurrence, whereas in the open group, the number of patients who underwent salvage transplantation was five (not significant). The authors referred to the role of surgical resection as a bridge to transplantation: LLS seems to be the best approach in this situation. In fact, the authors found 13.2% of severe complications (Clavien–Dindo III–V) after open surgery compared with only 2.8% after LLS [22]. Better postoperative results are also con-

Table 22.1 Medium and long-term follow-up after laparoscopic liver surgery (LLS) for colorectal metastases (CRM)

Author	Number of patients	Type of study	3-year disease-free survival	3-year overall survival	5-year disease-free survival	5-year overall survival
Nguyen et al. (2009) [6]	2,804; 115 CRM	Review	51%	80–87%	n.a.	n.a.
Kazaryan et al. (2010) [2]	1,77, 96 CRM	Single center	n.a.	n.a.	n.a.	46% (actuarial)
Welsh et al. (2010) [17]	1,152, 266 CRM	Single center	n.a.	69.7%	n.a.	44.2%
Castaing et al. (2009) [27]	60 CRM	Two center	n.a.	82%	n.a.	64%
Sasaki et al. (2009) [28]	39 CRM	Single center	n.a.	64%	n.a.	64%
Ito et al. (2009) [29]	13 CRM	Single center	n.a.	72%	n.a.	n.a.
Vibert et al. (2006) [30]	42 CRM	Single center	n.a.	87%	n.a.	n.a.

n.a., not applicable.

Table 22.2 Medium and long-term follow-up after laparoscopic liver surgery (LLS) for hepatocellular carcinoma (HCC)

Author	Number of patients	Age	Child–Pugh A	1-year overall survival	3-year overall survival	5-year overall survival
Dagher et al. (2010) [20]	163	65 (24–84) (median)	111 (68%)	92.6%	68.7	64.9
Aldrighetti et al. (2010) [21]	16	65 (10) (mean, SD)	9 (56.2%)	n.a.	Mean overall survival of 40.2 months	n.a.
Truant et al. (2011) [22]	36	60.6 (10.2) (mean, SD)	32 (88.9%)	n.a.	n.a.	70%
Tranchart et al. (2010) [23]	42	63.7 (13.1) (mean, SD)	30 (71.4%)	93.1%	74.4%	59.5%
Lai et al. (2009) [24]	25	59 (35–79) (median)	23 (92%)	n.a.	60%	n.a.
Lee et al. (2011) [25]	33	59 (36–85) (median)	33 (100%)	87.5%	87.5%	75%
Belli et al. (2011) [26]	65	63.2 (6.8) (mean, SD)	58 (89%)	95% (actuarial)	70% (actuarial) 55% (actuarial)	

SD, standard deviation; *n.a.*, not applicable.

firmed by others [18, 19]. Furthermore, Tranchart et al., comparing 42 patients undergoing LLS and 42 undergoing open surgery, found that blood loss ($p < 0.0001$), use of portal clamping ($p < 0.0001$), hospital stay ($p < 0.0001$), and cirrhotic decompensation ($p = 0.03$) were better in the LLS group [23]. Table 22.2 shows medium and long-term follow-up after LLS for HCC.

22.5 Conclusions

LLS is a safe, feasible, and efficient procedure for treating benign and malignant liver tumors, even in patients with prior abdominal surgery, and provides short-term benefits without compromising long-term oncological results. However, to achieve good oncological results, LLS should be reserved for expert surgical teams and proposed only for carefully selected patients. Although surgical margins, recurrence rate, and OS rates seem similar to or better than open surgery, controlled trials are needed to compare results of LLS and open surgery in specific groups of patients matched for lesion size, number and resection type.

References

1. Cherqui D (2003) Laparoscopic liver resection. Br J Surg 90:644–646
2. Kazaryan AM, Pavlik Marangos I, Rosseland AR et al (2010) Laparoscopic liver resection for malignant and benign lesions: ten-year Norwegian single-center experience. Arch Surg 145:34–40
3. Abu Hilal M, McPhail MJ, Zeidan B et al (2008) Laparoscopic versus open left lateral hepatic sectionectomy: A comparative study. Eur J Surg Oncol 34:1285-1288
4. Nguyen KT, Marsh JW, Tsung A et al (2011) Comparative benefits of laparoscopic vs open hepatic resection: a critical appraisal. Arch Surg 146:348–356
5. Mirnezami R, Mirnezami AH, Chandrakumaran K et al (2011) Short- and long-term outcomes after laparoscopic and open hepatic resection: systematic review and meta-analysis. HPB (Oxford) 13:295–308
6. Nguyen KT, Gamblin TC, Geller DA (2009) World review of laparoscopic liver resection-2,804 patients. Ann Surg 250:831–841
7. Pilgrim C, TO H, Ustatoff V et al (2009) Laparoscopic hepatectomy is a safe procedure for cancer patients. HBP Surg 11:247–251
8. Farges O, Jagot P, Kirstetter P et al (2010) Prospective assessment of the safety and benefit of laparoscopic liver resections. J Hepatobiliary Pancreat Surg 9:242–248
9. Ardito F, Tayar C, Laurent A et al (2007) Laparoscopic liver resection for benign disease. Arch Surg 142:1188–1193
10. Gigot JF, Glineur D, Santiago Azagra J et al (2002) Laparoscopic liver resection for malignant liver tumors: preliminary results of a multicenter European study. Ann Surg 236:90–97
11. Abu Hilal M, Underwood T, Zuccaro M et al (2010) Short- and medium-term results of totally laparoscopic resection for colorectal liver metastases. Br J Surg 927–933
12. Croome KP, Yamashita MH (2010) Laparoscopic vs open hepatic resection for benign and malignant tumors: an updated meta-analysis Arch Surg 145:1109–1118
13. Gustafson JD, Fox JP, Ouellette JR et al (2012) Open versus laparoscopic liver resection: looking beyond the immediate postoperative period. Surg Endosc 26:468–472

14. Vanounou T, Steel JL, Nguyen KT (2010) Comparing the clinical and economic impact of laparoscopic versus open liver resection. Ann Surg Oncol 17:998–1009
15. Tu JF, Jiang FZ, Zhu HL et al (2010) Laparoscopic vs open left hepatectomy for hepatolithiasis. World J Gastroenterol 16:2818–2823
16. Lai EC, Ngai TC, Yang GP et al (2010) Laparoscopic approach of surgical treatment for primary hepatolithiasis: a cohort study. Am J Surg 199:716–721
17. Welsh FKS, Tekkis PP, John TG, Rees M (2010) Open liver resection for colorectal metastases: better short- and long-term outcomes in patients potentially suitable for laparoscopic liver resection HPB (Oxford) 12:188–194
18. Nguyen KT, Geller DA (2010) Outcomes of laparoscopic hepatic resection for colorectal cancer metastases. J Surg Oncol 102:975–977
19. Poon RT, Fan ST, Lo CM et al (2001) Improving survival results after resection of hepatocellular carcinoma: a prospective study of 377 patients over 10 years. Ann Surg 234:63–70
20. Dagher I, Belli G, Fantini C et al (2010) Laparoscopic hepatectomy for hepatocellular carcinoma: a European experience. J Am Coll Surg 211:16–23
21. Aldrighetti L, Guzzetti E, Pulitano C et al (2010) Case-matched analysis of totally laparoscopic versus open liver resection for HCC: short and middle term results. J Surg Oncol 102:82–86
22. Truant S, Bouras AF, Hebbar M et al (2011) Laparoscopic resection vs. open liver resection for peripheral hepatocellular carcinoma in patients with chronic liver disease: a case-matched study. Surg Endosc 25:3668–3677
23. Tranchart H, Di Giuro G, Lainas P et al (2010) Laparoscopic resection for hepatocellular carcinoma: a matched-pair comparative study. Surg End 24:1170–1176
24. Lai EC, Tang CN, Ha JP et al (2009) Laparoscopic liver resection for hepatocellular carcinoma ten-year experience in a single center. Arch Surg 144:143–147
25. Lee KT, Chong CN, Wong J et al (2011) Long-term results of laparoscopic hepatectomy versus open hepatectomy for hepatocellular carcinoma: a case-matched analysis. World J Surg 35:2268–2274
26. Belli G, Fantini C, Bella A et al (2011) Laparoscopic liver resection for hepatocellular carcinoma in chirrhosis: long term results. Dig Surg 28:134–140
27. Castaing D, Vibert E, Ricca L et al (2009) Oncologic results of laparoscopic versus open hepatectomy for colorectal liver in two specialized centers. Ann Surg 250:849–855
28. Sasaki A, Nitta H, Otsuka K et al (2009) Ten-year experience of totally laparoscopic liver resection in a single institution. Br J Surg 96:274–279
29. Ito K, Ito H, Are C et al (2009) Laparoscopic versus open liver resection: a matched-pair case control study. J Gastrointest Surg 13:2276–2283
30. Vibert E, Perniceni T, Levard H et al (2006) Laparoscopic liver resection. Br J Surg 93:67–72

Part IV

Surgery: Techniques

Portal Vein Ligation

23

Giuseppe Aragiusto, Antonio Ceriello, Antonio Giuliani,
Carla Migliaccio and Fulvio Calise

23.1 Introduction

Portal vein ligation (PVL) is a standard procedure for achieving resectability in patients with an inadequate future remnant liver (FRL) prior to planning subsequent major hepatectomy. Its role in inducing FRL hypertrophy prior to major liver resection is clear [1]. Actually, PVL is the first step in a two-stage hepatectomy for treating initially unresectable liver metastases [1]. As reported in the literature, this procedure can be achieved safely without causing mortality [2]. Capussotti et al. showed how PVL is as effective as PV embolization in inducing hypertrophy of the remnant liver volume [2]. The possibility of using a laparoscopic approach seems to be favorable, even for achieving lower patient morbidity rates, particularly in the case of synchronous colorectal metastasis, avoiding the need to perform a further procedure (laparotomy/portal embolisation) [3, 4]. Finally, in a planned two-stage hepatectomy, laparoscopic PVL (LPVL) greatly reduces the presence of adhesions when it is time to perform the second surgical step [5].

23.2 Portal Pedicle Anatomy

The PV is about 8 cm long and is formed at the level of the second lumbar vertebra by the junction of the superior and inferior mesenteric veins and the splenic vein. It passes upward behind the superior part of the duodenum and then ascends in the right border of the lesser omentum to the right extremity

G. Aragiusto (✉)
Unit of Hepatobiliary Surgery and Liver Transplant Center, A. Cardarelli Hospital,
Naples, Italy
e-mail: giuseppe_aragiusto@fastwebnet.it

F. Calise and G. Casciola (eds.), *Minimally Invasive Surgery of the Liver,*
Updates in Surgery
DOI: 10.1007/978-88-470-2664-3_23, © Springer-Verlag Italia 2013

of the porta hepatis, where it divides into a right and a left branch, which accompany the corresponding branches of the hepatic artery into the liver parenchyma. In the lesser omentum, it is located behind and between the common bile duct and the hepatic artery, the former lying to the right of the latter. It is surrounded by the hepatic plexus of nerves and is accompanied by numerous lymphatic vessels and some lymph glands. The right branch of the PV enters the right lobe of the liver, but before doing so, it generally receives the cystic vein. The left branch, longer but of smaller caliber than the right, crosses the left sagittal fossa, branches to the caudate lobe, and then enters the left lobe of the liver. A frequent anatomical variation (11%) is the presence of a right artery originating from the superior mesenteric artery and entering the pedicle on the right side of the main portal trunk.

23.3 Surgical Technique

The patient is placed in the supine position with the lower limbs apart: the surgeon is placed between patient's legs, with one assistant on the left and one on the right side of the patient. Trocar placement is shown in Fig. 23.1. A minimum of four trocars is needed to perform the procedure, and should be spaced

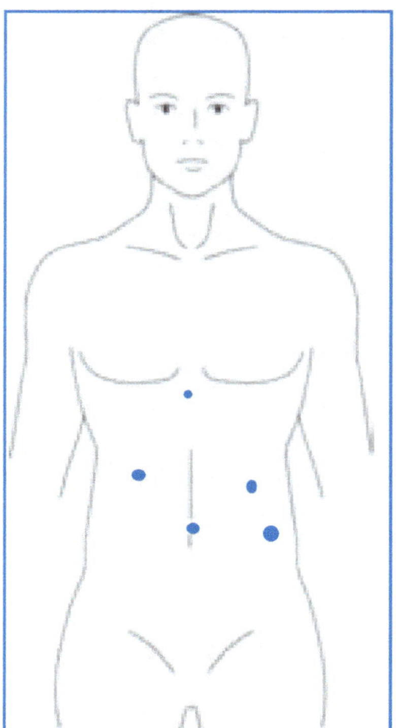

Fig. 23.1 Trocar placement

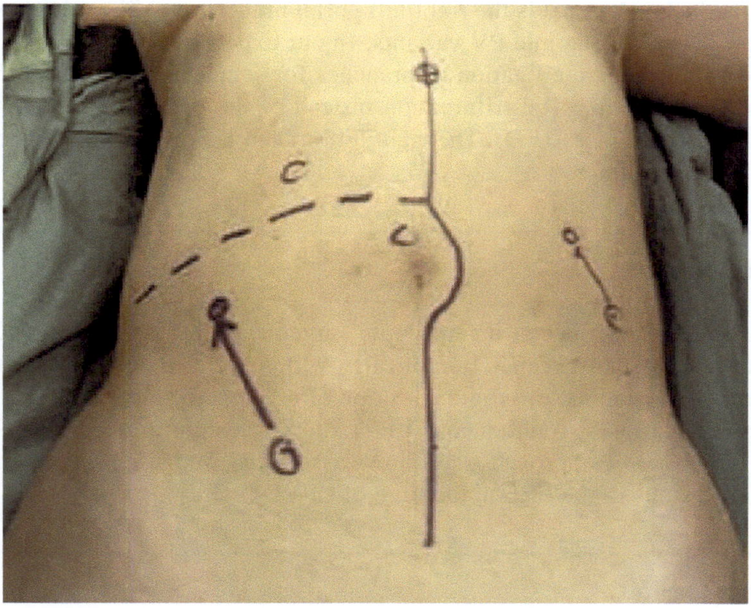

Fig. 23.2 Trocar placement for a left-sided or rectum tumor

out one from the other to avoid crossing over of the instruments. Special attention must be paid to small abdomens where the distance is reduced. In case of a simultaneous laparoscopic approach to colorectal disease and to the liver, trocar placement for a left-rectum-sided tumor is shown in Fig. 23.2. Pneumoperitoneum is obtained via open laparoscopy in the umbilical area and is maintained at 12 mmHg. The camera is placed at the level of the umbilical incision. The three remaining trocars are then placed: two 10- to 12-mm trocars are placed in the right and left paramedian area and the 5-mm trocar is placed in the epigastric area 2 cm under the xiphoid appendix. The screen is placed on the side of the operative table, close to the patient's head.

The first step after exploring the abdominal cavity is to encircle the pedicle to enable quick control of any eventual bleeding (see dedicated chapter in this volume). The portal triad is then dissected from the right side with the help of the Harmonic scalpel or bipolar forceps. Attention must always be paid to the presence of a right branch from the superior mesenteric artery, which can be checked using intraoperative ultrasound (US). The bile duct and right hepatic artery are lifted with a dissector to expose the main PV. The right PV is completely dissected and encircled with a vessel loop. Further dissection is performed in a cranial direction to identify the portal bifurcation. The right or left branch is then passed and encircled with another vessel loop. Intraoperative color Doppler US is routinely performed: the main trunk and the selected branch are sequentially clamped before portal branch ligation to ensure that the isolated vessel is the right/left portal branch, avoiding mistakes due to the presence of anatomic PV variants. In our experience, in two cases, we could perform a separate ligation of branches for segments 5 and 8 and 6 and 7 in the presence of portal trifurcation instead of the normal bifurcation (right and left branch) (see DVD). The right/left branch is now occluded with either clips or silk ligature.

23.4 Conclusions

In the hands of an experienced laparoscopic surgeon, LPVL is feasible and safe and induces adequate regeneration of FRL, with no related complications [2]. In patients requiring laparoscopic resection of colorectal cancer with synchronous liver metastases, simultaneous LPVL did not lead to increased morbidity. The laparoscopic approach also allows associated wedge resections of small anterior lesions in the FRL at the time of LPVL.

References

1. Jaeck D, Oussoultzoglou E, Rosso E et al (2004) A two-stage hepatectomy procedure combined with portal vein embolization to achieve curative resection for initially unresectable multiple and bilobar colorectal liver metastases. Ann Surg 240:1037–1051
2. Capussotti L, Muratore A, Baracchi F et al (2008) Portal vein ligation as an efficient method of increasing the future liver remnant volume in the surgical treatment of colorectal metastases. Arch Surg 143:978–982
3. Are C, Iacovitti S, Prete F et al (2008) Feasibility of laparoscopic portal vein ligation prior to major hepatectomy. HPB (Oxford) 10:229–233
4. Ayiomamitis GD, Low JK, Alkari B et al (2009) Role of laparoscopic right portal vein ligation in planning staged or major liver resection. J Laparoendosc Adv Surg Tech A 19:409–413
5. Machado MA, Makdissi FF, Surjan RC et al (2010) Two-stage laparoscopic liver resection for bilateral colorectal liver metastasis. Surg Endosc 24:2044–2047

Tumorectomy

24

Alberto Patriti, Giorgio Ercolani, Luciano Casciola
and Antonio Daniele Pinna

24.1 Introduction

Surgical treatment of colorectal liver metastases (CRLM) and hepatocellular carcinoma (HCC) is moving toward a parenchyma-sparing approach. The observation that surgical margin width is not definitively correlated with CRLM recurrence rate has encouraged favoring limited nonanatomic liver resections over major hepatectomies [1–3]. The optimal width of resection margin is unclear, with no clear minimum established. Pawlik and colleagues observed that the width of a negative margin did not affect survival, recurrence risk, or site of recurrence [4]. They concluded that a predicted margin width of <1 cm should not be used as exclusion to resection. A recent meta-analysis showed that a resection margin >1 cm is desirable, but disease-free survival is only slightly affected by a subcentimeter margin [5]. Emphasis on obtaining an R0 resection rather than striving for a minimal margin width was confirmed in the 2006 American Hepato-Pancreato-Biliary Association/ Society for Surgery of the Alimentary Tract/Society of Surgical Oncology (AHPBA/SSAT/SSO) Consensus Statement [6]. This approach has the advantage of reducing morbidity without changes in long-term results and offers the possibility of repeated hepatectomies in case of liver metastasis recurrence [7–9]. This trend has been improved by progresses in intraoperative ultrasound (US), which reduces the need for major hepatectomies even in demanding situations such as tumor invasion of the hepatic veins [10].

The concept of parenchymal preservation might be different in cases of HCC. Impaired liver function secondary to the underlying cirrhosis requires a

A. Patriti (✉)
Department of Surgery, Division of General, Minimally Invasive and Robotic Surgery,
ASL3 Umbria, San Matteo degli Infermi Hospital, Spoleto, Italy
e-mail: albertopatriti@gmail.com

F. Calise and G. Casciola (eds.), *Minimally Invasive Surgery of the Liver*,
Updates in Surgery
DOI: 10.1007/978-88-470-2664-3_24, © Springer-Verlag Italia 2013

balance between oncologic issues and parenchyma preservation. HCC spreads along portal veins and can give rise to satellite nodules up to 2 cm. Therefore, anatomic resection of parenchyma involved by the tumor, along with its feeding portal branch, is considered the gold standard treatment for HCC, and anatomic segmentectomy can be considered the more parenchymal-preserving oncologically effective procedure [11, 12]. However, in case of reduced liver function evaluated by indocyanine green clearance (IGC) or Model for End-Stage Liver Disease (MELD) score, subsegmentectomies can be applied, particularly in cases of superficial nodules not amenable to a percutaneous approach. This type of resection has been applied via the laparoscopic approach in patients with HCC on cirrhotic liver who are awaiting liver transplantation to control the tumor during the waiting period and to facilitate operative procedures at the time of transplantation [13]. In this chapter, we discuss the technique of nonanatomic resections, which can be applied for CRLM and HCC on cirrhosis. These considerations can be extended to endocrine tumors and noncolorectal, nonendocrine (NCRNNE) liver metastases [14].

24.2 Factors Affecting Surgical Strategy

Even though nonanatomic resections are commonly performed, even in low-volume centers, tumorectomies may present challenges that can hamper oncologic efficacy and patient safety. There are no standard techniques for tumorectomies as there are for anatomic resections, and there are no anatomic landmarks on the liver surface to drive the resection. Frequently, curved or angulated resection lines are required, and the nodule can be in close relation to a major liver vessel. Therefore, confidence with intraoperative US is imperative when approaching liver resection with a parenchymal-preserving attempt. As major portal pedicles are not usually dissected and ligated during a tumorectomy as they are in anatomic major hepatectomies, and the availability of a system for inflow occlusion (Pringle maneuver) is prudential in most cases, particularly if nodules are located in the right lobe, and are essential when dissection close to a major vascular structure is planned. Tumor proximity to the main vascular structure represents a possible limiting factor to carrying out a safe tumorectomy due to the risk of bleeding and biliary fistula, and in these situations, laparoscopy should be carefully evaluated.

- Position of the tumor is the primary consideration in surgical strategy for laparoscopic tumorectomy. Resections in the posterolateral segments are difficult due to inadequate exposure, poor operative field, and difficulty with parenchymal dissection. In studies investigating the laparoscopic approach to posterolateral liver malignancies, major resection is the prevalent procedure, in contrast to minor resections performed for lesions in the anterior segments [15]. This discrepancy can be ascribed to the rigid laparoscopic tools. In fact, segmentectomies and subsegmentectomies of the posterosuperior segments require curved or angulated section lines and are demanding procedures considering small degrees of freedom allowed by the instruments [15].

- Distance of the lesion from the liver surface is crucial due to the lack of tactile sensation in laparoscopic and – even more so – in robot-assisted surgery. For deeply located lesions, laparoscopic US is mandatory for locating the tumor and achieving a clear margin resection. In the second instance, lesion dimensions and echogenicity may contribute to hampering a margin-free tumorectomy or even tumor identification. Careful US exploration in an attempt to correlate intraoperative findings with preoperative images can help identify even small isoechoic lesions. An iPad (Apple Computer Inc., CA, USA) enveloped into a sterile plastic bag can be used to intraoperatively view preoperative images. Using the application OsiriX, an open-source Digital Imaging and Communications in Medicine (DICOM) viewer, it is possible to watch in real time the preoperative computed tomography (CT) and magnetic resonance (MR) images, scrolling through them while performing the laparoscopic US liver exploration.

24.3 Technical Details

For a safe laparoscopic tumorectomy, patient positioning and trocar placement should be individualized according to tumor location. Pneumoperitoneum induction with the Veress needle can help tailor trocar position, thus limiting the use of a periumbilical port to cases in which it is really necessary, as for tumor located in segments 4b and 5 and in the left lobe, and to avoid injury to a recannulated umbilical vein, which might be quite dangerous in cirrhotic patients. Nonanatomic liver resection of peripheric, superficial lesions located in anterior segments (3, 4b, 5, 6) is generally feasible and can be done with minimal morbidity and mortality rates. Pedicle clamping is optional or can be applied only in case of bleeding, and liver division can be performed with all the available transection devices; in these cases, using a cutting stapler can facilitate resection and markedly reduced operative time. Generally, three to four trocars disposed at the level of the umbilical line are adequate.

For lesions located in the posterolateral sector (upper segment 6; segments 7 and 8), the patient is rotated on the left flank to facilitate liver mobilization and inferior vena cava dissection, when necessary. The camera port and the left-sided trocars should be placed as close as possible to the right costal margin, whereas the right trocar can be inserted in the intercostal space between the 10th and 11th ribs along the scapular line. At this level, the risk of accidentally injuring the lung is very low, and direct access is provided to the posterolateral segments, as previously shown by Gumbs and Gayet in laparoscopic surgery and by our group for the robot-assisted approach [16, 17]. Due to the higher risk of bleeding, intermittent pedicle clamping is advisable when approaching posterolateral segments [17, 18].

Careful US exploration with demarcation on the Glisson capsule of the right hepatic vein can avoid major bleeding during parenchymal transection. For deeply located lesions or when the tumor is close to a major vessel, even in those located in anterior segments, the "corkscrew" technique can be useful: After identifying the lesion by inspection and intraoperative US, Glisson's capsule is marked with electrocautery 1- to 2-cm away from the tumor margin. According to tumor location, the marked area is anchored by stitches, with caution taken to prevent the needle from entering the tumor. The suture is held together by metallic clips, and upward traction is performed, facilitating parenchymal transection and accurate identification of vascular and biliary structures. Parenchymal transection is performed with the monopolar shears for the first liver layer (1 cm from the Glisson capsule) and then with the Kelly-clamp crushing technique or an ultrasonic dissector. Whenever necessary, metallic clips or stitches are applied to achieve vascular and biliary control. Control of the surgical margin should be always verified by intraoperative US during parenchymal transection [19]. Generally, small specimens can be extracted using an EndoBag through any port site; otherwise, enlarging the umbilical port or, rarely, a Pfannenstiel incision, could be necessary.

24.4 Robot-Assisted Nonanatomic Resections

Robot assistance is particularly useful for performing parenchymal-preserving resections, especially in the posterolateral segments and when the tumor is in contact with a portal branch or hepatic veins and when both are close to the tumor mass (Fig. 24.1) [17]. In fact, EndoWrist instruments allow fine movements and complex transection planes, reducing discomfort coming from the use of rigid tools. Principles of patient and trocar positions in conventional laparoscopic surgery are applicable also for the robot-assisted approach. For liver surgery, the robot is docked over the patient's head.

All liver resections should be guided by US performed by the on-table surgeon. The console surgeon can view the US screen in picture-in-picture modality, directing the dissection plane, which appears as an echogenic line between the cut surfaces. The parenchyma is usually transected with the Harmonic scalpel for straight-line resections. The Kelly-clamp crushing technique with the EndoWrist bipolar Precise forceps (Intuitive Surgical Systems, Sunnyvale, CA, USA) is preferred for curved and angulated section lines and tumor dissection close to a major liver vessel. Hemostasis of small vessels is obtained with monopolar or bipolar cautery. To secure larger vessels on the transection line, we use Hem-o-lok® clips or ligatures with Vicryl® or Prolene®. The hepatic veins are usually divided with the laparoscopic linear stapler or sutured with Prolene®. Biliostasis is assessed by observation and bile leaks controlled with sutures, as in open surgery.

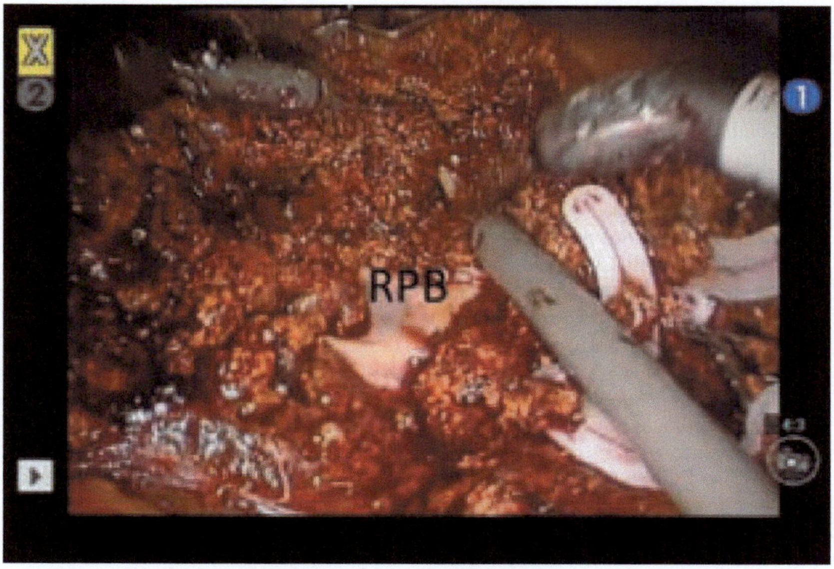

Fig. 24.1 Right portal branch (RPB): final field after tumorectomy of a colorectal liver metastasis close to the RPB

References

1. Elias D, Cavalcanti A, Sabourin JC et al (1998) Resection of liver metastases from colorectal cancer: the real impact of the surgical margin. Eur J Surg Oncol 24(3):174–179
2. Muratore A, Ribero D, Zimmitti G, Mellano A, Langella S, Capussotti L (2010) Resection margin and recurrence-free survival after liver resection of colorectal metastases. Ann Surg Oncol 17(5):1324–1329
3. Torzilli G, Montorsi M, Donadon M et al (2005) "Radical but conservative" is the main goal for ultrasonography-guided liver resection: prospective validation of this approach. J Am Coll Surg 201(4):517–528
4. Pawlik TM, Scoggins CR, Zorzi D et al (2005) Effect of surgical margin status on survival and site of recurrence after hepatic resection for colorectal metastases. Ann Surg 241(5):715–722, discussion 722–714
5. Cucchetti A, Ercolani G, Cescon M et al (2012) Impact of subcentimeter margin on outcome after hepatic resection for colorectal metastases: A meta-regression approach. Surgery 151(5):691–699. doi:10.1016/j.surg.2011.12.009
6. Charnsangavej C, Clary B, Fong Y, Grothey A, Pawlik TM, Choti MA (2006) Selection of patients for resection of hepatic colorectal metastases: expert consensus statement. Ann Surg Oncol 13(10):1261–1268. doi:10.1245/s10434-006-9023-y
7. Gold JS, Are C, Kornprat P et al (2008) Increased use of parenchymal-sparing surgery for bilateral liver metastases from colorectal cancer is associated with improved mortality without change in oncologic outcome: trends in treatment over time in 440 patients. Ann Surg 247(1):109–117. doi:10.1097/SLA.0b013e3181557e47; 00000658-200801000-00017 [pii]

8. Ferrero A, Viganò L, Lo Tesoriere R, Russolillo N, Sgotto E, Capussotti L (2009) Bisegmentectomies as alternative to right hepatectomy in the treatment of colorectal liver metastases. Hepatogastroenterology 56(94-95):1429–1435
9. Viganò L, Ferrero A, Sgotto E, Polastri R, Muratore A, Capussotti L (2005) Parenchyma sparing: evolution of the resective surgical approach of hepatic metastasis from the colorectum. Suppl Tumori 4(3):S35
10. Torzilli G, Montorsi M, Del Fabbro D, Palmisano A, Donadon M, Makuuchi M (2006) Ultrasonographically guided surgical approach to liver tumours involving the hepatic veins close to the caval confluence. British J Surg 93(10):1238–1246
11. Torzilli G, Procopio F, Cimino M et al (2010) Anatomical segmental and subsegmental resection of the liver for hepatocellular carcinoma: a new approach by means of ultrasound-guided vessel compression. Ann Surg 251(2):229–235
12. Makuuchi M, Hasegawa H, Yamazaki S (1981) Intraoperative ultrasonic examination for hepatectomy. Jpn J Clin Oncol (11):367–390
13. Laurent A, Tayar C, Andreoletti M, Lauzet JY, Merle JC, Cherqui D (2009) Laparoscopic liver resection facilitates salvage liver transplantation for hepatocellular carcinoma. Journal Hepatobiliary Pancreat Surg 16(3):310–314. doi:10.1007/s00534-009-0063-0
14. Goéré D, Elias D (2008) Resection of liver metastases from non-colorectal non-endocrine primary tumours. Eur J Surg Oncol 34(3):281–288. doi:10.1016/j.ejso.2007.07.205
15. Cho JY, Han HS, Yoon YS, Shin SH (2009) Outcomes of laparoscopic liver resection for lesions located in the right side of the liver. Arch Surg 144(1):25–29
16. Gumbs AA, Gayet B (2008) Video: the lateral laparoscopic approach to lesions in the posterior segments. J Gastrointest Surg 12(7):1154
17. Casciola L, Patriti A, Ceccarelli G, Bartoli A, Ceribelli C, Spaziani A (2011) Robot-assisted parenchymal-sparing liver surgery including lesions located in the posterosuperior segments. Surg Endosc 25(12):3815–3824. doi:10.1007/s00464-011-1796-9
18. Rotellar F, Pardo F, Bueno A, Marti-Cruchaga P, Zozaya G (2012) Extracorporeal tourniquet method for intermittent hepatic pedicle clamping during laparoscopic liver surgery: an easy, cheap, and effective technique. Langenbecks Arch Surg 397(3):481–485. doi:10.1007/s00423-011-0887-3
19. Makdissi FF, Surjan RC, Machado MA (2009) Laparoscopic enucleation of liver tumors. Corkscrew technique revisited. J Surg Oncol 99(3):166–168. doi:10.1002/jso.21206

Segmentectomies (Chapters 26-34): A Foreword

25

Fulvio Calise

25.1 Introduction

In 1957, Couinaud published his fundamental paper on the surgical anatomy of the liver [1], describing the eight segments, everyone with its pedicle: artery, portal vein, and bile duct. Since then, every liver surgeon has referred to it as a bible to be followed during routine surgical practice. Only with the experience with liver transplantation in the mid-1990s was splitting segment 1 into two sections suggested, identifying the portion of segment 1 on the right side of the caval vein as segment 9 [2]. This was only a minor modification of the original classification. Since the rapid spread of liver surgery in the 1970s, several authors have suggested technical tricks to remove the entire segment by sectioning the tributary pedicle [3–5].

Nowadays, intraoperative ultrasound (US)-guided vessel compression is applicable when performing anatomic segmental and subsegmental resections and coupling parenchyma sparing and tumor resection. Is it therefore mandatory, thanks to progress in US-guided liver resections, to resect the entire segment in order to be oncologically correct? Several reports [6, 7] show that the minimal margin for metastasis may go from the traditional 1 cm to 1 mm, or even predicting, as with Torzilli et al., R1 for metastases lying on a major vein [8, 9]. The situation is different for hepatocellular carcinoma (HCC) due to frequent satellite nodules; however, according to several reports, it seems there are no differences in oncological outcome between regulated (segmentectomy) and unregulated resection [10–12].

F. Calise (✉)
Hepatobiliary Surgery and Liver Transplant Center, A. Cardarelli Hospital
Naples, Italy
e-mail: fulvio.calise@libero.it

F. Calise and G. Casciola (eds.), *Minimally Invasive Surgery of the Liver,*
Updates in Surgery
DOI: 10.1007/978-88-470-2664-3_25, © Springer-Verlag Italia 2013

According to Dahiya et al. [10], the choice between minor and major resection for small (<5 cm), solitary HCC in cirrhotic patients is unclear. Evaluating the long-term disease-free survival and overall survival after minor (259) or major (114) hepatic resections for small solitary HCC in cirrhotic patients, the authors concluded that the severity of cirrhosis and tumor characteristics rather than resection type predict long-term survival. Matsui et al. [11] reported data on 365 patients who underwent hepatic resections without tumor surface exposure and 62 with the HCC surface at the cut stump with no surgical margins due to cancer adherence to the major hepatic vascular structures. No significant differences were registered between the two groups regarding recurrence and overall survival rates, recurrence rate at the cut stump, or number and location of intrahepatic recurrences. According to Torzilli et al. [12], parenchymal-sparing intraoperative US-guided finger compression should be part of the modern liver surgeon's armamentarium when performing conservative resection for malignant hepatic lesions.

25.2 Parenchymal Sparing and Minimally Invasive Surgery of the Liver: A Possible "Marriage"?

Minimally invasive surgery of the liver (MISL) has contributed to the support of formally uncompleted segmentectomies by simplifying the technical procedures of resection, particularly due to the difficulty of transferring into a laparoscopic approach all the sophisticated techniques carried out in open surgery. Whether or not a real parenchyma-sparing approach with minimal resection margin is applicable to a laparoscopic approach is yet under debate. Theoretically, a >1-cm resection margin may be difficult during laparoscopic liver resection (LLR) due to the loss of surgeon's tactile sense to precisely determine tumor margin and increased traction on the specimen leading to tearing away of parenchyma from the intended resection plane. These problems are most readily appreciated in laparoscopic wedge resection (WR), as transection requires to create a deep margin around the tumor; conversely, a laparoscopic anatomic segmental resection is a more standardized and well-defined procedure.

In the largest review of LLR series, reported by Nguyen et al. [13], the most common type of liver resection (45%) is wedge resection or segmentectomy (1,258/2,804). LLR achieves adequate oncologic treatment and does not seem to unequivocally affect resection margins. Certain techniques for parenchymal transection may alter the evaluation of adequate oncological resection because margins are artificially lowered by the stapler, commonly used in some centers during LLR, as the perilesional parenchyma is destroyed at the transection line [14, 15].

The impact of the extent of LLR on oncologic outcome is the topic of various studies. Several compare resection margins between LLR and traditional hepatectomy [16–19]; others compare them between anatomic and nonanatomic LLR. Abu Hilal et al. [20] focused on left lateral sectionectomy (LLS), showing there was no difference in resection margin between the laparoscopic (11 mm; 1.5–30 mm) and open (12 mm; 4–40 mm) approaches. Aldrighetti et al. [21] showed that the resection margin of laparoscopic LLS (1.1 ± 0.3 cm) is comparable with the open approach (1.3 ± 0.5 cm). McPhail et al. [22], in a collective review of five case–control series on LLS versus open surgery suggest that the laparoscopic approach did not compromise margin status. Kazaryan et al. [23] reported a comparative evaluation of 75 segmental LLR performed for malignant tumors localized in posterosuperior (1, 7, 8, 4a: 28 procedures) and anterolateral (2, 3, 4b, 5, 6: 47 procedures) segments. An infiltrated margin resection was detected in 5.3% of procedures: two cases in each group had positive resection margins. Furthermore, for one additional resection in each group, the resection margin was <1 mm. The minimal distance from the resection line to the tumor tissue was significantly shorter in the posterosuperior (median 3 mm) than in the anterolateral (median 8 mm) group.

25.3 Conclusion

The actual resective approach to hepatic cancer (primary or secondary) is oriented toward minimal parenchymal resection. This methodology is sustained by observation that surgical margin width is not correlated with cancer recurrence. Parenchyma-sparing resection reduces morbidity rates without changing long-term results and allows the possibility of redo liver resection in case of recurrence. With regard to segmentectomies, MISL has opened new frontiers that yet need to be supported by years and years of experience: however, our impression is that we are rediscussing dogma that dominated liver surgery for more than 30 years. This is, once more, a crucial challenge to enable us to better serve our patients.

References

1. Couinaud C (1957) Le foie, etude anatomiques et chirurgicales. Masson, Paris
2. Couinaud C (1994) Surgical approach to the dorsal section of the liver. Chirurgie 119(9):485–488
3. Iwamoto S, Sanefuji H, Okuda K (2003) Angiographic subsegmentectomy for the treatment of patients with small hepatocellular carcinoma. Cancer 97(4):1051–1056
4. Tanaka K, Shimada H, Matsumoto C et al (2008) Anatomic versus limited non-anatomic resection for solitary hepatocellular carcinoma. Surgery 143(5):607–615
5. Torzilli G, Procopio F, Cimino M et al (2010) Anatomical segmental and subsegmental resection of the liver for hepatocellular carcinoma: a new approach by means of ultrasound-guided vessel compression. Ann Surg 251:229–235
6. Pawlik TM, Scoggins CS, Zorzi D et al (2005) Effect of surgical margin status on survival and site of recurrence after hepatic resection for colorectal metastases. Ann Surg 241(5):715–724
7. Muratore A, Ribero D, Zimmitti G et al (2010) Resection margin and recurrence-free survival after liver resection of colorectal metastases. Ann Surg Oncol 17(5):1324–1329

8. Torzilli G, Montorsi M, Del Fabbro D et al (2006) Ultrasonographically guided surgical approach to liver tumours involving the hepatic veins close to the caval confluence. Br J Surg 93:1238–1246
9. Torzilli G, Botea F, Donadon M et al (2010) Minimesohepatectomy for colorectal liver metastasis invading the middle hepatic vein at the hepatocaval confluence. Ann Surg Oncol 17:483
10. Dahiya D, Wu TJ, Lee CF et al (2010) Minor versus major hepatic resection for small hepatocellular carcinoma (HCC) in cirrhotic patients: a 20-year experience. Surgery 147:676–685
11. Matsui Y, Terakawa N, Satoi S et al (2007) Postoperative outcomes in patients with hepatocellular carcinomas resected with exposure of the tumor surface: clinical role of the no-margin resection. Arch Surg 142:596–602
12. Torzilli G, Donadon M, Cimino M et al (2009) Systematic subsegmentectomy by ultrasound-guided finger compression for hepatocellular carcinoma in cirrhosis. Ann Surg Oncol 16(7):1843
13. Nguyen KT, Gamblin TC, Geller DA (2009) World review of laparoscopic liver resection-2,804 patients. Ann Surg 250(5)831–841
14. Kaneko H, Otsuka Y, Takagi S et al (2004) Hepatic resection using stapling devices. Am J Surg 187:280–284
15. Gumbs AA, Gayet B, Gagner M (2008) Laparoscopic liver resection: when to use the laparoscopic stapler device. HPB (Oxford) 10(4):296–303
16. Lee KF, Cheung YS, Chong CN, Tsang Y, Ng W, Ling E et al (2007) Laparoscopic vs. open hepatectomy for liver tumours: a case control study. Hong Kong Med J 13:442–448
17. Simillis C, Constantinides VA, Tekkis PP, Darzi A, Lovegrove R, Jiao L et al (2007) Laparoscopic vs. open hepatic resections for benign and malignant neoplasms: a meta-analysis. Surgery 141:203–211
18. Topal B, Fieuws S, Aerts R, Vandeweyer H, Penninckx F (2008) Laparoscopic vs. open liver resection of hepatic neoplasms: comparative analysis of short term results. Surg Endosc 22:2208–2213
19. Tsinberg M, Tellioglu G, Simpfendorfer CH et al (2009) Comparison of laparoscopic vs. open liver tumor resection: a case-controlled study. Surg Endosc 23:847–853
20. Abu Hilal MA, McPhail MJW, Zeidan B et al (2008) Laparoscopic vs. open left lateral sectionectomy: a comparative study. Eur J Surg Oncol 34:1285–1288
21. Aldrighetti L, Pulitano C, Catena M et al (2008) A prospective evaluation of laparoscopic vs. open left lateral sectionectomy. J Gastrointest Surg 12:457–462
22. McPhail MJ, Scibelli T, Abdelaziz M et al (2009) Laparoscopic vs. open left lateral hepatectomy. Expert Rev Gastroenterol Hepatol 3:345–351
23. Kazaryan AM, Røsok BI, Marangos IP et al (2011) Comparative evaluation of laparoscopic liver resection for posterosuperior and anterolateral segments. Surg Endosc 25(12):3881–3389

Segment 1: Laparoscopic Approach

26

Umberto Cillo, Riccardo Boetto, Alessandro Vitale and Daniele Neri

26.1 Introduction

The hepatic caudate lobe (S1), or Spiegel lobe, has been widely considered a "nonlaparoscopic" segment due to its particular anatomical location between the hilar plate and inferior vena cava, which technically restricts the use of a conventional laparoscopic approach when treating segment 1 primitive and metastatic lesions. Since the early 2000s, the increasing detail in understanding liver segmental anatomy, improved preoperative imaging and intraoperative anesthesiologic management, as well as improvements in laparoscopic surgical skills and equipment, have allowed a significant increase in the adoption of minimally invasive procedures. Initially confined to wedge resections and segmentectomies of the anterior liver (laparoscopic segments), more advanced minimally invasive liver resections, such as in left and right sections, are now extensively performed and attain acceptable morbidity and mortality rates, with 3- and 5-year survival rates reported for hepatocellular carcinoma (HCC) and colorectal metastases comparable with those of open procedures [1]. Even though extremely rare, isolated laparoscopic resection of hepatic segment 1 (S1) has also been reported in the context of technically dyshomogeneous series. With the exception of a couple of reports, there is substantial lack, however, of a systematic technical description of the procedure.

U. Cillo (✉)
Hepatobiliary Surgery Unit and Liver Transplant Center,
University of Padua,
Padua, Italy
e-mail: cillo@unipd.itt

F. Calise and G. Casciola (eds.), *Minimally Invasive Surgery of the Liver,*
Updates in Surgery
DOI: 10.1007/978-88-470-2664-3_26,

26.2 Anatomical Background

Hepatic S1 is usually divided into three portions according to Kumon's classification: the Spiegel lobe, the caudate process, and the paracaval portion. The Spiegel lobe is located underneath the lesser omentum on the left part of the inferior vena cava (IVC). The paracaval portion lies in front of the intrahepatic portion of the IVC, just to the right of the Spiegel lobe, and is surrounded by the right and middle hepatic veins. The caudate process is projected between the IVC and portal vein, just to the right of the paracaval portion. Kumon defined the paracaval portion as the liver parenchymal portion ventral to the IVC between the Spiegel lobe and the right hepatic lobe, adjacent to the middle hepatic vein ventrally; Coinaud also demonstrated and confirmed the existence of a paracaval portion, classifying this area as a separate S9 [2, 3]. The dorsal side of the IVC is sometimes covered by a parenchymal bridge of the S1 and often by a membranous structure called the IVC ligament. One or two thick (2–3 mm in diameter) veins, usually termed caudate veins, and several thin veins ensure S1 drainage, with some presenting as proper drainage for the S1 Spiegel lobe and others with common drainage with the paracaval portion of S1 and/or S4, 7, and 8. Thick veins enter the IVC, whereas thin veins merge with the IVC and middle and/or right hepatic veins. Almost half of the two to four S1 ducts merge with the posterior sectorial hepatic duct (B6 and 7) originating from S6 and 7, usually showing an epiportal course; the remaining S1 ducts join the left hepatic duct, formed by the joining of B2, 3, and 4 [4].

Arterial supply is represented by multiple small branches arising from the left (LHA) and right (RHA) hepatic arteries; singular branches from these arteries only are present in 35% and 12% of individuals, respectively. More frequently (53%), vascularization is guaranteed by both arterial branches. Artery for S1 does not designate a single vessel, but some large branches can be identified at angiography. Artery for S1 arising from the RHA courses posteriorly and medially and mainly supplies the lateral portion of S1 (the paracaval portion); Artery for S1 arising from the LHA, courses posteriorly, and mainly supplies the medial portion of S1 (caudate process, Spiegel lobe) [5].

Hepatic S1 presents an articulated and variable vascular supply. All three portions present vascular inflow from primary glissonian branches originating from the right and left portal vein, with the hilar bifurcation branch supplying not only the paracaval portion but also the left Spiegel lobe (29%) and the right caudate process (21%), allowing possible metastatic spreading of hepatic tumors to the entire liver [6].

26.3 Previous Clinical Experience

Experience in laparoscopic isolated resection of S1 is extremely limited, and data interpretation is confounded by the fact that S1 resection series are often included in wider sets of different laparoscopic approaches [1]. Therefore, it is impossible at the present time to draw definitive conclusions on safety and oncologic efficacy of such a procedure due to the lack of sufficiently large, focused studies. However, a few series report acceptable rates of adverse events associated with the approach, suggesting encouraging safety profiles.

Indeed, even though an isolated laparoscopic approach to lesions located in the posterosuperior segments (S1, 7, 8, and 4a) are significantly associated with longer operative time (331.4 vs. 258.5 min, $P = 0.009$) and intraoperative transfusion (47.2% vs. 25%, $P = 0.015$) compared with procedures on the anterolateral segments (S2, 3, 5, and 7), the approach was technically feasible and safe, with morbidities comparable with open surgery (19.4% vs. 16.3%, NS) in a recent monocentric series [7]. Dulucq and colleagues described the two cases to be reported in the literature and the surgical technique for isolated laparoscopic R0 resection of S1 for single colorectal metastasis using a six-trocar approach and the Pringle maneuver: operative time was 150 and 105 min, respectively, with no major complications or significant blood loss [8].

26.4 Technical Details

As for the other segmental laparoscopic hepatic resections, it is possible to identify at least three technical approaches: totally laparoscopic, hand-assisted, and laparoscopic-assisted open hybrid surgical techniques [1]. Before any further consideration regarding the most appropriate laparoscopic approach, however, it is important to recall that as far as the classic open, isolated resection of S1 is concerned, three common approaches have been described: left, right, and transparenchymal. The left approach is the most frequently described in laparoscopic resections, whereas the right approach, potentially indicated in the presence of masses located in the caudate process, is barely achievable by laparoscopy due to the difficult complete left rotation of the right hemiliver to expose the right row of spigelian veins. The anterior transparenchymal approach is theoretically applicable laparoscopically, primarily in association with right hepatectomy, even though S1 exposure with this technique is not always optimal.

With this in mind, it is clear that correct patient and trocar positioning is critical to patient safety and maximal comfort for the surgeon when performing minimally invasive S1 liver resection. In our experience, when a left approach is planned, the patient is positioned in the supine decubitus (30° anti-Trendelenburg), legs apart, with the surgeon standing between them and the assistant and the surgeon holding the camera standing on the patient's right and left side, respectively.

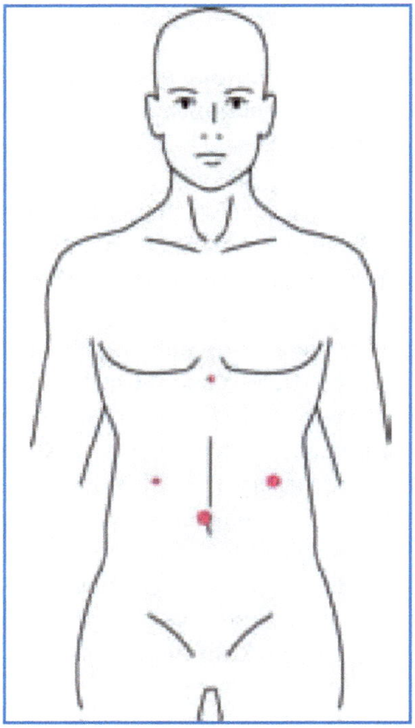

Fig. 26.1 Trocar positioning

Pneumoperitoneum (14 mmHg) is induced by positioning a supraumbilical 12-mm optical trocar (30°) using an open technique; 15-mm and 5-mm operative trocars are placed under direct vision in the left and right midclavicular lines, respectively, 5 cm above the transverse umbilical line. A 5-mm trocar is placed in the epigastrium, allowing inferior liver-surface retraction with adequate S1 and IVC exposure. Central venous pressure is maintained >5 mmHg (Fig. 26.1).

The peritoneal cavity is explored to rule out extrahepatic disease, and intraoperative liver ultrasonography is performed on the S1 before and after careful hepatoduodenal ligament dissection with a cautery hook (in search of the left hepatic accessory arteries) when looking for the presence of a lesion and possible IVC direct involvement.

The left lobe is not mobilized, and the liver is lifted upward with a 5-mm laparoscopic liver retractor, allowing glissonian pedicle interruption and hook cauterization of the peritoneal reflection of the IVC by gently grasping the spigelian lobe on the left side; blunt IVC dissection is continued, interrupted between clips, exposing the hepatic accessory veins. Once the spigelian lobe is completely released from the IVC, the glissonian pedicle is approached, with clip positioning on dorsal portal accessory veins. Indeed, we consider it safer to first mobilize the lobe, even though this interrupts venous outflow, to attain better control of the glissonian area.

Parenchymal transection is performed cephalad using radiofreqency forceps and bipolar cautery with intermittent water-drip irrigation until S1 resection is complete. The resection surface is meticulously checked for biliary leaks, and bleeding control is completed with bipolar cautery and water-drip irrigation. The tumor specimen is retrieved through the 12-mm supraumbilical port inside a dedicated laparoscopic plastic bag to prevent seeding. A small suction drain is positioned through the right 15-mm trocar at the bed of the hepatic resection. Pringle maneuver is not usually applied in our experience during parenchymal transection.

Due to the difficulty of this approach, laparoscopic S1 resection should be performed only in specialized centers by hepatobiliary surgeons with significant expertise in laparoscopic surgery.

References

1. Nguyen KT, Gamblin TC, Geller DA (2009) World review of laparoscopic liver resection-2,804 patients. Ann Surg 250(5):831–841
2. Kumon M (1985) Anatomy of the caudate lobe with special reference to the portal vein and bile duct. Acta Hepatol Jpn 26:1193–1199
3. Pulitanò C, Aldrighetti L (2008) The current role of laparoscopic liver resection for the treatment of liver tumors. Nat Clin Pract Gastroenterol Hepatol 5:648–654
4. Murakami G, Hata F (2002) Human liver caudate lobe and liver segment. Anat Sci Int 77(4):211–224
5. Furuta T, Maeda E, Akai H et al (2009) Hepatic segments and vasculature: projecting CT anatomy onto angiograms. Radiographics 29(7):1–22
6. Kaneko H (2005) Laparoscopic hepatectomy: indications and outcomes. J Hepatobiliary Pancreat Surg 12:438–443
7. Cho JY, Han HS, Yoon YS et al (2008) Feasibility of laparoscopic liver resection for tumors located in the posterosuperior segments of the liver, with a special reference to overcoming current limitations on tumor location. Surgery 144:32–38
8. Dulucq JL, Wintringer P, Stabilini C et al (2006) Isolated laparoscopic resection of the hepatic caudate lobe: surgical technique and report of 2 cases. Surg Laparosc Endosc Percut Tech 16:32–35

Segment 2: Laparoscopic and Robot-Assisted Approach

27

Alberto Patriti, Antonio Giuliani and Vincenzo Scuderi

27.1 Introduction

Segment 2 (S2) is the most posterior part of the anatomical left lobe and comprises S2 and S3 in the Couinaud description of liver segmental anatomy [1]. Liver tumors located in S2 are generally resected with a formal laparoscopic left lateral sectionectomy that is considered the approach of choice for tumors located in the left hepatic lobe [2]. In fact, the straight resection line along the falciform ligament is particularly feasible with all laparoscopic transection devices, including vascular linear staplers. Minimally invasive S2 segmentectomy is more challenging, requiring a high level of confidence with laparoscopic ultrasound (US) and creating a curved resection line not readily feasible in laparoscopic surgery. Indications for S2 segmentectomy are colorectal liver metastases and hepatocellular carcinoma without invasion of the left hepatic vein (LHV) and the surrounding S3 and 4.

27.2 S2 Anatomy

S2 is the upper part of the left hepatic lobe. According to the Couinaud classification, S2 is bordered medially by the falciform ligament and inferiorly by the LHV. Cranially, S2 is in contact with the diaphragm through a small bare area enclosed by the left triangular ligament. On the inferior surface, the medial border of S2 is the ligamentum venosum. Blood supply is maintained by a pedicle

A. Patriti (✉)
Department of Surgery, Division of General, Minimally Invasive and Robotic Surgery,
ASL3 Umbria, San Matteo degli Infermi Hospital, Spoleto, Italy
e-mail: albertopatriti@gmail.com

F. Calise and G. Casciola (eds.), *Minimally Invasive Surgery of the Liver,*
Updates in Surgery
DOI: 10.1007/978-88-470-2664-3_27, © Springer-Verlag Italia 2013

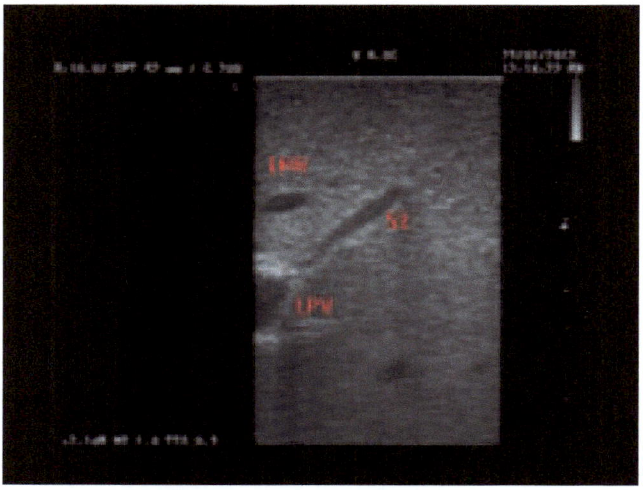

Fig. 27.1 Intraoperative laparoscopic ultrasound showing the portal branch for segment 2 (*S2*) arising from the umbilical portion of the left portal vein (*LPV*). The left hepatic vein (*LHV*) crosses the S2 branch at its origin

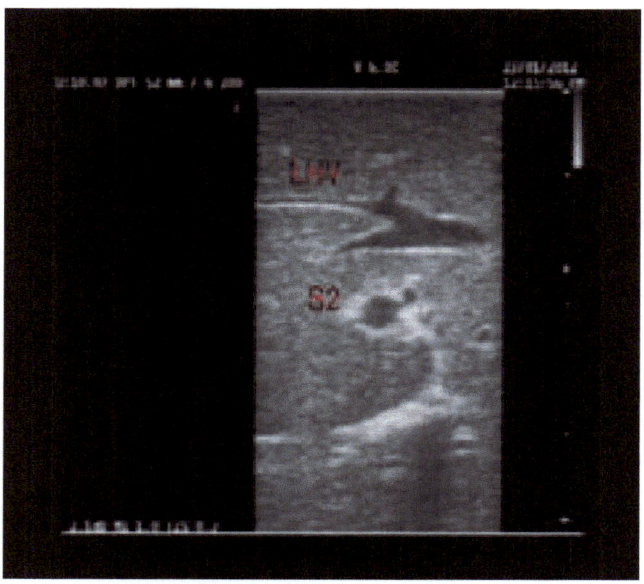

Fig. 27.2 Intraoperative laparoscopic ultrasound showing the left hepatic vein (*LHV*) crossing the portal pedicle for segment 2 (*S2*) and the ligamentum venosum (the *hyperechoic line* below S2)

(P2) arising directly from the left portal vein (LPV) in the first part of the umbilical portion. At US exploration P2 can be observed at the level of the angle where the LPV enters the liver parenchyma. In a transverse direction, P2 reaches the right triangular ligament crossing the LHV (Figs. 27.1 and 27.2)

27.3 Technical Details

27.3.1 Laparoscopic

The patient is placed in the supine position in a 30–35° reverse Trendelenburg position with the table tilted to the right side. The operator is placed between patient's legs with an assistant on each side. The camera in inserted in the umbilicus with an open approach. Three trocars (two 10–12 mm in the right and left paramedian positions, respectively, and one 5 mm in the epigastrium) are placed, with an accessory 10- to 12-mm trocar in the left flank when the Pringle maneuver is needed (Fig. 27.3). The camera used is a 30° laparoscope. An overall inspection of the cavity is made to evaluate the presence of gross pathologies, followed by a US scan. The first step is to pass the pedicle to facilitate the Pringle maneuver by gently inserting a Roticulator grasp behind the pedicle; a vessel loop is then passed and closed with two medium titanium clips. The round ligament is left in place, and the falciform ligament, the left triangular ligament, the left coronal ligament, the lesser omentum, and the splenic ligament are sectioned with the Harmonic scalpel. When the left lobe is completely mobilized, the transection plane is outlined on the liver capsule with monopolar diathermy, and another US scan is performed to control resec-

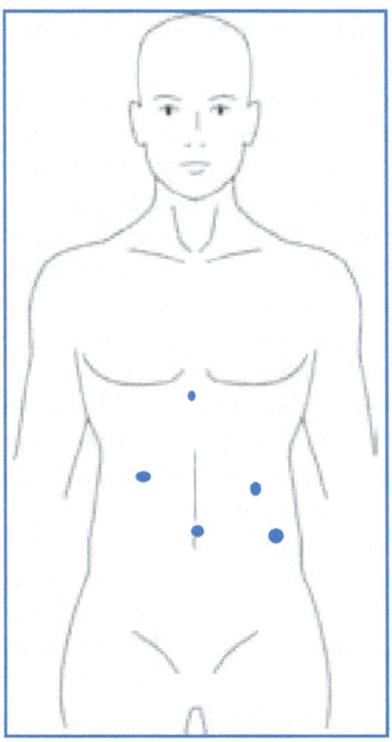

Fig. 27.3 Trocar positioning for laparoscopic resection of S2

tion margins. The transection is begun, without clamping the pedicle, with the Harmonic scalpel and the Cavitron Compact Ultrasonic Surgical Aspirator (CUSA) on one hand and the bipolar forceps on the other; in this phase, it is important to change the operative hand to attain the required position for the transection and to perform hemostasis on both transection surfaces. The assistant exposes S2 with the aspirator. Portal and arterial inflow to S2 is controlled via intrahepatic access with clips. During the transection, it is important to pay attention to the confluence of the LHV in the inferior vena cava to avoid creating lesions. The transection is completed using a vascular stapler to secure small branches of the LHV close to the hepatocaval confluence. The procedure is completed by applying fibrin glue to the transection surface and placing a tubular drain behind the pedicle.

27.3.2 Robot-Assisted

Laparoscopic S2 segmentectomy is difficult due to the curved resection line and technical difficulties controlling the inflow pedicle. For these reasons, we approach this segmentectomy using the four-arm da Vinci robotic system. The patient is placed in the 30° reverse Trendelenburg position with the robot over the head. Pneumoperitoneum is generally induced with the Veress needle inserted in the left upper quadrant and maintained during surgery at 10–12 mmHg. Abdominal pressure can be reduced during LHV dissection to avoid gas embolism. A 12-mm trocar is placed in the paraumbilical area for the robotic camera. Two additional 8-mm robotic ports are introduced into the right and left upper quadrant. In resections performed with the fourth arm, an 8-mm trocar is placed in the left subcostal area five fingerbreadths from the port placed in the right upper quadrant. One or two accessory trocars can be placed in the right and left lower quadrant for use by the on-table assistant for suction and retraction (Fig. 27.4). The operation begins with umbilical and falciform ligament takedown and division of the right triangular ligament, exposing the hepatocaval confluence of the LHV. The left and fourth robotic arms are used to maintain traction on the ligaments that are cut using the monopolar scissor of the right robotic arm. Using careful US exploration, the course of the LHV is marked on the liver surface, delineating the border between S2 and S3. The crossing point between P2 and the LHV is marked as well. From this point to the superior border of the liver, the transection line can proceed straight to avoid accidental injury to the hepatocaval confluence. In this manner, the transection line will assume a bidirectional shape, with two intersecting lines forming an obtuse angle at the crossing point between the LHV and P2. Two 2-0 stay sutures are placed at both sides of the origin of the LHV at the level of the liver's inferior border. The suture in S3 is held by the assistant and that in S2 by the fourth robotic arm and will be used to retract liver surfaces during transection. A tourniquet is passed around the liver pedicle, as described in Chap. 15, for

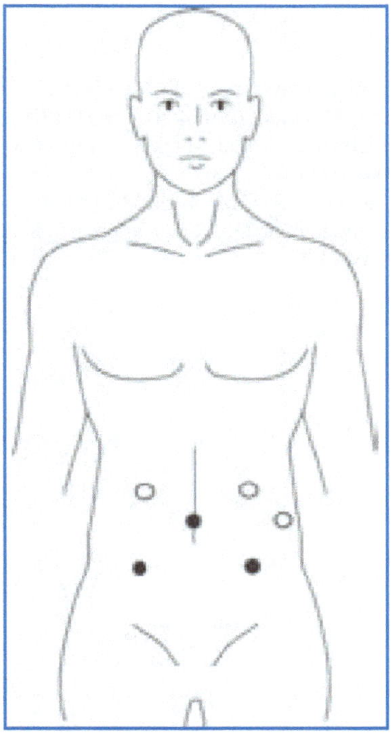

Fig. 27.4 Trocar positioning for robot-assisted resection of S2

intermittent Pringle maneuver. Liver transection is carried out with the Precise bipolar forceps on the left arm and the monopolar scissor on the right arm in the fashion of the Kelly-clamp crushing technique. Small vessels are coagulated while the LHV and P2 branches are clipped or ligated. If properly performed, the transection line runs along the left border of the LHV, reaching underneath the P2. Liver parenchyma around the portal pedicle must be crushed until it is completely exposed. Otherwise, its oblique direction makes it difficult to properly clip or ligate the entire pedicle, and multiple fires are necessary to secure the vessel. After the pedicle is divided, liver transection proceeds in a nearly straight line. Application of a vascular stapler could be useful in the last 2–3 cm of the resection plane to secure small branches of the LHV close to the hepatocaval confluence and conclude the resection.

Machado et al. suggest a different approach to S2. They describe an intrahepatic glissonian access to P2, performing two incisions laterally to the origin of P2 at the level of the inferior liver surface. The pedicle is secured with a vascular linear stapler, and transection is carried out following the demarcation between S2 and S3 without using the Pringle maneuver. The technique is reported to be safe and feasible, even in laparoscopic surgery [3].

References

1. Couinaud C (1956) Contribution of anatomical research to liver surgery. Fr Med 19(5):5–12
2. Aldrighetti L, Pulitanò C, Catena M, Arru M, Guzzetti E, Casati M, Comotti L, Ferla G (2008) A prospective evaluation of laparoscopic versus open left lateral hepatic sectionectomy. J Gastrointest Surg 12(3):457–462. doi:10.1007/s11605-007-0244-6
3. Machado MA, Makdissi FF, Surjan RC, Herman P, Teixeira AR, C Machado MC (2009) Laparoscopic resection of left liver segments using the intrahepatic Glissonian approach. Surg Endosc 23(11):2615–2619. doi:10.1007/s00464-009-0423-5

Segment 3: Laparoscopic Approach

28

Marco Filauro, Claudio Bagarolo, Giulio Angelini,
Fiorenza Belli and Andrea Barberis

28.1 Introduction

Segment 3 (S3) of the liver is considered highly suitable for laparoscopic resection [1, 2] due to its anterior position (favorable access without liver mobilization), the possibility of accessing the segmental pedicle using fixed and constant anatomical landmarks, and the possibility of direct visualization and ultrasound (US) scanning of the lesion(s) to be resected.

28.2 Indications

Indications to isolated resection [3] of S3 are:

- Benign diseases: giant/symptomatic hemangiomas, symptomatic focal nodular hyperplasia (FNH) (compressive effect to adjacent organs)
- Border line: giant/symptomatic/complicated adenoma
- Histologically or radiologically uncertain diseases (mainly to define differential diagnosis with malignancy)
- Malignant diseases: hepatocellular carcinoma (HCC), secondaries, when segmental resection allows a radical (R0) operation [4].

28.3 Preoperative Care and Instruments

The patient is carefully worked up for accurate indication to S3 resection and laparoscopic feasibility: preoperative US (with contrast-enhanced US, if indi-

M. Filauro (✉)
General and Hepatobiliary Pancreatic Surgery, Galliera Hospitals
Genoa, Italy
e-mail: marco.filauro@galliera.it

F. Calise and G. Casciola (eds.), *Minimally Invasive Surgery of the Liver*,
Updates in Surgery
DOI: 10.1007/978-88-470-2664-3_28,

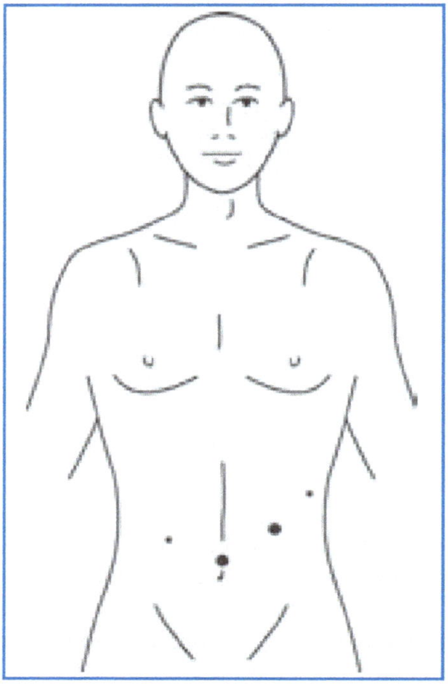

Fig. 28.1 Trocar positioning

cated), abdominal multislice computed tomography (CT) with contrast agent, and liver magnetic resonance imaging (MRI) with a specific contrast agent are performed with appropriate and individually selected indications.

After anesthesiologic evaluation, the same preoperative schedule as that used in the laparotomic approach is followed, considering the intraoperative possibility of converting laparoscopy to open surgery. For this reasons, we prefer to use central venous access and epidural catheter for blended anesthesia.

The patient is positioned supine on the operating table, with legs spread and fixed with appropriate padding to permit lateral and/or caudal movement during operation. A laparoscopic video unit is placed at the patient's head: a double screen is advisable to facilitate viewing by the entire surgical team. Four trocars are inserted (Fig. 28.1):

- Hasson trocar on the supraumbilical midline
- A 5-mm trocar on the midclavicular right line
- A 10- to 12-mm trocar on the midclavicular left line
- A 5-mm trocar on the left flank (or alternatively in the left xiphoid region).

This schema could be modified considering patient size, intraperitoneal adhesions, liver volume, and other variables. The patient is positioned in the anti-Trendelenburg position at a degree that depends upon the intra-abdominal situation. The following instruments must be available on the scrub-nurse's table:

- A 30° laparoscope
- Laparoscopic ultrasonography probe (10 mm)
- Johann atraumatic graspers (2)
- Right-angle dissector (5 mm)
- Monopolar coagulation device
- Bipolar coagulation grasper (5mm)
- US shears (5 mm) and/or radiofrequency (RF) coagulation grasper (5 mm)
- Laparoscopic Cavitron Compact Ultrasonic Surgical Aspirator (CUSA) (as an option, if available)
- Laparoscopic retractor (5 mm)
- Linear laparoscopic stapler
- Irrigation and suction device (5 mm)
- Clips (medium-large) applier (10 mm)
- Retrieval bag.

28.4 Surgical Technique

Trocars are positioned (Fig. 28.1) after creation of a pneumoperitoneum at 12 mmHg; adhesions (if present) are treated, and a general abdominal exploration is carried out to evaluate concomitant diseases not previously demonstrated (peritoneal implants, fluid collections, etc). The liver is widely exposed and parenchymal status is evaluated for firmness, nodules appearing on the surface, and abnormal perihepatic venous vessel (portal hypertension) (Fig. 28.2). Locoregional lymph nodes are also evaluated. Segment 3 lesion(s) are explored and evaluated by instrumental "palpation".

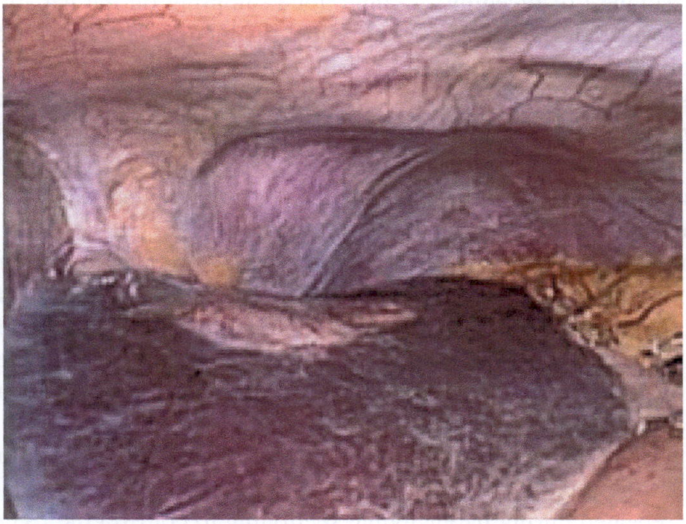

Fig. 28.2 Lesion overview

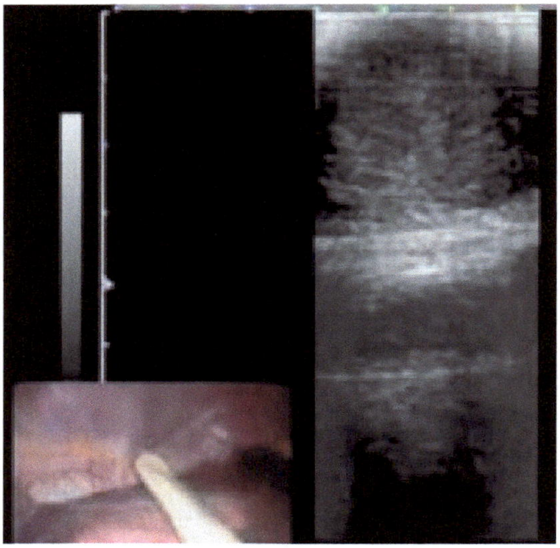

Fig. 28.3 Intraoperative laparoscopic ultrasound

US probes are now inserted through 10- to 12-mm trocar, and liver scanning – laparoscopic ultrasound (LUS) – is systematically performed (S6, 7; S5, 8; S4, 1; S2, 3); the hepatic pedicle is also scanned. LUS confirms lesion location and number, demonstrates the presence of further lesion(s), and facilitates US-guided needle biopsy on suspected nodules (Fig. 28.3). US-guided tattooing using electrocautery is carried out to identify the resection margin on the liver surface, as well as the anatomical landmark of S3. The hepatic pedicle is isolated and encircled with a soft rubber tourniquet to facilitate intracorporeal Pringle maneuver (in case of bleeding). Lymph node sampling on the hepatic pedicle can be performed at this time. The umbilical ligament is not sectioned in order to maintain the left lobe; the falciform ligament is not opened.

Parenchymal section begins at the lateral margin of the umbilical ligament on the marked line, which is identified ultrasonically: dissection is achieved by the operating surgeon with tissue crushing using US shears (Fig. 28.4) (with the right hand) or by CUSA (optional); hemostasis is attained using the same US shears and bipolar grasper (with the left hand).

The second surgeon inserts and uses the suction device or Johann grasper in the left flank (xiphoid) trocar to keep the left liver on a lateral margin and facilitate opening the parenchyma during dissection and to clean the sectioned line. Parenchymal sectioning proceeds cranially and allows easy access to the pedicle of S3: the pedicle can be clamped to demonstrate segmentary ischemia and then clipped and sectioned.

Sectioning can also be achieved using the linear stapler (Fig. 28.5) [5] with a vascular cartridge; the section line must be checked to exclude biliary leakage. LUS must be repeated during parenchymal sectioning to confirm the right line is being followed and that lesion(s) is adequately located in the lateral side.

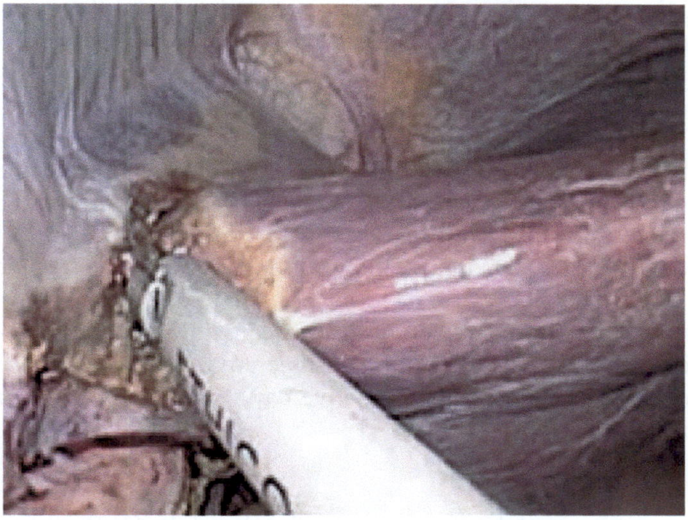

Fig. 28.4 Parenchymal section with ultrasonic (US) shears

Fig. 28.5 Division of segmental pedicle with linear stapler

When the upper cranial landmark of resection is reached, hemostatic material (fibrillar oxidized cellulose) can be placed in the sectioned line; the lateral section can then be started. In this part of operation, the US shears are moved in the left-flank trocar and the bipolar grasper in the left midclavicular trocar. During parenchymal sectioning, well-isolated and accurately identified vessels are closed using US shears and/or clips: anatomical resection must be always provided. LUS must be repeated during this stage. When the two lines of the parenchymal section are joined, resection is completed, and the specimen is inserted in a removal bag. The Pringle maneuver is usually not necessary if accurate, anatomical dissection and hemostasis are achieved.

Hemostasis must be checked after reduction of abdominal pressure (pneumoperitoneum must be lowered to 2–3 mmHg) and improved by selected bipolar coagulation; fibrin glue and fibrillar oxidized cellulose are applied to the liver surface. One tubular or Jackson–Pratt drainage tube is positioned on the resected liver surface. The specimen can be extracted through the umbilical incision (adequately enlarged) or, if too bulky, through a small suprapubic incision.

The postoperative period does not require specific care: the patient is mobilized early, and food intake begins on the first postoperative day. The drainage tube can be removed on the second postoperative day if no bleeding and/or bile leakage occurs.

Acknowledgements A contribution to this chapter is gratefully acknowledged to Dr. Stefano Cappato.

References

1. Rao A, Rao G, Ahmed I (2012) Laparoscopic or open liver resection? Let systematic review decide it. Am J Surg [Epub ahead of print, January 13, PMID: 22245507]
2. Kazaryan AM, Pavlik Marangos I, Rosseland AR et al (2010) Laparoscopic liver resection for malignant and benign lesions: ten-year Norwegian single-center experience. Arch Surg 145:34–40
3. Buell JF, Cherqui D, Geller DA et al (2009) The international position on laparoscopic liver surgery: The Louisville Statement, 2008. Ann Surg 250:825–830
4. Abu Hilal M, Di Fabio F, Abu Salameh M et al (2012) Oncological efficiency analysis of laparoscopic liver resection for primary and metastatic cancer: a single-center UK experience. Arch Surg 147:42–48
5. Gumbs AA, Gayet B, Gagner M (2008) Laparoscopic liver resection: when to use the laparoscopic stapler device. HPB (Oxford) 10:296–303

Segment 4: Laparoscopic Approach

29

Giuseppe Maria Ettorre, Giovanni Vennarecci,
Pasquale Lepiane, Andrea Laurenzi and Roberto Santoro

29.1 Introduction

Segment 4 (S4) laparoscopic liver resection (LLR) is demanding for two main reasons: laparoscopic liver surgeons more frequently perform either isolated S4 resection (i.e., segmentectomy, wedge resection) or in association with central or left hepatectomy; for transplant teams involved in living-related pediatric living transplantation, S4 resection represents a crucial part of laparoscopic left lateral sectionectomy because the cutting plane crosses S4 and can be moved according to the volume of the harvested liver (left lateral sectionectomy, left hepatectomy) [1–3].

29.2 S4 Anatomy

S4 is part of the left liver. It represents the left paramedian sector in association with S3, according to Couinaud classification [4], and the left medial section, according to Brisbane classification. Its anatomical limits are the main portal scissura medially (Cantlie's line), which contains the middle hepatic vein; the umbilical fissure laterally, which contains in up to 60% of cases a hepatic scissural vein; and the anterior surface of S1 posteriorly. This posterior limit can be conceived of as a plane composed of the left portal branch confluence, the origin of the middle hepatic vein on the inferior vena cava, and the Arantius' ligament. S4 can be further divided into two subsegments: 4a (superior–posterior) and 4b (inferior–anterior) according to the transverse scissura.

G. M. Ettorre (✉)
General Surgery and Transplantation, San Camillo Hospital
Rome, Italy
e-mail: gmettorre@scamilloforlanini.rm.it

F. Calise and G. Casciola (eds.), *Minimally Invasive Surgery of the Liver*,
Updates in Surgery
DOI: 10.1007/978-88-470-2664-3_29, © Springer-Verlag Italia 2013

Portal vein supply to S4 develops late in embryogenesis, and for this reason, the vascular branches can be multiple and variable. Usually, portal flow is provided by a cluster of three to thirteen veins that stem from the umbilical part of the left main portal branch, with one stronger vein that heads toward subsegment 4b. Rarely, it is possible to find a venous branch that comes directly from the left main portal branch [5].

The arterial supply is usually provided by an arterial branch arising from the left hepatic artery. In 10% of cases, the S4 artery originates from the right hepatic artery. In a few cases, there can be a double arterial vascular supply originating from both right and left hepatic arteries [6].

The biliary drainage consists of one or two bile ducts, which join the left hepatic duct. The S4 bile duct can be either close to the common hepatic duct confluence or join directly the hepatic confluence or the common hepatic duct.

Suprahepatic drainage is mainly provided by two to three branches that drain into the left surface of the middle hepatic vein, and there are also some branches that drain either in the scissural vein or in the left hepatic vein. The scissural vein drains S4 and a small part of S3 and enters the left hepatic vein before the common trunk [5, 6].

29.3 Technical Details

S4 can be laparoscopically approached either as anatomical segmental resection or limited resection as subsegmentectomy or wedge. The S4 laparoscopic approach is also considered in major hepatectomies, such as central (S4, 5, 8) or left hepatectomy (S2, 3, 4). Our purpose in this chapter is to describe the anatomical S4 resection.

Patient position for S4 LLR is supine, with the legs opened. The surgeon stands between the patient's legs and the first and second assistants on the patient's left. The laparoscopic tower is placed on the right side of the patient's head. Trocar positions are shown in Fig. 29.1. A 30° optical device with flexible extremity is preferred, particularly when performing a subsegment 4a resection.

As the hepatic resection can go through a deep plane, the Pringle maneuver is always prepared for. For a subsegment 4b LLR, we avoid any liver mobilization. For lesions located in subsegment 4a, surgical dissection of the left triangular, round, and falciform ligament can be considered on a case-by-case basis.

Considering that S4 is in the central region of the liver, LLR must be safe and should avoid any type of injuries to the liver, particularly to the left hepatic pedicle, gallbladder, middle hepatic vein, and P2–3.

The procedure usually begins with an intraoperative ultrasound (US) of the liver to determine exact lesion location and vascular structures surrounding it (i.e., portal pedicles, hepatic veins). The cutting surfaces are then marked with monopolar diathermy. The first step is to section the parenchymal bridge – if

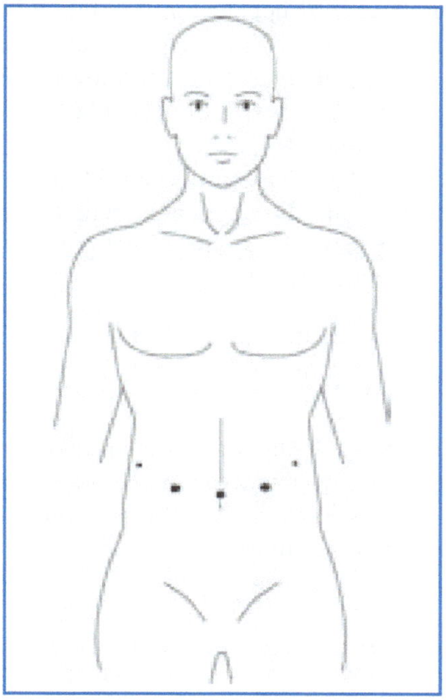

Fig. 29.1 Trocar positions for S4 laparoscopic liver resection

present – which connects the S4 to the left lobe on the inferior surface of the liver. This sectioning allows wide and clear vision, P4 identification, and the origin of the Arantius' ligament. All parenchymal dissections can be performed with US devices [Ultracision Ethicon, Cavitron Compact Ultrasonic Surgical Aspirator (CUSA), Olympus]. To seal 1-to-3-mm vessels, we usually utilize vascular clips, Hem-o-lok clips (Teleflex Medical), and UltraCision ultrasonic dissector (Ethicon). In case of bigger vascular structures (i.e., portal pedicles or hepatic veins), we use Hem-o-lok clips or a laparoscopic stapler.

For a laparoscopic anatomic segmentectomy, the operative field can be exposed by traction from the gallbladder toward the patient's right side and traction of the round ligament toward the left side. Cutting surfaces run parallel to the falciform ligament on the left side and to the middle hepatic vein (Cantlie's line) on the right side. These two cutting surfaces continue to join in the suprahepatic region. The deep section surface is parallel to a plane composed of the left portal branch and Arantius' ligament.

In S4 wedge resections, it is not mandatory to respect the anatomical planes, and a traction stitch on the specimen can be useful to expose the cutting surface. At the end of the S4 resection, it is preferable to avoid cholecystectomy in patients in whom a further liver resection may be necessary (i.e., colorectal metastases, neuroendocrine tumors), whereas it is mandatory in all other cases, particularly if the gallbladder has been injured during traction maneuvers or when a transcystic bile leak test has been performed.

One of the main risks in S4 resection or central hepatectomy is the wide hepatic surface achieved at the end of the procedure, which requires accurate hemostasis and biliostasis. Hemostasis can be achieved with the monopolar diathermy or hemostatic devices on the cutting surfaces. A drain is placed on the operative field in case of anatomical resection.

Among 102 LLRs performed in our unit, S4 was approached 21 times: three during left hepatectomy, three during anatomic segmental resections, eleven during wedge resections, three during multiple wedge resections, and one during left lateral sectionectomy extended to S4.

29.4 Conclusion

In conclusion, the laparoscopic approach to S4 is demanding; it is involved in as many as 20% of cases in dedicated hepatic surgery units. Knowledge of the difficult S4 anatomy allows specialized surgical teams to avoid major complications and to approach more difficult hepatic resections.

Acknowledgements A contribution to this chapter is gratefully acknowledged to Dr. Marco Colasanti.

References

1. Soubrane O, Cherqui D, Scatton O et al (2006) Laparoscopic left lateral sectionectomy in living donors: safety and reproducibility of the technique in a single center. Ann Surg 244:815–820
2. Sommacale D, Farges O, Ettorre GM et al (2000). In situ split liver transplantation for two adults recipient. Transplantation 15;69:1005–1007
3. The terminology of committee of the IHPBA (2000) The Brisbane 2000 terminology of hepatic anatomy and resections. HPB (Oxford) 2:333–339
4. Couinaud C (1957) Le foie etudes anatomiques et chirurgicales. Masson & Cie, Paris
5. Gadzijev EM, Ravnik D (1996) Atlas of applied internal liver anatomy. Springer, Wien, New York
6. Couinaud C (1993) Un scandale: segment IV et transplantation du foie. J Chir 130:443–446

Segment 4b: Laparoscopic Approach 30

Felice Giuliante, Francesco Ardito, Maria Vellone
and Gennaro Nuzzo

30.1 Introduction

Laparoscopic liver resection (LLR) of segment 4b (S4b) involves removing the inferior portion of S4. S4 is divided into two portions: 4a (superior portion) and 4b (inferior portion) [1]. The main indication for isolated anatomic liver resection of S4b is the presence of a tumor, which can be resected with adequate free surgical margins; absence of involvement of the left portal pedicle; no extension through the umbilical scissure on the left; and no involvement of S5 on the right (Fig. 30.1). Limited resections, such as isolated resection of S4b, should always be attempted in order to preserve as much healthy liver parenchyma as possible to decrease the risk of postoperative liver failure after more extended resections and allow re-resection in case of recurrence.

30.2 S4b Anatomy

The anatomic landmarks of S4b are the falciform ligament on the left; the left half of the gallbladder bed on the right and the peripheral route of the middle hepatic vein (MHV); the left portal pedicle superiorly; and the hilar plate inferiorly [1]. Arterial and portal blood supply of S4b arises from the left portal pedicle, which curves forward at the left border of the hilus along the umbilical fissure and terminates 1–2 cm from the anterior edge of the liver at the level of the Rex recessus, where the round ligament is joined anteriorly. Here, at the intrahepatic portion of the left portal pedicle, is where pedicles to S2, 3, and 4 originate before branching into their respective segments [1]. This point

F. Giuliante (✉)
Hepatobiliary Surgery Unit, Università Cattolica del Sacro Cuore
Rome, Italy
e-mail: fgiuliante@rm.unicatt.it

F. Calise and G. Casciola (eds.), *Minimally Invasive Surgery of the Liver*,
Updates in Surgery
DOI: 10.1007/978-88-470-2664-3_30, © Springer-Verlag Italia 2013

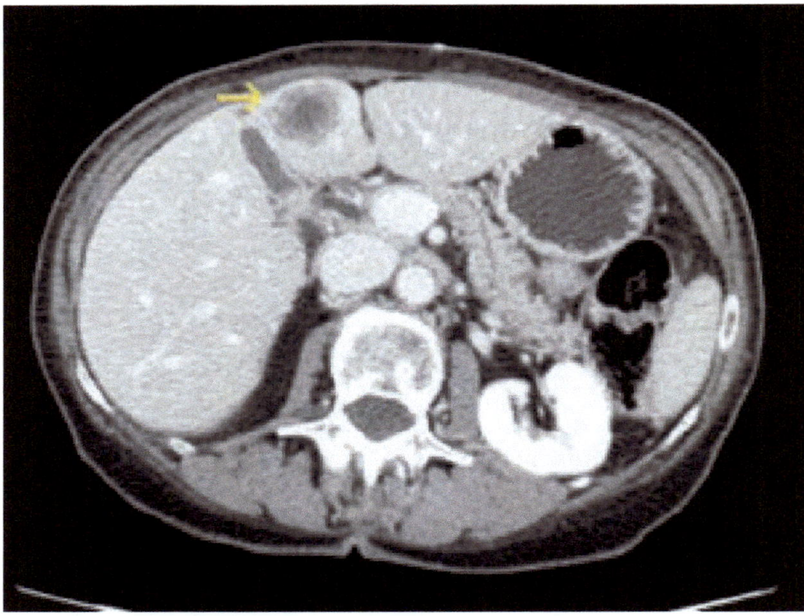

Fig. 30.1 An 83-year-old female patient with a 3-cm colorectal liver metastasis in segment 4b (*yellow arrow*)

is a clinically important ultrasound (US) landmark. From the left side, a branch arises to the anterior portion of the left lobe, P3 (Fig. 30.2a). Another branch arises from the right side to the portion of the liver situated to the right of the umbilical fissure, referred to as P4 (Fig. 30.2b, c).

30.3 Technical Details

Liver resection of S4b has two transection planes: on the left along the umbilical scissure, and on the right along the gallbladder bed. The left transection plane is an anatomic plane and easy to follow during parenchymal transection because it is located along the falciform ligament. The right transection plane has no external landmarks.

Two laparoscopic surgical techniques for liver resection of S4b have been described in the literature: the US-guided LLR with intraparenchymal ligation of P4b [2–5], and the intrahepatic glissonian approach [6]. In the first technique, P4b is identified and ligated during the left-side parenchymal transection. The transection plane between S4b and S5 is determined by the ischemic demarcation line, which can be a good guide for the plane of parenchymal transection. Of course, the intraoperative US should also be used to localize the MHV because the right transection plane is along the right side of the vein.

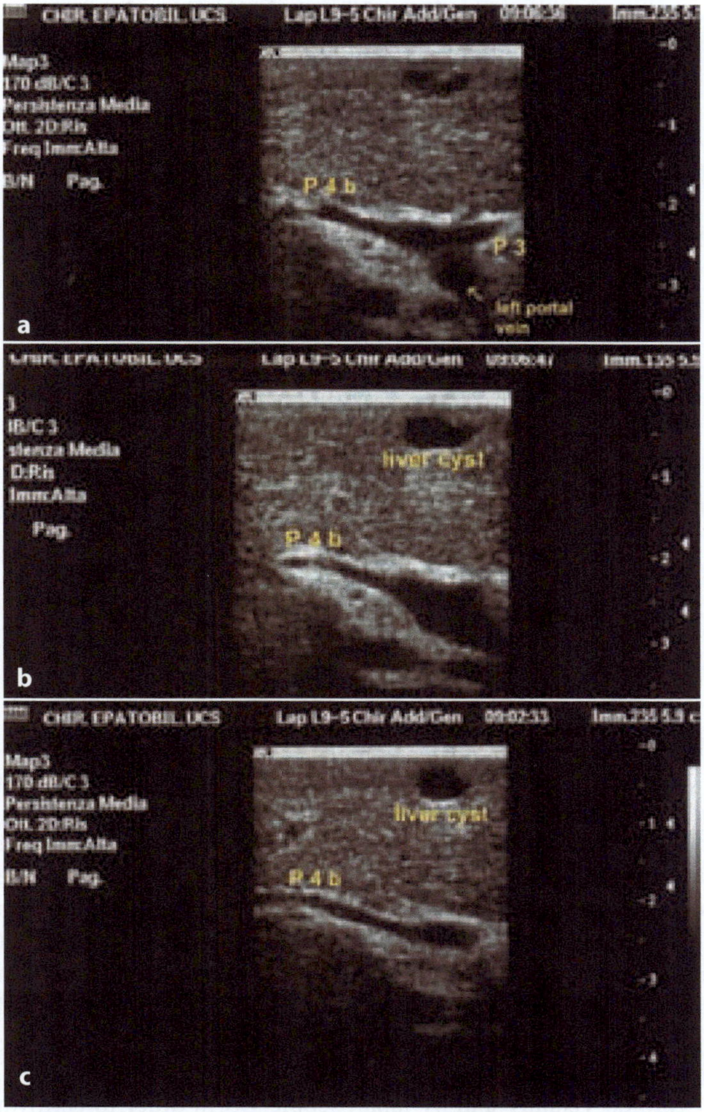

Fig. 30.2 Intraoperative laparoscopic ultrasound. Pedicles to segments 4b and 3 arising from the left portal vein (**a**). Longitudinal view of the pedicle to segment 4b (**b**, **c**)

The intrahepatic glissonian approach, described by Machado et al. [6], is based on clamping around P4b, including the parenchymal tissue around the pedicle. This technique is based on small incisions according to anatomic landmarks. The round ligament is retracted upward, exposing the umbilical fissure between S3 and 4. Using the round ligament as a guide, on its right side, it is possible to identify the anterior aspect of P4b. Two small incisions are made,

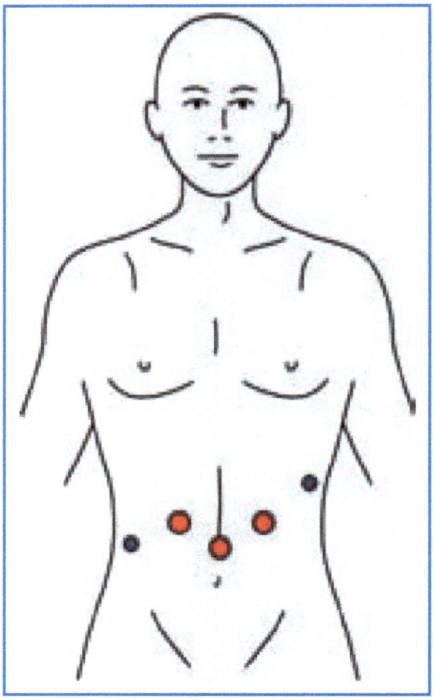

Fig. 30.3 Trocar position. Three 12-mm trocars (*red*) and two 5-mm trocars (*blue*) are used

one in front of the hilum and one at the base of the round ligament. In this way, a large vascular clamp is introduced and P4b is clamped. The demarcation area includes S4b and represents the guide for the anatomic resection.

Under general anesthesia, the patient is placed in the supine position. The surgeon is between the patient's legs, with one assistant on each side of the patient. Five trocars are usually inserted. The procedure is performed with pressure-controlled carbon dioxide (CO_2) pneumoperitoneum, maintained at 12 mmHg. Using an open technique, a 12-mm trocar is placed about 3 cm above the umbilicus. Through this port, a 30° laparoscope is introduced and four additional trocars are placed at the sites, as shown in Fig. 30.3 (two 12-mm and two 5-mm trocars).

The liver is explored visually and by laparoscopic US in order to confirm the location of the tumor, to assess its relationship with P4b and the hilar plate and with the radicals of the MHV, and to ensure adequate resection margin. The round ligament is divided. The portal pedicle is systematically encircled with a tape to allow intermittent pedicle clamping (15-min clamping and 5-min release periods), if required. Cholecystectomy is usually performed. Liver transection is usually performed using bipolar multifunctional shears and forceps (Aesculap; B. Braun, Melsungen, Germany). Small vascular or biliary pedicles are divided after bipolar coagulation or between absorbable clips. During parenchymal transection, it is fundamental to repeatedly perform laparoscopic US to ensure an adequate resection margin. P4b is isolated intraparenchymally and divided between Hem-o-lok nonabsorbable clips (Weck Teleflex Medical, Research Triangle Park, NC, USA).

The resected specimen may be removed into an endoscopic bag through the extended incision of the 12-mm trocar, through the Pfannenstiel incision, or through a previous appendectomy incision. Abdominal drainage is usually left in place.

30.4 Conclusion

LLR of S4b is a feasible operative procedure in selected patients. The use of laparoscopic US is mandatory to guide liver resection and ensure an adequate resection margin. Careful intraparenchymal dissection of the glissonian pedicle makes LLR as feasible as open liver resection.

References

1. Bismuth H (1982) Surgical anatomy and anatomical surgery of the liver. World J Surg 6:3–9
2. Ahn KS, Han HS, Yoon YS, Cho JY (2011) Laparoscopic anatomic S4 segmentectomy for hepatocellular carcinoma. Surg Laparosc Endosc Percutan Tech 21:e183–186
3. Bryant R, Laurent A, Tayar C, Cherqui D (2009) Laparoscopic liver resection-understanding its role in current practice: the Henri Mondor Hospital experience. Ann Surg 250:103–111
4. Koffron AJ, Auffenberg G, Kung R, Abecassis M (2007) Evaluation of 300 minimally invasive liver resections at a single institution: less is more. Ann Surg 246:385–392; discussion 392–394
5. Ardito F, Tayar C, Laurent A, Karoui M, Loriau J, Cherqui D (2007) Laparoscopic liver resection for benign disease. Arch Surg 142:1188–1193; discussion 1193
6. Machado MA, Makdissi FF, Surjan RC, Herman P, Teixeira AR, C Machado MC (2009) Laparoscopic resection of left liver segments using the intrahepatic Glissonian approach. Surg Endosc 23:2615–2619

Segment 5: Laparoscopic Approach

31

Lorenzo Capussotti, Alessandro Ferrero, Luca Viganò and Roberto Lo Tesoriere

31.1 Introduction

Since the early experiences of laparoscopic liver surgery (LLS), segment 5 (S5) resections have been depicted [1]. The segment's anterior position allows easy access, manipulation, transection, and bleeding control. Even if no series specifically analyze outcomes of S5 laparoscopic resections, their feasibility and efficacy have been clearly demonstrated by available series that include quite a large number of these procedures [2–4]. In this chapter, we analyze the surgical anatomy and laparoscopic resection of S5.

31.2 S5 Anatomy

The surgical liver anatomy is still today based on the liver segmental classification established by Couinaud in 1957. This classification separated the liver into eight different segments, each one having its own glissonian or portal pedicle (portal, arterial, and biliary branches) and shares its hepatic vein drainage with adjacent segments. According to this classification, the right portal pedicle divides secondarily into two branches: the right anterior pedicle, feeding S5 and 8; and the right posterior pedicle, feeding S6 and 7. Actually, S5 is fed not by a single pedicle but usually by several branches (from two to five), which arise from the caudal side of the right anterior pedicle before the origin of the two main S8 branches (Fig. 31.1). The S5 pedicles

L. Capussotti (✉)
Department of HPB and Digestive Surgery, Ospedale Mauriziano Umberto I, Turin, Italy
e-mail: lcapussotti@mauriziano.it

F. Calise and G. Casciola (eds.), *Minimally Invasive Surgery of the Liver*, Updates in Surgery
DOI: 10.1007/978-88-470-2664-3_31, © Springer-Verlag Italia 2013

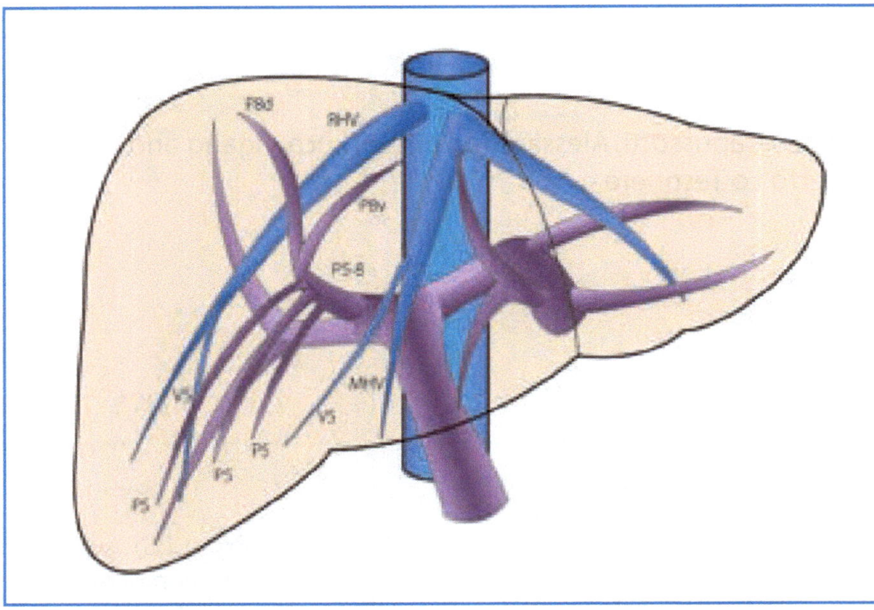

Fig. 31.1 Segment 5 (S5) anatomy. *P5–8*, right anterior portal branch; *P8d*, S8 dorsal portal branch; *P8v*, S8 ventral portal branch; *P5*, S5 portal branches; *MHV*, middle hepatic vein, *RHV*, right hepatic vein; *V5*, hepatic veins draining S5

arise within the liver, and their ligature can be performed only inside the parenchyma by following the right anterior portal pedicle.

Hepatic venous outflow of S5 is drained into both the middle and the right hepatic veins, which can be considered the lateral landmarks of the parenchyma of S5. Two to four veins arising in S5 drain into the middle hepatic vein, and a similar number of veins drain into the right hepatic vein. Many studies of the right liver venous drainage have been carried out in centers performing living-donor liver transplantation. The Kyoto group, using 3D reconstructed images of the hepatic vascular anatomy, divided the anterior sector morphologically into two types: a right-hepatic-vein-dominant graft in which the territory draining into the middle hepatic vein is <40%, and the middle-hepatic-vein-dominant graft [5].

31.3 Surgical Technique

The technique for S5 segmentectomy is described here, but in practice, almost all published S5 laparoscopic procedures are wedge resections. Complete state-of-the-art equipment is recommended, including two monitors. Although some groups use 0° laparoscopes [2, 3], 30° laparoscopes are usually preferred.

The patient is placed in the supine position, in mild reverse Trendelenburg

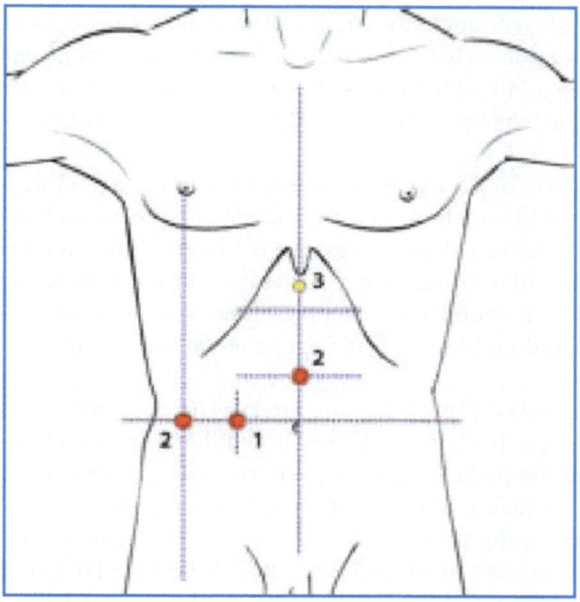

Fig. 31.2 Trocar position. *Red* 10- to 12-mm port; *yellow* 5-mm port; *1*, laparoscope port; *2*, surgeon right and left operative port; *3*, assistant port

position, with lower limbs apart. The right arm is along the body; the left arm is abducted for vein and artery cannulation. The surgeon stands between patient's legs with one assistant on each side. Some authors prefer the patient in the supine position, with the surgeon standing on at the patient's side and the assistant on the opposite side [6].

Carbon dioxide (CO_2) pneumoperitoneum is created using a Veress needle or the open technique. We prefer the open access, which is mandatory in the presence of previous laparotomies. Electronic monitoring of intra-abdominal pressure is required, and it should be maintained <14 mmHg. Port-site positioning varies among authors, but the one preferred in our experience is shown in Fig. 31.2.

Hand-assisted laparoscopy has been proposed to render laparoscopic liver resection (LLR) safer and more accessible than open resection [6]. The technique may help in abdominal exploration, liver mobilization, parenchymal transection, and bleeding control. In our experience, hand assistance is rarely needed and is usually not beneficial in S5 resections.

The first surgical step is always exploration of the liver and abdominal cavity, checking for ascites, signs of portal hypertension, carcinomatosis, and primary tumor recurrence, according to the patient's disease. Frozen sections of any peritoneal deposits are recommended. Liver volume and quality (steatosis or cirrhosis) and tumor(s) characteristics are evaluated. Subglissonian infracentimetric tumors missed by preoperative imaging can be also be revealed. Laparoscopic ultrasonography (LUS) is mandatory to complete the patient's evaluation. It allows study of liver anatomy and its variations, locates known lesions, defines tumor connections with portal pedicles and hepatic veins, and screens the parenchyma for additional tumors. Several studies demonstrate that intraoperative US might change surgical strategy in up to 20–25% of cases [7, 8].

Once laparoscopic resection of S5 has been decided upon, the round ligament can be divided and used for traction. In cirrhotic patients, the round ligament should be spared to preserve collateral vessels. No liver mobilization is required before resection. In order to improve visualization of the transection plane, the gallbladder is partially mobilized along its left margin and then used for traction. It will be removed en bloc with the specimen at the end of transection.

The lesser omentum is checked for left hepatic artery. The porta hepatis is encircled with an umbilical tape. It has been demonstrated that liver resections can be safely performed without pedicle clamping [9]. Moreover, pneumoperitoneum itself significantly decreases blood inflow to the liver. Salvage intermittent clamping (15-min clamping followed by a 5-min interval) can be performed whenever necessary in case of bleeding during transection [10]. The procedure is safe and well tolerated [11] but is being used less and less in laparoscopic practice [12]. If needed, pedicle clamping is performed by inserting a tourniquet into the abdominal cavity through a port. The tape is then pulled and the tourniquet fixed by a forceps inserted through an additional port.

Before starting parenchymal transection, the boundaries of S5 must be identified (Fig. 31.3). No superficial landmarks are evident except for the gallbladder bed on the left side. In open surgery, S5 can be identified by US, blue die injection in P5 pedicles, or pedicle digital compression. The last two options have not been reproduced during laparoscopy. LUS is the only way to define S5 boundaries and is achieved as follows: the middle hepatic vein lies along the gallbladder bed on left side; the right hepatic vein is on the right side; the origin of P5 pedicles from the right portal branch is on the upper part. The boundaries are marked by cautery incision of Glisson's capsule.

Parenchymal transection is then carried out, starting along the left side. Different transection techniques are available and are described elsewhere in this book. LUS is often performed during transection to identify encountered vascular structures and to check adequacy of surgical margins (Fig. 31.4). Hepatic veins draining S5 into the middle hepatic vein are encountered and divided by applying clips. The transection is continued up to the origin of S5 pedicles, which are identified and sectioned between clips. The ischemic area corresponding to S5 is checked. The transection is now started along the right side. In this phase, the transection device can be handled by the right or the left hand, according to the surgeon's preferences, and on a case-by-case evaluation basis. The hepatic veins draining S5 into the right hepatic vein are divided, and the resection is completed.

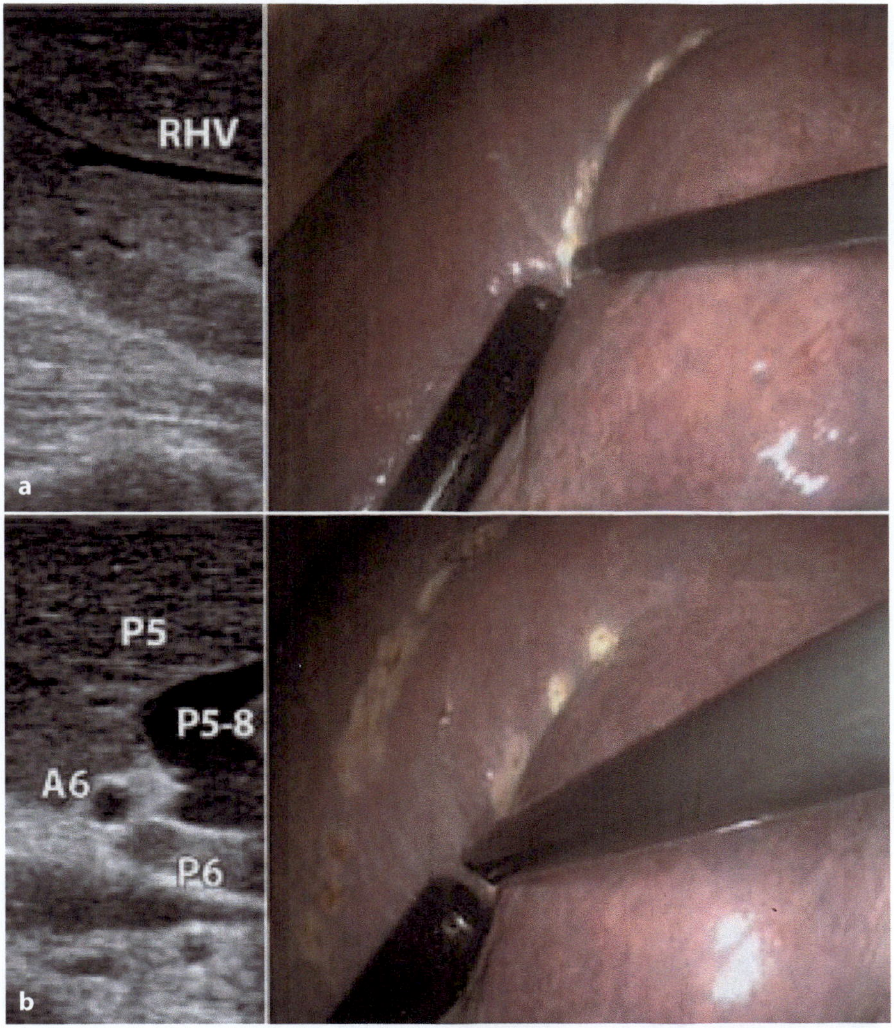

Fig. 31.3 Laparoscopic ultrasound (LUS) identification of segment 5 (S5) landmarks. **a** Right hepatic vein (RHV), defining the right boundary of S5, is marked under LUS guidance. **b** Main S5 pedicle (*P5*) is marked under LUS guidance. S6 artery (*A6*) and portal branch (*P6*) and right anterior portal pedicle (*P5–8*) are visible

In all cases, the specimen is placed in a plastic bag and extracted through a separate incision, either along a new suprapubic horizontal incision or a previous abdominal scar. Fragmentation must be avoided. After specimen extraction, the incision is sutured and the pneumoperitoneum is again created. Hemostasis is checked, as is the presence of any bile leak. According to surgeon preference, the raw cut surface can be treated by any hemostatic agent, and an abdominal drainage can be used (Fig. 31.5).

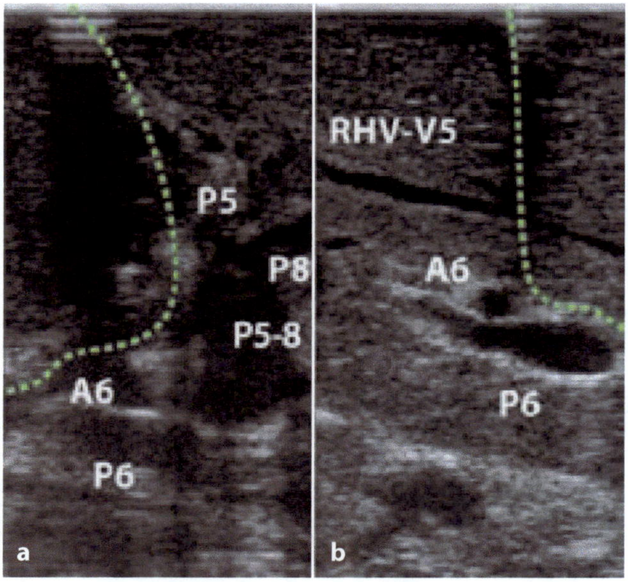

Fig. 31.4 Laparoscopic ultrasound (LUS) control of transection line.
a The planned section line is checked on the left side of the cut surface. It runs across segment 5 (*S5*) pedicle (*P5*) and along S6 pedicle (*P6*). *A6* S6 artery.
b The planned section line is checked on the right side of the section plane. It runs across a hepatic vein draining S5 into the right hepatic vein (*RHV-V5*) and along S6 pedicle (*P6, A6*)

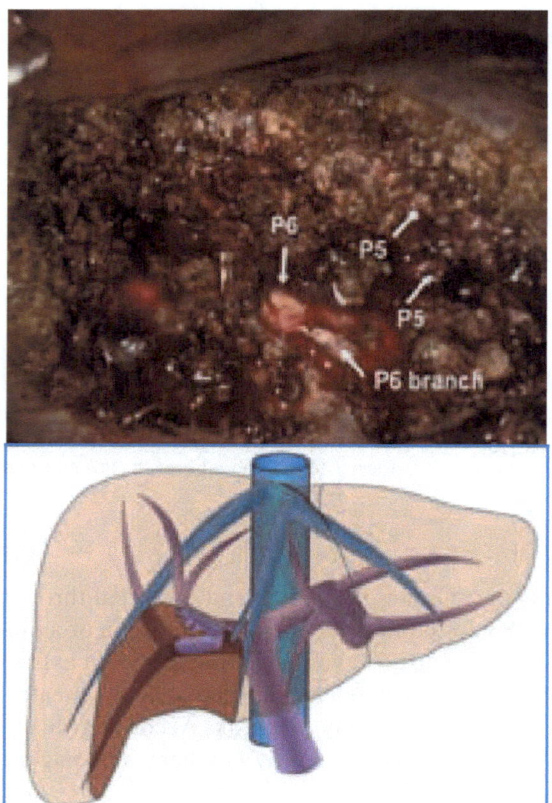

Fig. 31.5 Cut surface of the liver. Segment 5 (S5) resection is completed. S6 portal branch (*P6 branch*) is exposed on the cut surface. Stumps of S5 pedicles (*P5*) and of a small S6 pedicle (*P6*) are visible

References

1. Cherqui D, Husson E, Hammoud R et al (2000) Laparoscopic liver resections: a feasibility study in 30 patients. Ann Surg 232(6):753–762
2. Dagher I, Proske JM, Carloni A et al (2007) Laparoscopic liver resection: results for 70 patients. Surg Endosc 21(4):619–624
3. Vibert E, Perniceni T, Levard H et al (2006) Laparoscopic liver resection. Br J Surg 93(1):67–72
4. Viganò L, Tayar C, Laurent A, Cherqui D (2009) Laparoscopic liver resection: a systematic review. J Hepatobiliary Pancreat Surg 16(4):410–421
5. Yu PF, Wu J, Zheng SS (2007) Management of the middle hepatic vein and its tributaries in right lobe living donor liver transplantation. Hepatobiliary Pancreat Dis Int 6:358–363
6. Koffron AJ, Auffenberg G, Kung R, Abecassis M (2007) Evaluation of 300 minimally invasive liver resections at a single institution: less is more. Ann Surg 246(3):385–392
7. John TG, Greig JD, Crosbie JL et al (1994) Superior staging of liver tumors with laparoscopy and laparoscopic ultrasound. Ann Surg 220(6):711–719
8. Foroutani A, Garland AM, Berber E et al (2000) Laparoscopic ultrasound vs triphasic computed tomography for detecting liver tumors. Arch Surg 135:933–938
9. Capussotti L, Muratore A, Ferrero A et al (2006) Randomized clinical trial of liver resection with and without hepatic pedicle clamping. Br J Surg 93(6):685–689
10. Viganò L, Jaffary SA, Ferrero A et al (2011) Liver resection without pedicle clamping: feasibility and need for "salvage clamping". Looking for the right clamping policy. Analysis of 512 consecutive resections. J Gastrointest Surg 15(10):1820–1828
11. Decailliot F, Cherqui D, Leroux B et al (2001) Effects of portal triad clamping on haemodynamic conditions during laparoscopic liver resection. Br J Anaesth 87(3):493–496
12. Viganò L, Laurent A, Tayar C et al (2009) The learning curve in laparoscopic liver resection: improved feasibility and reproducibility. Ann Surg 250(5):772–782

Segment 6: Laparoscopic Approach

32

Umberto Cillo, Marina Polacco, Francesco D'Amico
and Enrico Gringeri

EXTRA MATERIALS

32.1 Introduction

Totally laparoscopic liver resection (LLR), first described by Gagner et al. in 1992 [1], now appears to be safe and effective procedure to treat lesions located on left lateral (2 and 3), middle (4b), and anterior right (5, 6) segments. Tumors localized in the posterior and superior segments of the liver or masses requiring major liver resections (right hepatectomy, left hepatectomy, extended right or left hepatectomy) are still considered less frequent indications for the minimally invasive approach and should be reserved to centers with significant experience in laparoscopic liver surgery (LLS), open hepatic surgery, and intraoperative ultrasound (US) [2]. Indications for LLS, with particular reference to resection of segment 6 (S6), remain basically the same as in open surgery and, technical feasibility of S6 resection using a minimally invasive approach is increasing with surgeon skill [2]. On this basis, a liver lesion located on S6 should be routinely evaluated for LLR in high-volume centers.

32.2 S6 Anatomy

According to Couinaud's division into self-contained units, each liver segment can be resected without damaging the others. For the liver to remain viable, resections must proceed along the vessels that define the peripheries of these segments, with resection lines roughly parallel to the hepatic veins. S6 is the right posterolateral anterior/inferior segment. It is limited medially by

U. Cillo (✉)
Hepatobiliary Surgery Unit and Liver Transplant Center, University of Padua, Padua, Italy
e-mail: cillo@unipd.it

F. Calise and G. Casciola (eds.), *Minimally Invasive Surgery of the Liver*, Updates in Surgery
DOI: 10.1007/978-88-470-2664-3_32, © Springer-Verlag Italia 2013

the right scissure posteriorly from the S7 fissure (detectable only by US examination). Laterally it is in contact with the abdominal wall and the lower portion with the hepatic colonic flexure. Its blood supply is guaranteed by the right posterior branch of the right artery. Literature reports describe some variations in portal-branches anatomy: in most patients, the right posterior trunk has a constant pattern of division and is directed dorsolaterally in a more or less horizontal plane. Covey and Brown demonstrated, in a cohort of 200 computed tomography (CT) portograms, that about 35% of patients had variant portal-vein anatomy; in particular, 6% of P6 originates as a separate branch of the right portal vein, and in 1%, P7 arises as well as a separate branch of the right portal vein [3]. Some reports also describe the presence of a posterior trunk in the absence of a common right portal trunk, or a separate origin of segmental branches from the right portal trunk. The right hepatic vein shows marked variability, described as early confluence or presence of multiple right hepatic veins and sometimes as an accessory inferior hepatic vein draining directly into the inferior vena cava (IVC) from S6 (right inferior) [4]. Biliary ducts from S6 and 7 usually drain into the right posterior hepatic duct. Rarely, the right posterior hepatic duct drains directly into the left hepatic duct (13–19%) or into the common hepatic duct (5%) [5].

32.3 Clinical Experience

The literature does not report focused series of S6 totally operated by LLR. It is therefore impossible to draw definitive clinical conclusions and to make technical recommendations about a procedure that has not been fully studied. In fact, although the safety and feasibility of S6 LLR, in terms of clinical and oncologic results, have been suggested by various reports, to date, the technique lacks a systematic analysis of associated clinical patterns and detailed step-by-step technical description [6].

32.4 Technical Details

S5 and 6 are the most easily accessible and suitable locations for LLR on the right side. However, whereas laparoscopic procedures for left liver are well described and standardized, the approach to S6 is still matter of some degree of discussion.

As with the other segmental laparoscopic hepatic resections, it is possible to identify at least three technical approaches: totally laparoscopic, hand-assisted, and laparoscopic-assisted open "hybrid" surgical techniques. There remains, however, a debate about patient positioning: we prefer a left lateral decubitus or semilateral decubitus position that ranges from 45° to 70° with the aid of a beanbag mattress (as described by Belli et al. [7]). The patient is placed in a secure position with safety straps and tape. Nerve-compression injuries are prevented by identifying and coating all pressure points. The patient is tilted in mild anti-Trendelenburg position to help with exposure, and the right arm is elevated with an arm rest. The surgeon and first assistant usually stand in front of the patient on the left side, with the second assistant in front of them. Some authors suggest adopting the standard supine position with the patient's legs spread open.

According to a previous description, four trocars may be positioned along a semicircular line, with the concavity facing the right subcostal margin: two 12-mm trocars (a, b) for introducing the laparoscope, the endoscopic stapler device, and the intraoperative US probe; and two 5-mm surgical trocars (c, d). The first 12-mm trocar (a) is placed, using the open technique (Hasson technique), in the midline at one third of the xiphoumbilical distance [7, 8]. The carbon dioxide (CO_2) pneumoperitoneum is then induced, with an intra-abdominal pressure to be maintained at 12–14 mmHg to prevent risk of gas embolism. The remaining three trocars all positioned under direct vision: the second 12-mm trocar (b) is positioned laterally to the umbilicus transrectally; one 5-mm trocar (c) is positioned in the right flank inferior and slightly posterior to the margin of the 11th rib; the second 5-mm trocar (d) is placed on the left, subcostally [9].

To obtain adequate vascular control at the level of the S6 portal pedicle during parenchymal transection and to better facilitate a round-shaped transection plane, we propose a different trocar positioning, as shown in Fig. 32.1. The right subcostal trocar (D) allows retraction of the liver's inferior surface and a greater vascular control at the end of the transection procedure, when it is necessary to control the glissonian pedicle.

A cold-light-source 30° laparoscope is used. The absence of carcinosis and/or extrahepatic localization of disease is confirmed by a standard exploratory laparoscopy. Nodule localization, number, volume, and vascular relationship are determined by intraoperative US. The next step is to approach the hepatoduodenal ligament to prepare for the eventual Pringle maneuver. Incision of the pars lucida of the lesser omentum and isolation of the hepatic pedicle is then performed, and the hilum is subsequently encircled by a tape. We do not recommend routinely performing a cholecystectomy during S6 LLR.

A liver retractor is used to gently expose the right hepatic lobe and begin mobilization of the right hemiliver. The triangular ligament is divided by hook cautery, and the dissection is carried on to the diaphragm medially toward the IVC. US examination is used to establish the exact transaction plane and position of the right hepatic vein. Both positions are then traced on the liver surface by monopolar cautery.

Liver parenchymal transection begins by using monopolar cautery for the glissonian surface and the very superficial layers. More deeply, it is performed by US dissection [Cavitron Compact Ultrasonic Surgical Aspirator (CUSA Excel), Tyco Healthcare] or a US Harmonic scalpel (Ethicon Ultracision Harmonic ACE, Ethicon, Somerville, NJ, USA). At our center, we more fre-

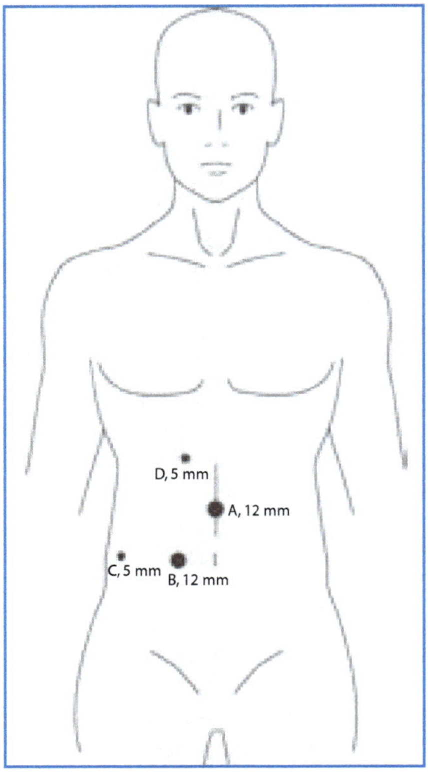

Fig. 32.1 Trocar positioning for S6 laparoscopic liver resection

quently adopt a 5-mm sharp-tip radiofrequency (RF) device (LigaSure Vessel Sealing System, Valleylab, Tyco, UK), which provides a nice balance between a gentle dissection and good capability to control small- and medium-size intraparenchymal vessels.

We consider it advisable to proceed with the parenchymal transection on both superior and inferior surfaces of S6 and progressively deepening the transection on both sides. This approach favors a progressive and safe approach to the more significant S6 vessels, in particular, achieving a step-by-step skeletonization of P6. S6 vascular hilum, indeed, is usually situated deep in the segment, below the right hepatic vein. Near the end of parenchymal transection, the glissonian pedicle is divided by positioning a linear vascular endostapler (Endo GIA™ Ultra Universal Stapler, Covidien, Mansfield, MA, USA). The major venous S6 outflow (V6) may be encountered either ventromedially to the glissonian pedicle in the more frequent presentation, or more dorsally if a dominant right inferior vein is present. In both cases, it may be controlled and divided after clip positioning.

After parenchymal transection is completed, accurate hemostasis and biliostasis must be achieved. The transected surface can be sealed with fibrin glue (Tisseel®; Baxter, Vienna, Austria) to reduce the incidence of biliary fistulas and improve hemostasis.

The liver specimen is extracted using a retrieval bag and removed, without fragmentation (EndoBag™ Specimen Retrieval, Covidien), to minimize the risk of neoplastic cells seeding at trocar levels. This is attained after enlarging one of the trocar sites or by performing a Pfannenstiel incision.

The use of abdominal drainage after LLR is still under debate; an 18-F rubber tube can be used for this purpose and removed on postoperative day 1 or 2 in absence of complications.

References

1. Gagner M, Rheault M, Dubuc J (1992) Laparoscopic partial hepatectomy for liver tumor. Surg Endosc 6:99
2. Topal B, Fieuws S, Aerts R, Vandeweyer H, Penninckx F (2008) Laparoscopic versus open liver resection of hepatic neoplasms: comparative analysis of short-term results. Surg Endosc 22(10):2208–2013
3. Covey AM, Brown KT (2007) Prevalence of right portal vein branching anomalies. AJR Am J Roentgenol 188(5):W492
4. Van Leewen MS, Noordzij J, Arancha Fernandez M, Hennipman A, Feldberg MAM, Dillon EH (1994) Portal venous and segmental anatomy of the right hemiliver. AJR 163:1395–1404
5. Catalano OA, Singh AH, Uppot RN, Hahn PF, Ferrone CR, Sahani DV (2008) Vascular and biliary variants in the liver: implications for liver surgery. Radiographics 28(2):359–378
6. Rao A, Rao G, Ahmed I (2012) Laparoscopic or open liver resection? Let systematic review decide it. Am J Surg [Epub ahead of print, January 13, PMID: 22245507]
7. Belli G, Fantini C, D'Agostino A, Cioffi L, Limongelli P, Russo G, Belli A (2008) Laparoscopic segment VI liver resection using a left lateral decubitus position: a personal modified technique. J Gastrointest Surg 12(12):2221–2226
8. Han HS, Cho JY, Yoon YS (2009) Techniques for performing laparoscopic liver resection in various hepatic locations. J Hepatobiliary Pancreat Surg 16(4):427–432
9. Cherqui D, Chouillard E, Laurent A, Tayar C (2006) Laparoscopic hepatectomy. Elsevier Masson SAS, Paris, France

Segment 7: Robot-Assisted Approach

33

Alberto Patriti, Graziano Ceccarelli and Luciano Casciola

33.1 Introduction

Laparoscopic liver resection (LLR) was initially indicated for benign and peripherally located lesions, suggesting concern about the oncologic effectiveness of these procedures and a fear of massive bleeding when approaching posterior and superior liver segments [1]. A right hepatectomy remains the preferred choice for deeply located lesions in segments 7 (S7) and 8 by most laparoscopic liver surgeons to avoid curved and angulated transection lines close to the right hepatic vein (RHV) and the inferior vena cava (IVC) [2]. However, in the last few years, the number of minimally invasive complex liver resections has increased, and even segmentectomies of the posterosuperior (PS) segments have been successfully performed [2–4]. Thus, laparoscopic segmental and subsegmental resection of S7 are an important step toward the fulfillment, even in minimally invasive surgery, of the principles of parenchyma preservation that represent the actual trend in the treatment of both colorectal liver metastases and hepatocellular carcinoma [5].

33.2 S7 Anatomy

S7 is the upper part of the right posterior sector. According to the classification of Couinaud, S7 is delimited medially by the RHV, and the blood supply is maintained by one of the two branches of the common trunk for the posterior sector of the right portal pedicle. More recent anatomical studies describe

A. Patriti (✉)
Department of Surgery, Division of General, Minimally Invasive and Robotic Surgery,
ASL3 Umbria, San Matteo degli Infermi Hospital, Spoleto, Italy
e-mail: albertopatriti@gmail.com

F. Calise and G. Casciola (eds.), *Minimally Invasive Surgery of the Liver*,
Updates in Surgery
DOI: 10.1007/978-88-470-2664-3_33, © Springer-Verlag Italia 2013

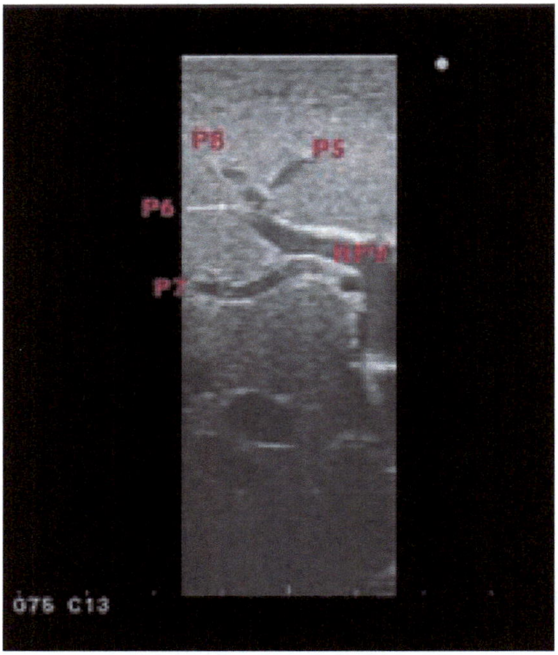

Fig. 33.1 Intraoperative laparoscopic ultrasound showing the right portal vein (*RPV*) branching off P7 and the branch for the anterior sector. The latter trifurcates in a small P6 and in P5 and P8

a complex vascular anatomy of the posterior sector, knowledge of which could be useful during ultrasound (US) exploration and liver resection. Four vascular pattern of the posterior portal vein are described. The "arch-like" pattern is characterized by a common posterior trunk with a curved course upward and backward, providing several inferior branches for S6 (P6). The trunk for S7 (P7) is a direct continuation of this posterior trunk. In the "simple bifurcation" pattern, the posterior sectorial trunk branches off into the P6 and P7 trunk. The "trifurcation" pattern is characterized by an intermediate branch that supplies an area between S6 and S7. The last group comprises variations involving the right anterior sectorial branch. The most common variations, accounting for 33.5% of 197 cadaveric dissections, are the tri-furcation forming P6 and P7 and the anterior sectorial trunk (Fig. 33.1) [6]. Therefore, identifying the inferior border of S7 with S6 according to the portal distribution could be demanding and requires careful preoperative workup and accurate intraoperative US exploration.

The RHV can only be used as the medial border of S7 close to the hepatocaval confluence, where a distinct trunk is present. More distally, tributaries of the RHV pass between subsegmental branches of the posterior sector, making medial definition of S7 and S6 impractical. In nearly 20% of cases, the RHV is poorly developed, and S7 is drained by one or more accessory veins directly into the IVC. In this situation, a thick accessory vein draining S7 is present and can be used as the medial border of S7.

33.3 Surgical Technique

S7 segmentectomy is challenging due to its deep location and contiguity to the RHV and IVC. Due to the high risk of bleeding, intermittent pedicle clamping and low central venous pressure anesthesia are advisable when approaching S7 [7, 8].

To date, the most frequently used laparoscopic approaches to S7 are lateral and anterior. Both techniques require a 30° reverse-Trendelenburg position and patient rotation on the left flank in order to facilitate liver mobilization and IVC dissection. Due to the atypical patient position, pneumoperitoneum is generally induced with the Veress needle inserted in the right upper quadrant and maintained during surgery at 10–12 mmHg. Reductions in abdominal pressures can be applied during RHV and IVC dissection or in case they tear to avoid gas embolism.

Gumbs and Gayet described the lateral approach in 2008 using conventional laparoscopy. The camera port and left operative trocar are inserted transcostally, whereas the third and forth operative trocars are placed along the right costal margin. The main advantage of this approach is the frontal view to S7 and the right hepatocaval confluence. The right triangular and coronal ligaments are divided with the Harmonic scalpel or the monopolar scissors. Caval dissection proceeds in a right to medial direction, clipping and dividing the accessory veins and right posterior liver ligament and exposing the right hepatocaval confluence. Under US exploration, the borders of S7 are marked with monopolar forceps on the liver surface, and parenchymal transection can be carried out using the Harmonic scalpel and the bipolar forceps. Anatomical S7 segmentectomy implies two right-angled resection lines; therefore, the main transection device can be moved from the transcostal operative trocar for the inferior border of S7 and the subcostal trocar for the medial border parallel to the RHV. The main advantage of this approach is the frontal view to S7 and the ergonomic position of the operative trocars, which allow excellent access to S7, even with the traditional laparoscopic tools. Our personal criticism to this approach is the impossibility of safely encircling the liver pedicle to apply the Pringle maneuver.

The anterior approach was described by our group in 2011 and requires the use of the articulated arms of the da Vinci robotic system, which is docked over the patient's head. The camera port and the left-sided trocars are placed as close as possible to the right costal margin, whereas the right trocar is inserted in the intercostal space between the 10th and 11th rib along the scapular line. At this level, the risk of accidental injury to the lung is very low, and direct access to the posterolateral segments is provided [7]. Two accessory trocars can be placed along the middle and anterior axillary line or in the umbilical area (Figs. 33.2 and 33.3).

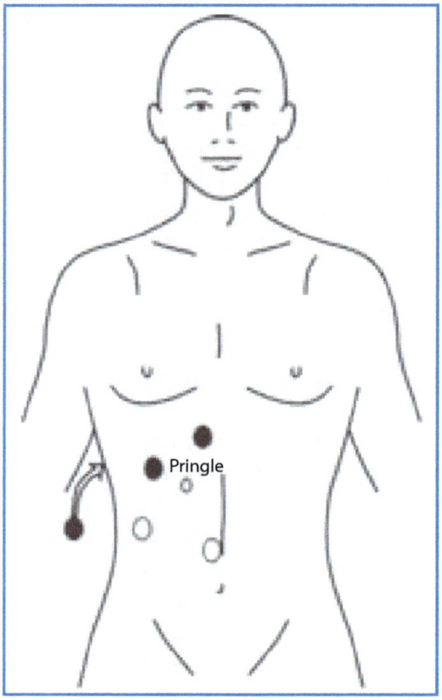

Fig. 33.2 Trocar disposition to robotically approach segment 7 (S7). *Black dots* robotic trocars. *White circles* 12-mm accessory trocars. *Pringle* position to exteriorize the tube for extracorporeal Pringle maneuver

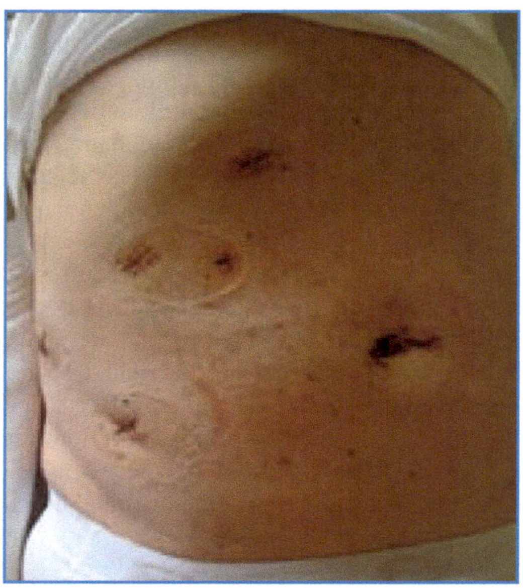

Fig. 33.3 Postoperative aesthetic outcome of a robot-assisted resection in segment 7 (S7)

Surgery starts by encircling the liver pedicle for the Pringle maneuver. In case of a large right lobe, the camera can be inserted in the accessory trocar in the anterior axillary line or in the umbilical area for better vision. The device for inflow occlusion is composed of a 20-F chest tube, an umbilical tape, and a plug used to occlude the Foley catheter. The chest tube is inserted percutaneously in the epigastric region. The umbilical tape is passed around the hepatoduodenal ligament with the use of an EndoWrist Cadiere forceps (Intuitive Surgical Inc, Sunnyvale, CA, USA). Under direct visualization of the Winslow foramen, the Cadiere forceps easily encircles the hepatoduodenal ligament, breaking out the pars flaccida of the lesser omentum. The umbilical tape is then exteriorized through the chest tube with the use of a 5-mm laparoscopic forceps. The chest tube is finally closed with the plug in order to avoid air loss. When inflow occlusion is needed, the on-table surgeon removes the plug and blindly pulls the umbilical tape with the left hand while pushing the tube with the right hand. When the desired tape tension is achieved, the chest tube is closed with the plug. Once the device for inflow occlusion is arranged, the camera can be introduced in its original subcostal port. Monopolar scissors are inserted in the right robotic arm and the bipolar forceps in the left. The right triangular and coronal ligaments are divided with the monopolar scissors. Caval dissection proceeds in a downward to upward direction, clipping and dividing the accessory veins and the right posterior liver ligament, exposing the right hepatocaval confluence. The accessory veins are clipped with plastic locking clips by the on-table assistant and divided by the console surgeon using the EndoWrist robotic scissors on the right robotic arm. Surgical margins are marked on a routine basis using the laparoscopic US used by the on-table assistant. Parenchymal dissection is carried out under intermittent inflow occlusions using the bipolar forceps in the fashion of the Kelly-clamp crushing technique. Liver dissection starts from the inferior border of S7 and, once the RHV is reached, proceeds upward along the main RHV trunk. The specimen is generally removed through an enlarged port-site protected by an EndoBag.

The main advantage of this approach is the safety provided by the inflow occlusion and the robot's EndoWrist movements that allow fine dissection along the intraparenchymal portion of the RHV and curved resection lines without changes in instrument positions.

References

1. Buell JF, Cherqui D, Geller DA et al (2009) The international position on laparoscopic liver surgery: The Louisville Statement, 2008. Ann Surg 250(5):825–830
2. Nguyen KT, Gamblin TC, Geller DA (2009) World review of laparoscopic liver resection-2,804 patients. Ann Surg 250(5):831–841. doi:10.1097/SLA.0b013e3181b0c4df
3. Cho JY, Han HS, Yoon YS, Shin SH (2008) Feasibility of laparoscopic liver resection for tumors located in the posterosuperior segments of the liver, with a special reference to overcoming current limitations on tumor location. Surgery 144(1):32–38. doi:10.1016/j.surg.2008.03.020

4. Gumbs AA, Gayet B (2008) Video: the lateral laparoscopic approach to lesions in the posterior segments. J Gastrointest Surg 12(7):1154. doi:10.1007/s11605-007-0455-x
5. Torzilli G, Montorsi M, Donadon M, Palmisano A, Del Fabbro D, Gambetti A, Olivari N, Makuuchi M (2005) "Radical but conservative" is the main goal for ultrasonography-guided liver resection: prospective validation of this approach. J Am Colle Surg 201(4):517-528. doi:10.1016/j.jamcollsurg.2005.04.026
6. Hata F, Hirata K, Murakami G, Mukaiya M (1999) Identification of segments VI and VII of the liver based on the ramification patterns of the intrahepatic portal and hepatic veins. Clin Anat 12(4):229–244
7. Casciola L, Patriti A, Ceccarelli G, Bartoli A, Ceribelli C, Spaziani A (2011) Robot-assisted parenchymal-sparing liver surgery including lesions located in the posterosuperior segments. Surg Endosc 25(12):3815–3824. doi:10.1007/s00464-011-1796-9
8. Rotellar F, Pardo F, Bueno A, Marti-Cruchaga P, Zozaya G (2012) Extracorporeal tourniquet method for intermittent hepatic pedicle clamping during laparoscopic liver surgery: an easy, cheap, and effective technique. Langenbecks Arch Surg 397(3):481–485. doi:10.1007/s00423-011-0887-3

Segment 8: Robot-Assisted Approach

34

Alberto Patriti, Graziano Ceccarelli, Alessandro Spaziani and Luciano Casciola

34.1 Introduction

Laparoscopic resection of segment 8 (S8) is considered particularly demanding due to its vascular anatomy and location in the "dome" of the liver [1]. As for S7, a right hepatectomy remains the preferred technique of most laparoscopic liver surgeons for resecting lesions in S8 in order to avoid complex transection lines close to the right (RHV) and middle (MHV) hepatic veins [2]. Therefore, laparoscopic S8 segmentectomy remains purely anecdotal and reserved to centers with significant experience in laparoscopic (LLS) and open (OLS) liver surgery. There are no studies designed to investigate the best minimally invasive approach to S8. However, in the few series in which posterolateral lesions were approached laparoscopically, resection was associated with higher blood loss, operative time, and conversion rate in respect to resection of the anterior segments [3]. Introduction of the da Vinci Robotic System (Intuitive Surgical Inc, Sunnyvale, CA, USA), with its three articulated operative arms, has made liver transection easier, even for nonlinear resections. Therefore, it is conceivable that the robot could be useful also to improve the safety of minimally invasive S8 resection, even though evidence for a routine robot-assisted S8 segmentectomy is still lacking [4].

34.2 S8 Anatomy

S8 is the anterior–superior area of the liver. According to the classification of Couinaud, S8 is delimited medially by the MHV and laterally by the RHV, but

A. Patriti (✉)
Department of Surgery, Division of General, Minimally Invasive and Robotic Surgery, ASL3 Umbria, San Matteo degli Infermi Hospital, Spoleto, Italy
e-mail: albertopatriti@gmail.com

F. Calise and G. Casciola (eds.), *Minimally Invasive Surgery of the Liver*, Updates in Surgery
DOI: 10.1007/978-88-470-2664-3_34, © Springer-Verlag Italia 2013

more recent studies revealed a discrepancy between the "classical" functional anatomy of S8 and the relationships between the portal branch to S8 (P8) and the longitudinal scissurae along the two hepatic veins. Blood supply to S8 is maintained by one branch arising from the common trunk for the anterior sector of the right portal pedicle. Takayasu et al. described the division of P8 into four subsegmental branches, that appear, counterclockwise, as: (a) ventral, (b) dorsolateral, (c) dorsal, and (d) medial (Figs. 34.1, 34.2) [5]. P8d is the cranially directed subsegmental branch that runs through the space between the two hepatic veins laterally and S1 dorsally. As described by Cho et al., P8c is directed cranially and gives rise to branches that cross the RHV, distributing over the entire dorsocranial area, including the upper part of the area classified by Couinaud as S7 [6]. Therefore, limits of S8 go beyond the RHV, and different types of subsegmentectomies are possible due to its peculiar vascular anatomy [7].

34.3 Surgical Technique

Due to the high risk of bleeding, intermittent pedicle clamping and low central venous pressure anesthesia are advisable when approaching S8 [8].

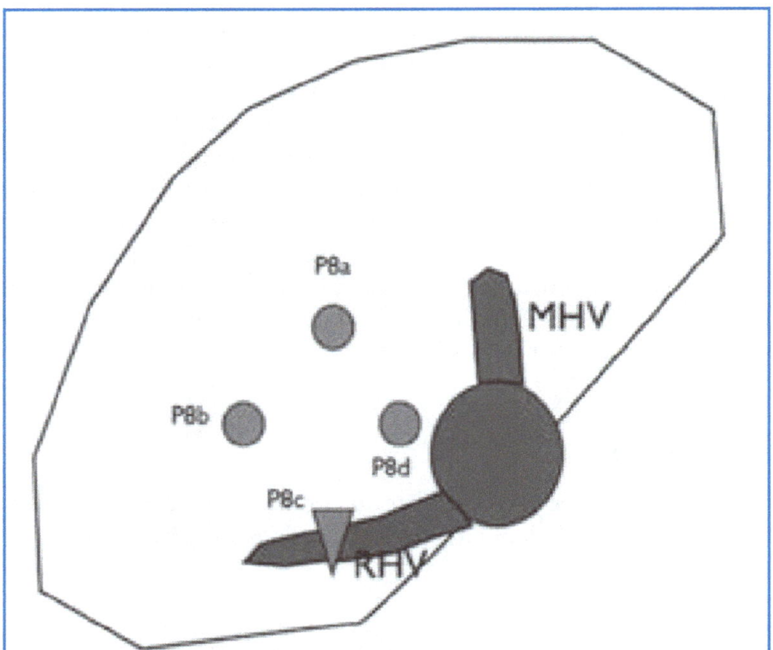

Fig. 34.1 Transverse section of the liver at the level of the hepatocaval confluence. *MHV*, middle hepatic vein; *RHV*, right hepatic vein; *P8*, portal branches of segment 8

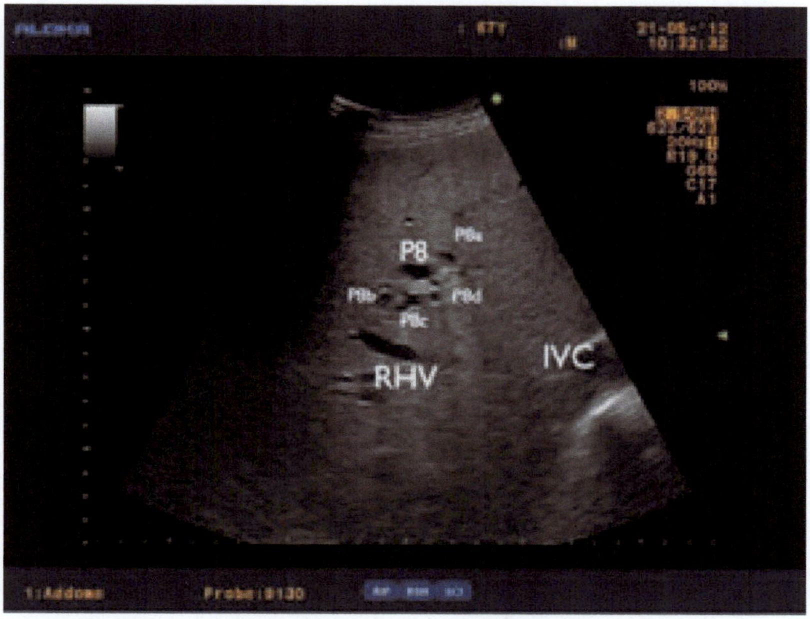

Fig. 34.2 Echographic image showing the transverse section of the liver. *P8,* portal branches of segment 8; *RHV,* right hepatic vein; *IVC,* inferior vena cava

The patient is placed in the supine position with the lower limbs apart on a split leg table. The on-table surgeon stands between the patient's legs. The scrub nurse and surgical instruments are positioned lateral to the left leg. A reverse Trendelenburg position is used, and the table is tilted to the left, as necessary, to take advantage of gravity and liver weight to improve exposure. The robot cart is docked over the patient's head.

Pneumoperitoneum is generally induced with the Veress needle inserted in the left upper quadrant and maintained during surgery at 10-12 mmHg. Reductions in abdominal pressures can be applied during RHV and MHV dissection or if they tear, to avoid gas embolism.

The camera port is inserted supraumbilically. Two 8-mm trocars, one in the left and one in the right upper quadrant are used for the robotic arms, and two accessory trocars placed in the umbilical line are used for suction and retraction by the on-table assistant (Fig. 34.3). Through the same accessory ports, the on-table assistant can perform the intraoperative ultrasound (US). We suggest using only 12-mm trocars for this purpose. Two robotic instruments are used: a bipolar Precise forceps and a monopolar scissor.

The round ligament is divided between clips by the robotic articulated scissors on the right arm. The falciform ligament is transected close to the abdominal wall along its length to the confluence of the hepatic veins and inferior vena cava. This confluence is then carefully dissected in order to identify inser-

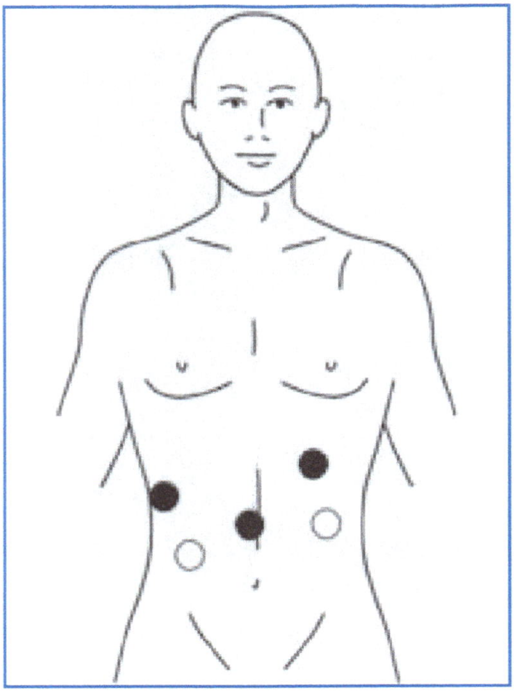

Fig. 34.3 Trocar placement. *Black dots,* robotic trocars; *white circles,* accessory trocars

tion of the RHV and common trunk of the MHV and left (LHV) hepatic vein. To expose the bare area of the liver, liver mobilization begins by cutting the right triangular ligament while the assistant lifts up and gradually rotates the right lobe to the left. The coronal ligament is then divided until reaching the confluence of the RHV, which is already partially dissected. With US exploration and guidance the borders of S8 are marked on the liver surface with monopolar forceps, and parenchymal transection can be carried out using the Kelly-clamp crushing technique under intermittent pedicle clamping. During division of the inferior boundary, the right anterior sectional pedicle and the S8 pedicle should be sought. Applying Pringle maneuver to obtain a bloodless operative field is critical and facilitates identification of these pedicles within the liver, minimizing erroneous ligation of the S5 pedicle. Dissection along the two hepatic veins is carried out with a gentle Kelly-clamp crushing technique, clipping small branches with titanium clips and larger structures with plastic locking clips. The main advantage of using the robot is the possibility to ligate and suture small defects on the vessel wall. The transection plane is finally inspected for bleeding and biliary leaks, which can be selectively repaired with 5-0 sutures.

References

1. Buell JF, Cherqui D, Geller DA et al (2009) The international position on laparoscopic liver surgery: The Louisville Statement, 2008. Ann Surg 250(5):825–830
2. Nguyen KT, Gamblin TC, Geller DA (2009) World review of laparoscopic liver resection-2,804 patients. Ann Surg 250(5):831–841. doi:10.1097/SLA.0b013e3181b0c4df
3. Cho JY, Han HS, Yoon YS, Shin SH (2008) Feasibility of laparoscopic liver resection for tumors located in the posterosuperior segments of the liver, with a special reference to overcoming current limitations on tumor location. Surgery 144(1):32–38. doi:S0039-6060(08)00215-8 [pii] 10.1016/j.surg.2008.03.020
4. Casciola L, Patriti A, Ceccarelli G, Bartoli A, Ceribelli C, Spaziani A (2011) Robot-assisted parenchymal-sparing liver surgery including lesions located in the posterosuperior segments. Surg Endosc 25(12):3815–3824. doi:10.1007/s00464-011-1796-9
5. Takayasu K, Moriyama N, Muramatsu Y, Shima Y, Goto H, Yamada T (1985) Intrahepatic portal vein branches studied by percutaneous transhepatic portography. Radiology 154(1):31–36
6. Cho A, Okazumi S, Takayama W, Takeda A, Iwasaki K, Sasagawa S, Natsume T, Kono T, Kondo S, Ochiai T, Ryu M (2000) Anatomy of the right anterosuperior area (segment 8) of the liver: evaluation with helical CT during arterial portography. Radiology 214(2):491–495
7. Torzilli G, Procopio F, Palmisano A, Donadon M, Del Fabbro D, Marconi M, Scifo G, Montorsi M (2011) Total or partial anatomical resection of segment 8 using the ultrasound-guided finger compression technique. HPB (Oxford) 13(8):586–591.
8. Rotellar F, Pardo F, Bueno A, Marti-Cruchaga P, Zozaya G (2012) Extracorporeal tourniquet method for intermittent hepatic pedicle clamping during laparoscopic liver surgery: an easy, cheap, and effective technique. Langenbecks Arch Surg 397(3):481–485.

Left Lateral Sectionectomy: Laparoscopic Approach

35

Luca Aldrighetti, Francesca Ratti, Federica Cipriani,
Michele Paganelli and Gianfranco Ferla

35.1 Introduction

Interest in laparoscopic liver surgery (LLS) has increased since first being described [1], thanks to the possibility of reducing postoperative pain and disability, shortening hospital stay and time required for functional recovery, the growing experience in both laparoscopy and hepatic surgery, continual technological progress, and patient awareness about the benefits of this approach [2, 3]. Many series are available in the literature regarding laparoscopic resections of any liver segment, and even major hepatectomies are perfomed in selected cases [4]. Regardless, the laparoscopic approach is at present suggested as the gold standard only for left lateral sectionectomy [5] (according to Brisbane classification [6]) or left lobectomy, intending resection of segments 2 and 3 (S2, S3) (according to Couinaud). Experience gained in recent years has improved results of this type of surgery, even thanks in part to the learning-curve effect [7], and has led to continual research of minimal invasiveness through the Laparo-Endoscopic Single Site (LESS) surgery, which is now used also for liver surgery, especially for left lobectomy [8, 9].

35.2 Surgical Technique

For left lateral sectionectomy, the patient is placed in the French position, with the first surgeon standing between the patient's legs and one assistant on each side of the patient. A four-trocar configuration is used (one 5-mm trocar

L. Aldrighetti (✉)
Department of Surgery, Hepatobiliary Surgery Unit,
San Raffaele Hospital, Vita-Salute San Raffaele University,
Milan, Italy
e-mail: aldrighetti.luca@hsr.it

F. Calise and G. Casciola (eds.), *Minimally Invasive Surgery of the Liver,*
Updates in Surgery
DOI: 10.1007/978-88-470-2664-3_35, © Springer-Verlag Italia 2013

and three 12-mm trocars), positioned in a rhomboid configuration: one of these, inserted above the umbilicus, houses the 30° laparoscope. With an open laparoscopy technique, the first trocar is placed above the umbilicus, and continuous carbon dioxide (CO_2) pneumoperitoneum is induced to a pressure of 12 mmHg.

The liver is examined by direct vision, and intraoperative ultrasonography (US) is preliminarily performed to confirm lesion nature, number, and size and define their relationship with intrahepatic vascular structures. The line of intended resection is outlined with electrocautery marks on the left side of the falciform ligament. The transection plane is conducted from the convex surface (starting at the left side of the falciform ligament) to the ligamentum venosum and from the free margin of the liver to the diaphragm. Liver mobilization by sectioning the round, falciform, and triangular ligaments is never performed at the beginning of the intervention, usually conducted in open surgery, because ligaments suspend the liver without the need for other instruments and allow easy wide opening of the liver during the transection phase. Several techniques and technologies are available to transect the parenchyma and are chosen according to the preference of each surgeon, as no clear advantage of one technique over the others has been definitively demonstrated so far. In our center, a system integrating two surgical instruments – the ultrasonic coagulating cutter and the conventional ultrasonic dissector – is used: both tools are activated by ultrasonic energy produced by a single generator. However, the two hand pieces are independent and cannot be used at the same time. Ultrasonic dissection leaves arteries, veins, and bile ducts intact, which can be sealed and divided using the ultrasonic coagulator for minor structures or clips for larger ones. Bipolar coagulation with tiny-tip forceps is also extensively used to seal small blood vessels and treat the raw surface of the liver during transection. The intraparenchymal vascular anatomy is easily defined using the ultrasonic dissector, so that a decision pertaining to hemostatic technique can be based on vessel size. Ultrasonic dissection is highly effective in normal liver, as the presence of the soft and well-hydrated parenchyma enhances the effects of cavitation, dissolution, and aspiration of liver tissue, leaving intact collagen-rich structures (blood vessels and bile ducts). In cirrhotic livers, the prevalence of fibrotic tissue significantly lowers the efficacy of US dissection, so that the crushing technique with fine forceps is preferentially used or direct sealing and division of liver parenchyma is performed by energy-activated devices (i.e., radiofrequency, US). Parenchymal transection by direct application of several cartridges of mechanical staplers may also be used in cirrhotic livers. Portobiliary pedicles for the left lobe (S2 and 3) are reached through a transparenchymal approach, avoiding primary extrahepatic isolation of vessels. The glissonian plane, including the portobiliary pedicles for S2 and 3, is dissected free and then transected under direct vision with an articulated linear stapler charged with vascular cartridges. Two cartridges are usually needed to complete glissonian plane transection. Only in selected cases it is possible to separately dissect and identify the single portobiliary pedicles for S2 and 3, and caution is advised not to exceed in this attempt, as it may be challenging to repair any vascular or biliary lesion during this maneuver, thus leading to conversion to an open surgery. Caution must also be taken not to damage the hepatic vein with the edge of the stapler at the time of glissonian pedicle transection.

The left hepatic vein crosses the transection plane at the edge of the dihedral angle created between the two raw liver surfaces and stands on a more superficial plane than the glissonian pedicles. Its position should be checked by direct vision before applying the linear stapler to transect portobiliary pedicles. After glissonian pedicles are divided, alternating use of ultrasonic dissection and coagulating cutter or bipolar forceps is continued until the left hepatic vein is well recognized and exposed. An extreme dissection of the hepatic vein may lead to damage, so its preparation should be barely enough to place the linear stapler, avoiding strictures at the confluence with the middle hepatic vein. Transection of the left hepatic vein usually completes the hepatic resection.

Left triangular and coronary ligaments are finally divided. The resected specimen is then placed in a retrieval bag and removed, without fragmentation, through the umbilical port incision, extending the incision for larger specimens or, in selected cases, through a Pfannenstiel incision. A single, flat Jackson-Pratt drain is optionally placed in the posterior aspect of the resection bed through a port site. The Pringle maneuver is not routinely performed in laparoscopic left lobectomy [10, 11].

The LESS approach implies using an access device consisting of a plastic disk connected to a plastic ring, which is inserted into the abdomen by a plastic sheath. An introducer device allows single-port introduction. The outer surface of the device has three or four ports, with gel-capped access sites to maintain pneumoperitoneum, and an insufflation port. An additional 12-mm traditional port may be placed to allow dissector and linear stapler introduction just at the end of parenchymal transection.

35.3 Benefits of Laparoscopy

Since first being described, the laparoscopic approach to left lateral sectionectomy has been gaining increasing popularity, especially when compared with laparoscopic resections for liver lesions located in the right lobe [2]. Currently, laparoscopy is suggested as the gold standard approach, even though it requires good experience both in laparoscopic and liver surgery [5, 12]. Its benefits are well recognized: in general, faster postoperative functional recovery, less pain and therefore fewer analgesics requirements, and comparable oncological outcomes and incidence of postoperative complications, as demonstrated by several studies in the literature comparing results in open (OLS) and LLS [13, 14]. Some studies also report reduced blood loss and need for transfusions in patients undergoing laparoscopic left lateral sectionectomy. The advantages of laparoscopic versus open approach for left lateral sectionectomy have been confirmed by systematic reviews [2, 15]. Literature data of studies comparing laparoscopic and open left lateral sectionectomy are summarized in Table 35.1.

Currently, the open approach for left lateral sectionectomy may be reserved to patients who require synchronous resection of other lesions located in "non-laparoscopic" segments or to patients with major adhesions resulting from previous abdominal surgery. Previous surgery must not be considered an absolute exclusion criterion for the laparoscopic approach, as patients with prior colorectal surgery, cholecystectomy, or even liver surgery may be suitable for laparoscopic left lateral sectionectomy. The possibility of performing laparoscopic left lateral sectionectomy, however, must not influence – or, even worse, create the indication for – liver resection, so the pool of candidates for surgery must not be expanded simply because of the widespread application of minimally invasive techniques, even if some particular patient subsets may receive the finest benefits from this approach.

In cirrhotic patients undergoing resection for hepatocellular carcinoma (HCC), laparoscopy allows preservation of wall portosystemic shunts and the round ligament, and perioperative increases in portal pressure with the subsequent risks of bleeding and ascites may be reduced. Indeed, a recent meta-analysis [16] comparing results of laparoscopic and open approaches for surgical resection of HCC showed a significant advantage of laparoscopy in terms of postoperative complications as a consequence of a lower incidence of ascites.

Medium- and long-term outcomes are comparable with open surgery in terms of disease-free and overall survival, as demonstrated by previous work from our group [17]. The potential advantage of laparoscopic liver resection as a "bridge treatment" before orthotopic liver transplantation has to be validated yet; however, such an advantage may lie in fewer adhesions, with no negative impact on the transplantation procedure [18].

Patients affected by liver metastases from colorectal cancer (colorectal liver metastases – CLM) often require repeated liver resections, as recurrence after the first resection is frequent (30% of patients present isolated liver recurrence). Second, third, and even forth liver resections for CLM are actually reported in the literature [19]. The laparoscopic approach may lower the incidence and severity of postoperative adherence syndrome so that in patients with CLM who need reresection for recurrent disease, easier access to the liver area may be obtained, without limiting even the possibility of a repeat laparoscopic resection, if technically indicated.

Not least, the aesthetic result, traditionally considered a secondary outcome of oncological surgery, may need to be re-evaluated under the perspective of quality of life impact in the setting of modern oncology and in specific subgroups of patients, including left-lobe living donors [18] and resections for benign disease (adenomas, cystadenomas, hemangiomas, focal nodular hyperplasia) [20], even if satisfaction regarding the cosmetic outcome is not yet specifically investigated in literature series.

Table 35.1 Comparative studies between laparoscopic and open left hepatic lobectomy

	Abu Hilal et al. (2008) [21]			Aldrighetti et al. (2008) [13]			Lesurtel et al. (2003) [22]			Tang et al. (2005) [23]			Soubrane et al. (2006) [24]		
	LLLS	OLLS	*P* value	LLLS	OLLS	*P* value	LLLS	OLLS	*P* value	LLLS	OLLS	*P* value	LLLS	OLLS	*P* value
Participants	24	20	-	20	20	-	18	20	-	10	7	-	16	14	-
Sex(M:F)	7:17	10:10	NS	12:8	13:7	NS	7:11	5:15	NS	2:8	3:4	NS	10:6	9:5	0.07
Medianage, years	65	65	NS	63	66	NS	55	47	NS	52	57	NS	29	32	0.10
Indications			0.403472	Not assessed		-			NS	Not assessed		-	No assessed		-
CRM	14	11					1	1							
HCC	1	2					4	4							
FNH	2	4					3	3							
Others	4	7					10	10							
Median operative time, mins	180	155	NS	260	220	NS	202	145	<0.01	232	150	0.007	320	244	<0.005
Morbidity, %	13	25	NS	10	25	NS	11	15	NS	20	57	NS	19	36	NS
Median blood loss, ml	80	470	0.002	165	214	0.001	236	429	<0.05	350	400	NS	19	200	<0.005
Median hospital stay, days	3,5	7	<0.001	4.50	5.80	0.003	8	10	NS	8	14	0.019	7.5	8.1	NS
Median resection margin, mm	12	11	NS	Not assessed		-	Not assessed		-	Not assessed		-	Not assessed		-

LLLS, laparoscopic left lateral sectionectomy; *OLLS*, open left lateral sectionectomy.

The LESS approach represents an attempt to further minimize surgical invasiveness. The typical triangulation of laparoscopy, facilitated by traditional instruments and the laparoscope, is lacking, and the in-line vision gives the operator the impression of impaired vision. However, the main difficulty encountered by the surgeon is the conflict between hands outside the abdomen and among instruments into the abdomen. The LESS approach has already been described for many surgical procedures, such as cholecystectomy, colon resection, nephrectomy, and recently also for liver resection [8].

The actual benefits on clinical outcome due to a smaller incision are not yet clear. Significant differences cannot be expected to emerge as strong as those when comparing open and laparoscopic procedures, as the biological scenario is almost the same. Nevertheless, recent work by our group shows that the LESS approach in liver resection is feasible, safe, and at least not inferior to the standard laparoscopic resection in terms of complications and clinical outcome [9].

35.4 Conclusion

Laparoscopic approach for left lateral sectionectomy has gained wide acceptance, thanks to its safety and efficacy: many series from the literature are available and demonstrate significant advantages in terms of short term outcome when compared to open access, without impairing long term results in the treatment of both benign and malignant liver lesions. Laparoscopic approach is now considered a gold standard for resection of S2 and S3, even though requiring a good expertise in both laparoscopic and hepatic surgery to complete the learning curve.

References

1. Azagra JS, Goergen M, Jacobs C, Crema E (1994) Laparoscopic left lateral segmentectomy (left hepatic lobectomy). Endosurgery 1:7–9
2. Pulitanò C, Aldrighetti L (2008) The current role of laparoscopic liver resection for the treatment of liver tumors. Nat Clin Pract Gastroenterol Hepatol 5:648–654
3. Reddy SK, Tsung A, Geller A (2011) Laparoscopic liver resection. World J Surg 35:1478–1486
4. Buell JF, Cherqui D, Geller DA, O'Rourke N et al; World Consensus Conference on Laparoscopic Surgery (2009) The international position on laparoscopic liver surgery: The Louisville statement, 2008. Ann Surg 250(5):825–830
5. Azagra JS, Goergen M, Brondello S, Calmes MO, Philippe P, Schmitz B (2009) Laparoscopic liver sectionectomy 2 and 3 (LLS 2 and 3): towards the "gold standard". J Hepatobiliary Pancreat Surg 16:422–426
6. Terminology Committee of the International Hepato-Pancreato-Biliary Association (2000) The IHPBA Brisbane 2000 terminology of liver anatomy and resections. HPB (Oxford) 2:333–339
7. Viganò L, Laurent A, Tayar C, Tomatis M, Ponti A, Cherqui D (2009) The learning curve in laparoscopic liver resection. Improved feasibility and reproducibility. Ann Surg 250:772–782
8. Aldrighetti L, Guzzetti E, Ferla G (2011) Laparoscopic hepatic left lateral sectionectomy using the LaparoEndoscopic Single Site approach: evolution of minimally invasive liver surgery. J Hepatobiliary Pancreat Sci 18:103–105

9. Aldrighetti L, Ratti F, Catena M, Pulitanò C, Cipriani F, Ferla G (2012) Laparoendoscopic single site (LESS) surgery for left-lateral hepatic sectionectomy as an alternative to traditional laparoscopy: case-matched analysis from a single center. Surg Endosc 26(7):2016–2022
10. Pulitanò C, Catena M, Arru M, Guzzetti E, Comotti L, Ferla G, Aldrighetti L (2008) Laparoscopic liver resection without portal clamping: a prospective evaluation. Surg Endosc 22:2196–2200
11. Aldrighetti L, Pulitanò C, Arru M, Catena M, Guzzetti E, Casati M, Ferla G (2008) Ultrasonic-mediated laparoscopic liver transection. Am J Surg 195:270–272
12. Chang S, Laurent A, Tayar C, Karoui M, Cherqui D (2007) Laparoscopy as a routine approach for left lateral sectionectomy. Brit J Surg 94:58–63
13. Aldrighetti L, Pulitanò C, Catena M, Arru M, Guzzetti E, Casati M, Comotti L, Ferla G (2008) A prospective evaluation of laparoscopic versus open left lateral hepatic sectionectomy. J Gastrointest Surg 12:457–462
14. Zhou YM, Shao WY, Zhao YF, Xu DH, Li B (2011) Meta-analysis of laparoscopic versus open resection for hepatocellular carcinoma. Dig Dis Sci 56:1937–1943
15. Viganò L, Tayar C, Laurent A, Cherqui D (2009) Laparoscopic liver resection: a systematic review. J Hepatobiliary Pancreat Surg 16(4):410–421
16. Aldrighetti L, Guzzetti E, Pulitanò C, Cipriani F, Catena M, Paganelli M, Ferla G (2010) Case-matched analysis of totally laparoscopic versus open liver resection for HCC: short and middle term results. J Surg Oncol 102:82–86
17. Zhou YM, Shao WY, Zhao YF, Xu DH, Li B (2011) Meta-analysis of laparoscopic versus open resection for hepatocellular carcinoma. Dig Dis Sci 56:1937–1943
18. Troisi R, Van Huysse J, Berrevoet F, Vandenbossche B, Sainz-Barriga M, Vinci A, Ricciardi S, Bocchetti T, Rogiers X, de Hemptinne B (2011) Evolution of laparoscopic left lateral sectionectomy without the Pringle manuevre: through resection of benign and malignant tumors to living liver donation. Surg Endosc 25:79–87
19. Adam R, Pascal G, Azoulay D, Tanaka K, Castaign D, Bismuth H (2003) Liver resection for colorectal metastases: the third hepatectomy. Ann Surg 238(6):871–883
20. Ardito F, Tayar C, Laurent A, Karoui M, Loriau J, Cherqui D (2007) Laparoscopic liver resection for benign disease. Arch Surg 142(12):1188–1193
21. Abu Hilal M, Mc Phail MJ, Zeidan B et al (2008) Laparoscopic versus open left lateral hepatic sectionectomy: A comparative study. Eur J Surg Oncol 34:1285-8
22. Lesurtel M, Cherqui D, Laurent A, Tayar C, Fagniez PL (2003) Laparoscopic versus open left lateral hepatic lobectomy: a case-control study. J Am Coll Surg 196:236-42
23. Tang CN, Tai CK, Ha JP, Siu WT, Tsui KK, Li MK (2005) Laparoscopy versus open left lateral segmentectomy for recurrent pyogenic cholangitis. Surg Endosc 19:1232-6
24. Soubrane O, Cherqui D, Scatton O et al (2006) Laparoscopic left lateral sectionectomy in living donors: safety and reproducibility of the technique in a single center. Ann Surg 244:815-20

Right Hepatectomy: Laparoscopic Approach

36

Michael D. Kluger and Daniel Cherqui

36.1 Introduction

Laparoscopic liver surgery (LLS) has become increasingly common [1, 2]. A recent international position paper maintains that LLS is a safe and effective approach to the management of surgical liver disease in selected patients in the hands of trained surgeons [3]. However, most procedures are limited resections, and only 9% of nearly 3,000 cases reported in the international literature are right hepatectomies [4], as it remains a challenging procedure.

After a brief discussion of patient selection and necessary devices, we herein describe the technical aspects of the laparoscopic anterior approach for right hepatectomy in four stages: (1) patient positioning, port placement, peritoneal cavity and liver inspection, and ultrasonography; (2) approach, pedicle control, anterior liver mobilization, hilar-plate dissection; (3) marking the liver, parenchymal dissection, hemostasis, and bile-duct ligation; (4) specimen extraction, cholangiography, and drainage.

36.2 Patient Selection and Evaluation

Lesion location and, to a lesser extent, size are the most important determinants of when laparoscopic resection is appropriate. In the case of right hepatectomy, we recommend lesions that are not intact with the hilum, main hepatic veins, or the inferior vena cava. We consider large tumors (i.e. > 8 cm) relative contraindications to a laparoscopic approach. Formal contraindications

D. Cherqui (✉)
Section of Hepatobiliary Surgery and Liver Transplantation
New York Presbyterian Hospital – Weill Cornell Medical Center,
New York, USA
e-mail: dac7027@med.cornell.edu

F. Calise and G. Casciola (eds.), *Minimally Invasive Surgery of the Liver,*
Updates in Surgery
DOI: 10.1007/978-88-470-2664-3_36,

remain gallbladder cancer and hilar cholangiocarcinoma, the need for complete vascular occlusion, and whether oncologic principles could be better served via laparotomy. High-quality imaging with vascular reconstruction is necessary to understand the patient's intrahepatic arterial, portal, and biliary anatomy.

36.3 Essential Operating Room Equipment

36.3.1 The Laparoscopic Operating Room

- All laparoscopic equipment must be state of the art and in good working order
- Adjustable, remote-controlled electric split-leg table
- Monitors placed lateral to each shoulder of the patient, with a hanging monitor above the patient's head
- One to two carbon dioxide insufflators maintaining a pneumoperitoneum of 12 mmHg
- Ultrasound with B and D modes and a high-frequency laparoscopic transducer
- Energy vessel-sealing device
- Ultrasonic dissector device
- Set of readily available conventional open instruments.

36.3.2 Necessary Laparoscopic Instruments

- A 10-mm 30° laparoscope
- Atraumatic bowel graspers
- Curved and right-angle dissectors
- Scissors
- Needle drivers
- Liver retractor
- Bipolar diathermy forceps
- Monopolar diathermy hook
- Articulated vascular linear stapler (30-mm and 45-mm vascular cartridges)
- Suction irrigator
- Umbilical tape and a 4- to 6-cm 16-F rubber tube to serve as a tourniquet
- Titanium clip applier, small and medium
- Plastic locking clip applier, medium and large
- Endoscopic bag
- 5-0 monofilament sutures cut to 20–30 cm.

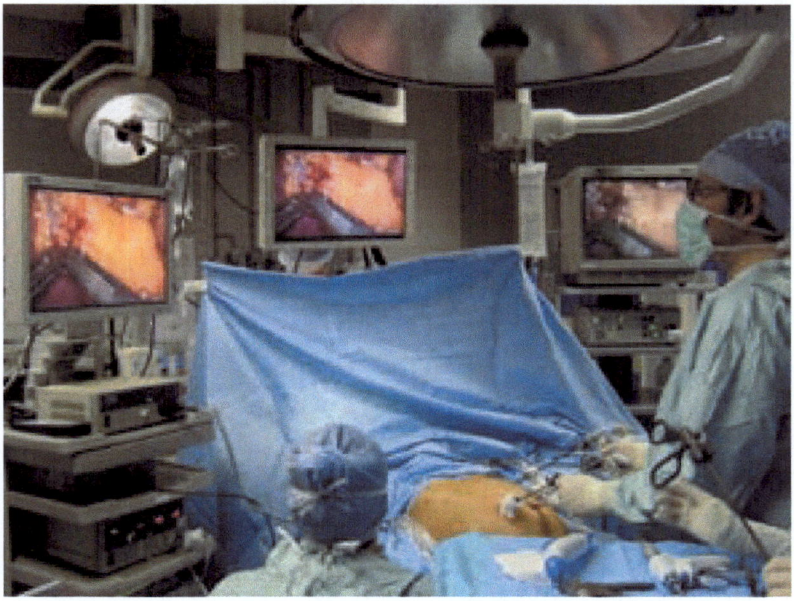

Fig. 36.1 View of the recommended operating room setup from the patient's right foot

36.4 The Procedure

36.4.1 Stage 1

36.4.1.1 Patient Positioning

The patient is placed in the supine position with the lower limbs apart on a split-leg table. The right arm is padded and tucked at the side. The surgeon stands between the patient's legs, with an assistant seated at each side. The scrub nurse and instruments are positioned lateral to the patient's right leg (Fig. 36.1). A Mayo stand positioned over the patient's right leg holds the most commonly utilized instruments. Reverse Trendelenburg position is used, tilting the table laterally as necessary to take advantage of gravity and the weight of the liver to improve exposure.

36.4.1.2 Port Placement

Five trocars are placed (Fig. 36.2):

- 1 × 12-mm port placed supraumbilically for the camera
- 2 × 12-mm paramedian ports are the working ports
- 2 × 5-mm lateral ports to be used for retraction.

The open technique is used to insert the camera port, and the remaining four ports are placed under direct vision generally following the profile of the liver in a curve from right to left.

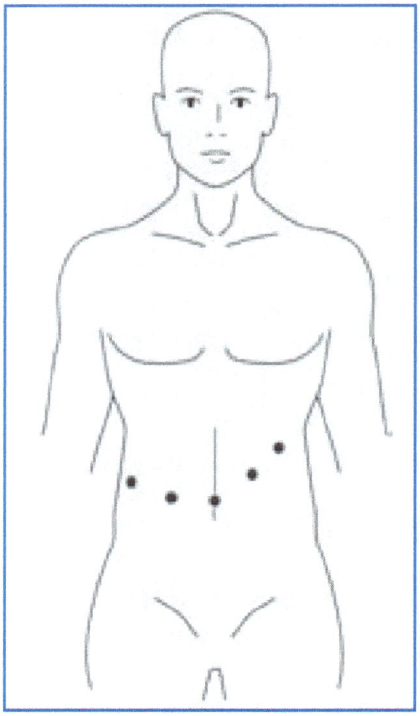

Fig. 36.2 Trocar placement for laparoscopic right hepatectomy

36.4.1.3 Inspection of the Peritoneal Cavity and Liver, and Ultrasonography

A thorough inspection of the peritoneal cavity for gross pathology is performed. Laparoscopic ultrasonography is performed to confirm lesion location and assess vascular anatomy.

36.4.2 Stage 2

36.4.2.1 Approach

A laparoscopic right hepatectomy can be performed from an anterior approach (i.e., transection without posterior mobilization of the right liver) or the conventional approach. The latter entails mobilization of the right liver, dissection off the inferior vena cava, and extrahepatic control of the right hepatic vein prior to transection; the use of a hand port may be useful in this setting. The hilar dissection is the same for both approaches. The pure laparoscopic anterior approach is presented here.

36.4.2.2 Anterior Liver Mobilization

The round ligament is divided using a vessel-sealing device. The falciform ligament is transected close to the abdominal wall along its length to the confluence of the hepatic veins and inferior vena cava. This confluence is then carefully dissected in order to identify the insertion of the right hepatic vein and common trunk of the middle and left hepatic veins.

36.4.2.3 Pedicle Control

A tourniquet can be constructed in case a Pringle maneuver is necessitated. A blunt bowel grasper is gently passed through the most lateral right 5-mm port under the hepatoduodenal ligament and through the pars flaccida. An umbilical tape is then handed to the bowel grasper – encircling the hepatoduodenal ligament – and extracorporeally inserted through a 16-F rubber tube and returned to the abdomen. To clamp, a laparoscopic instrument grasps the ends of the tape, pulling it taught, while a laparoscopic locking-clip applier pushes the tubing toward the pedicle and applies a clip. We do not routinely use pedicle clamping for right hepatectomies, as the specimen is devascularized prior to transection.

36.4.2.4 Hilar-Plate Dissection

The gallbladder and distal cystic duct can be used as "handles" to improve exposure of the transection plane after division of the cystic duct and cystic artery. The gallbladder is retracted to the patient's right, and the distal cystic duct is retracted to the patient's left, exposing the short course of the extrahepatic right pedicle. The right hepatic artery is circumferentially dissected for a distance of 1–2 cm and encircled with a tape to retract it away and protect the other hilar structures. After visualization of the contralateral arterial branch, the right hepatic artery is divided between plastic locking clips. If present, a replaced right hepatic artery is ligated and divided.

Right hepatic artery division greatly improves visualization of the right portal vein. It will often be necessary to ligate a small vein feeding segment 1 (S1) from the right portal vein with an electrosurgical device to safely dissect the right portal branch. This is now dissected circumferentially along its short extrahepatic course and a tape passed around it. The bifurcation and left portal branch must be clearly visualized. When transecting the right portal vein with a linear stapler, the tape should be pulled to the patient's left so as to displace the bifurcation to the left and lengthen the right branch. When possible, intraparenchymal dissection to the first-order branches of the hepatic artery and portal veins limits the risk of damaging the hilar structures.

36.4.3 Stage 3

36.4.3.1 Marking the Liver

Based on vascular demarcation, the transection plane is outlined on the liver capsule with monopolar diathermy.

36.4.3.2 Parenchymal Dissection, Hemostasis, and Bile-Duct Ligation

For the peripheral transection (i.e., first 2 cm in depth), the Harmonic scalpel or LigaSure are efficient instruments. Deeper in the liver parenchyma, larger vascular structures are encountered. These can be damaged by blind dissection, especially fragile hepatic venous branches, and we recommend using an ultrasonic dissector to identify the structures.

Vascular and biliary structures <3 mm are ligated and transected using bipolar diathermy or vessel-sealing devices. Larger structures are ligated using plastic locking clips. Laparoscopic staplers with 2.5 mm depth vascular loads are used for the glissonian pedicle and right hepatic vein.

Parenchymal transection is started at the inferior edge of the liver in the anterior-to-posterior and caudad-to-cephalad directions along the demarcation line. As progress is made cranially, the proximal hepatic veins draining S5 and S8 toward the middle hepatic vein are exposed, clipped, and divided.

Hilar-plate dissection and division of the right intraparenchymal bile ducts allows the transection plane to be opened widely for progression of parenchymal transection. At the hilar plate, the right bile duct is circumferentially dissected until the anterior and posterior branches are visualized, at which point the duct can be divided with locking clips or the stapler. It will be necessary to next divide the connective tissue junction between the right lobe and S1 with diathermy to fully expose the retrohepatic cava.

As parenchymal transection progresses posteriorly and cephalad, the posterior capsule of the liver is divided with sharp dissection and monopolar diathermy along the anterior surface of the cava while systematically clipping and dividing the small bridging veins.

The right hepatic vein is identified at its insertion into the vena cava and circumferentially dissected. An umbilical tape is placed around the vessel and retracted to the right in order to elongate the vein prior to transection with an articulating linear stapler. This prevents impingement on the inferior vena cava.

The specimen is now retracted to the right as the hepatocaval ligament is transected with multiple clips or a single firing of the linear stapler. Finally, the attachments of the specimen to the diaphragm are freed with electrosurgical instruments.

36.4.4 Stage 4

36.4.4.1 Specimen Extraction

In the virgin abdomen, specimens are removed through a Pfannenstiel incision. Alternatively, specimens can be removed by opening pre-existing incisions. Regardless of the chosen incision, the abdominal wall is incised to the fascia, and a 15-mm trocar is inserted at the center of the incision under direct laparoscopic vision. A large-capacity endoscopic bag is introduced and the specimen retained. The specimen is then retracted against the fascia, and the fascia and peritoneum are opened only as much as necessary for retrieval. The fascial layers are reapproximated, and pneumoperitoneum is reintroduced. The operative site is lavaged and examined for hemostasis and biliary-tract integrity.

36.4.4.2 Cholangiography, Drainage, and Closure

Intraoperative cholangiography is not routinely performed; however, a cholangiogram catheter can be inserted through the cystic duct for methylene blue injection or radiographic cholangiography, if necessary. Because of the large dead space created, a 10-F closed-circuit suction drain is placed in the subphrenic space. The falciform ligament is reattached to the diaphragm with monofilament sutures to avoid torsion of the left lobe. The skin and port site incisions are closed with absorbable subcuticular sutures.

References

1. Cherqui D, Husson E, Hammoud R et al (2000) Laparoscopic liver resections: a feasibility study in 30 patients. Ann Surg 232(6):753–762
2. Bryant R, Laurent A, Tayar C, Cherqui D (2009) Laparoscopic liver resection-understanding its role in current practice: the Henri Mondor Hospital experience. Ann Surg 250(1):103–111
3. Buell JF, Cherqui D, Geller DA et al (2008) The international position on laparoscopic liver surgery: The Louisville Statement. Ann Surg 250(5):825–830
4. Nguyen KT, Gamblin TC, Geller DA (2009) World review of laparoscopic liver resection-2,804 patients. Ann Surg 250(5):831–841

Right Hepatectomy: Robot-Assisted Approach

37

David Calatayud and Pier Cristoforo Giulianotti

37.1 Introduction

Liver surgery has advanced more than any other type of surgery since the early 1980s. This development is partly due to advances in technology and surgical instrumentation, as well as in anesthetic management of patients. In fact, indications for extended liver surgery interventions being performed today were thought of as unthinkable practice just a few decades ago.

In 1957, Couinaud [1] described the functional anatomy of the liver based on the distribution of portal pedicles and hepatic veins. This distribution of liver segments is still vital when planning liver surgery using the classification of Brisbane [2]. The open approach continues as the predominant one in liver surgery, despite great enthusiasm following the development and growth of laparoscopy since the early 1990s. In fact, in some high-volume centers with skilled surgeons, laparoscopy is now the method of choice when performing left lateral sectionectomy [resection of segments 2 and 3 (S2 and 3)], or when lesions are located in anterior segments [3–7]. However, major resections (removal of three or more segments) are typically performed using the open approach.

In our experience, the robot-assisted approach is distinctly different from laparoscopy for this type of surgery. Following a single-surgeon experience of 47 major resections, 31 of which were right hepatectomies (Table 37.1) [8, 9] we found major liver resection using the robot to be not only feasible but also advantageous, creating the ideal gateway for minimally invasive surgery of the liver (MISL).

D. Calatayud (✉)
Division of General, Minimally Invasive, and Robotic Surgery,
Department of Surgery, University of Illinois at Chicago,
Chicago, IL, USA
e-mail: dcalatay@clinic.ub.es

F. Calise and G. Casciola (eds.), *Minimally Invasive Surgery of the Liver*,
Updates in Surgery
DOI: 10.1007/978-88-470-2664-3_37, © Springer-Verlag Italia 2013

Table 37.1 Major liver resections performed using the robotic approach (single-surgeon experience)

Procedure	No.
Right hepatectomy	31
Left hepatectomy	5
Extended right hepatectomy	4
Trisegmentectomy or multiple (>3) liver segment resection	4

Of all hepatic resections performed, right hepatectomy is the most commonly performed worldwide, and its technique is well standardized. In this chapter, we describe the surgical steps of this technique using the robot-assisted approach.

37.2 Surgical Technique

37.2.1 Patient and Trocar Positions

The patient is placed in a mild reverse-Trendelenburg position, with left lateral decubitus. Pneumoperitoneum is made using a Veress needle in the left upper quadrant, maintaining a pressure of 15 mmHg. Four trocars are placed for the robot: one in the right pararectal position for the camera, two 8-mm trocars in the right flank and left pararectal, respectively, and another 8-mm trocar for the fourth arm in the left flank. Two laparoscopic trocars are also added, to be used by the assistant. These are placed in the periumbilical location and between the camera trocar and the left-arm trocar (Fig. 37.1).

37.2.2 Intraoperative Ultrasound

Intraoperative ultrasound (US) is performed routinely, which allows delineation of the lesion to be resected, its relationship to vascular structures to be evaluated, and additional lesions to be detected. The entire liver should be explored, always using the same steps (see Chap. 13).

37.2.3 Hepatic Hilum Dissection

After performing a cholecystectomy, the hepatic hilum is dissected using the monopolar hook with the right arm of the robot and the bipolar forceps with the left. The first element to dissect is the right hepatic artery, sectioning between ligatures. Once this is complete, the right portal vein is dissected, with special attention given to the additional branches of varying size (from small/medium

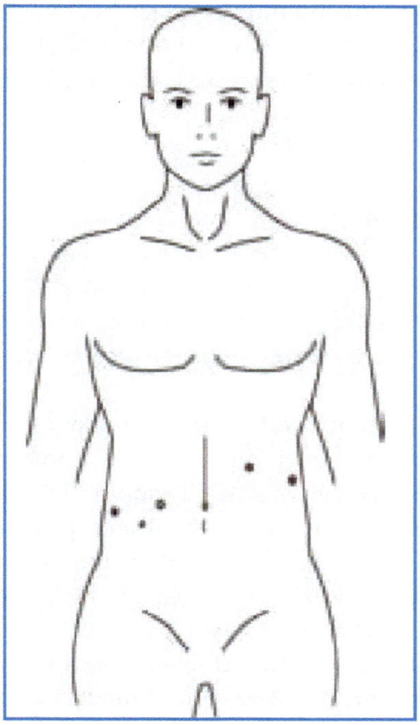

Fig. 37.1 Trocar positioning. *Green* robotic camera; *red* robotic trocars for instruments; *yellow* laparoscopic trocars (assistant trocars)

branches to the caudate lobe up to a possible portal trifurcation). After sectioning the right portal vein between hemoclips, we prefer to place a security stitch using 5-0 polypropylene in the right portal vein proximal stump.

Biliary tree dissection and evaluation requires special attention because sectioning will be made either at this time or later during the procedure, depending upon the location of the biliary bifurcation (extrahepatic, more or less proximal). Modern visualization technology with indocyanine green clearance (IGC) fluorescence enables extrahepatic biliary tree evaluation without requiring any invasive procedure; this is due to the physical characteristics of IGC, which allow unique hepatic metabolism and biliary excretion.

37.2.4 Hepatocaval Dissection (Piggyback Technique)

Use of the fourth arm is instrumental in completing this step because it helps separate the liver to the top and left. We use gauze between the instrument and the liver to prevent injury. The first step is to section the right triangular ligament using the hook. The Harmonic scalpel can also be used if the ligament is thickened. Next, we section the right triangular ligament to the right hepatic vein (RHV) then dissect the right hepatic lobe from the retroperitoneum above the right kidney up to the vena cava. At this point, we dissect the space between the vena cava and liver (piggyback), sectioning between hemoclips on the venous branches. For safety reasons, we also apply 5-0 polypropylene stitches on the side of the vena cava. A hepatocaval dissection is then made up to the RHV drainage.

37.2.5 Hepatic Parenchymal Transection

Prior to liver transection, we prepare the hilum for a Pringle maneuver if it becomes necessary. We use tape and a silicone tube to introduce both ends of the tape and apply two clips. In our experience, the Pringle maneuver has been required in one case only.

Ischemic delineation of transection along Cantlie's line [10] is marked following sectioning of vascular vessels of the right hepatic lobe. The monopolar hook can be used to mark this line and open the capsule of Glisson, although it is not mandatory. Prior to starting the transection, we apply 4-0 polypropylene stitches on each side of the Cantlie line in the edge of the liver. Transection is performed using the HARMONIC scalpel, closing gradually as monopolar energy is applied. Transection is performed layer by layer, from superficial to deep, opening the liver as with a book. This is made possible by using the polypropylene stitches initially applied. When vascular branches that cross between the right and left lobes are encountered, we ligate them. It is preferable to apply the hemoclips before sectioning. However, if this is not possible, or if the veins are broken, bipolar cautery can be used in smaller vessels and 4-0 polypropylene stitches in larger vessels. This maneuver is possible due to the articulation ability and degrees of freedom of movement of the robotic instruments.

Upon reaching the area of the RHV, a mechanical stapler is used to section the vein and complete the transection. Occasionally, depending upon patient and local conditions, the endoscopic stapler is needed prior to reaching the RHV. Once the transection is complete, we check for hemostasis and bile leakage on the transected surface; hemostatic glue is applied, if needed. The surgical specimen is then extracted in an EndoBag through a Pfannenstiel minilaparotomy incision. Surgical hemostasis and trocar orifices are again evaluated (Fig. 37.2).

37.3 Discussion

The robot-assisted surgical procedure follows the identical steps as those for the standardized open approach. This may be part of the reason for the success of robot-assisted surgery, as it requires no change to the standard technique while offering to the hepatic surgeon advantages that are not found with the laparoscopic approach. It is important to note, however, that because using the robotic approach is not standardized for major liver surgery, it represents a

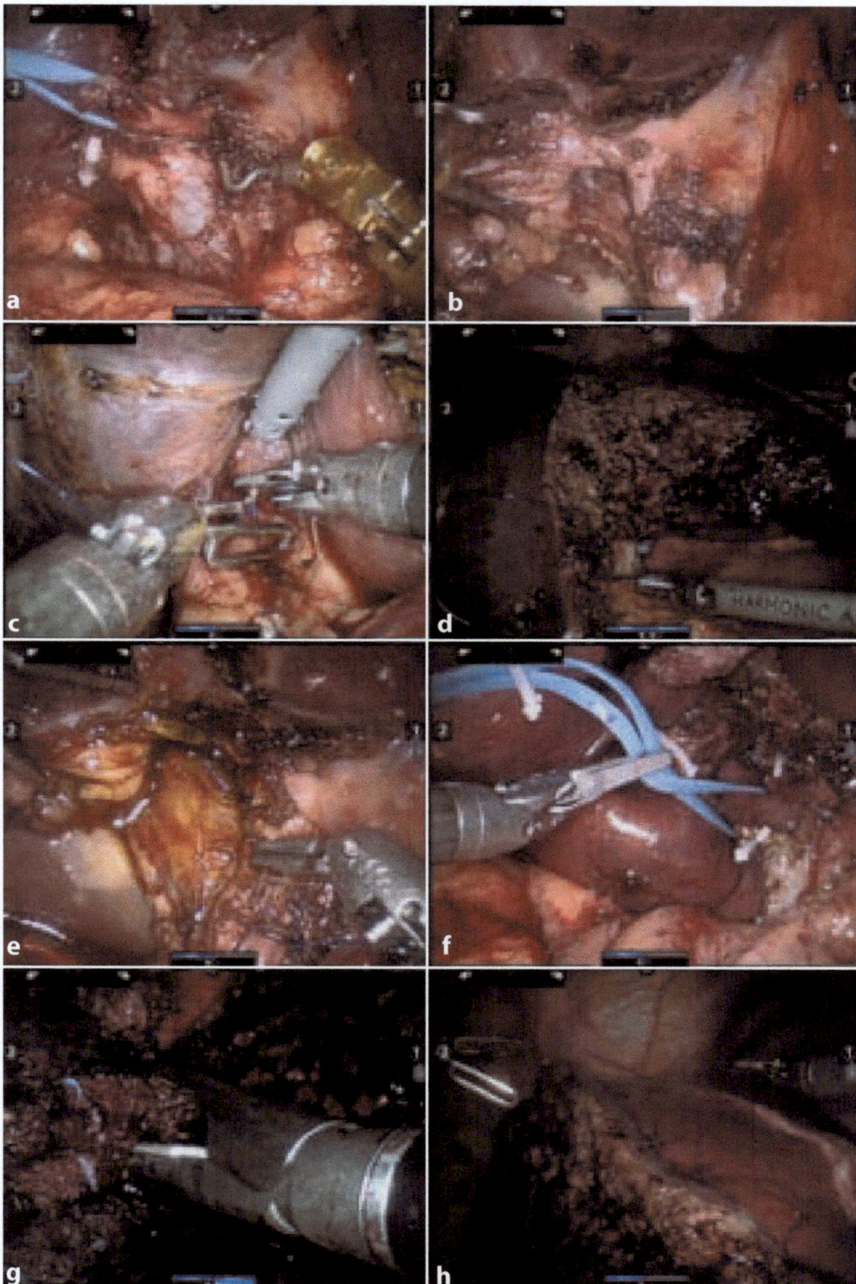

Fig. 37.2 Robot-assisted right hepatectomy. **a** Right hepatic artery dissection; **b** biliary bifurcation dissection; **c** right hepatic duct section; **d** right portal vein dissection; **e** hepatocaval dissection (piggyback); **f** liver parenchymal transection; **g** right hepatic vein section; **h** end of the surgery, remnant left hepatic lobe

major change in the field of MISL when compared with the laparoscopic approach. However, in the hands of skilled liver surgeons, this approach can help to significantly expand indications for liver surgery in a minimally invasive environment.

References

1. Couinaud C (1957) Le foie: etudes anatomiques et chirurgicales. Masson, Paris
2. Strasberg SM, Belghiti J, Clavien PA et al (2000) The Brisbane 2000 terminology of liver anatomy and resections. HPB Surg 2:333–339
3. Buell JF, Cherqui D, Geller DA et al (2009) The international position on laparoscopic liver surgery: The Louisville Statement, 2008. World Consensus Conference on Laparoscopic Surgery. Ann Surg 250(5):825–830
4. Shafaee Z, Kazaryan AM, Marvin MR et al (2011) Is laparoscopic repeat hepatectomy feasible? A tri-institutional analysis. J Am Coll Surg 212(2):171–179
5. Cherqui D, Belghiti J (2009) Hepatic surgery. What progress? What future? Gastroenterol Clin Biol 33(8-9):896–902
6. Nguyen KT, Gamblin TC, Geller DA (2009) World review of laparoscopic liver resection-2,804 patients. Ann Surg 250(5):831–841
7. Gayet B, Cavaliere D, Vibert E et al (2007) Totally laparoscopic right hepatectomy. Am J Surg 194(5):685–689
8. Giulianotti PC, Sbrana F, Coratti A et al (2011) Totally robotic right hepatectomy: surgical technique and outcomes. Arch Surg 146(7):844–850
9. Giulianotti PC, Coratti A, Sbrana F et al (2011) Robotic liver surgery: results for 70 resections. Surgery 149(1):29–39
10. Cantlie J (1897) On a new arrangement of the right and left lobes of the liver. Proceedings – Anatomical Society of Great Britain and Ireland 32:4–9

Left Hepatectomy: Laparoscopic Approach

38

Fulvio Calise, Piero Atelli, Antonio Giuliani, Vincenzo Scuderi and Aldo Rocca

38.1 Introduction

Since the first laparoscopic anatomical hepatectomy, reported by Azagra et al. in 1996 [1], major laparoscopic hepatectomies have become increasingly frequent: in a 2009 world review of almost 3,000 laparoscopic liver resections (LLR), 7 % were left hepatectomies [2]. Several studies show the safety and feasibility of major LLR in the hands of surgeons who are expert in both liver and laparoscopic surgery [3, 4]. The Louisville Conference in 2008 established that there is no formal contraindication to minimally invasive surgery of the liver (MISL), even for major hepatectomies, provided that established guidelines are attentively followed [5].

In this respect, we recommend not to operate lesions involving major vessels and, anyway, with a size over 7-8 cm. Currently there is a formal contraindication to operate Klatskin tumors.

The advantages of MISL are well known: reduced blood loss, morbidity, and hospital stay [6]. For the left liver in particular, the laparoscopic camera allows magnificent exposition of left branches of the artery and portal vein and on the left of the liver ligaments.

38.2 Left Liver Anatomy

The left liver comprises segments (S) 1, 2, 3, and 4 according to the Couinaud classification. These segments receive the arterial inflow from the left branch

F. Calise (✉)
Hepatobiliary Surgery and Liver Transplant Center, A. Cardarelli Hospital
Naples, Italy
e-mail: fulvio.calise@libero.it

F. Calise and G. Casciola (eds.), *Minimally Invasive Surgery of the Liver*, Updates in Surgery
DOI: 10.1007/978-88-470-2664-3_38, © Springer-Verlag Italia 2013

of the hepatic artery: in 8% of cases, a supplementary left artery arises from the gastric artery. The laparoscopic view allows easy identification of this anatomical variant. Venous inflow is assured by the left branch of the portal vein, providing tributaries on the left side of S1 before entering the liver. The left biliary duct is usually longer than the right one. Outflow is provided by the median (MHV) and left (LHV) hepatic veins, which generally merge before entering the vena cava.

38.3 Essential Operating Room Equipment

38.3.1 The Laparoscopic Operating Room

- An adjustable table
- Two monitors
- One or two carbon dioxide (CO_2) insufflators maintaining pneumoperitoneum of 12 mmHg
- Ultrasound with B and D modes, and a high-frequency laparoscopic transducer
- Energy vessel-sealing device
- Ultrasonic dissector device
- Readily available set of conventional open surgery instruments for possible rapid conversion.

38.3.2 Necessary Laparoscopic Instruments

- A 10- mm 30° laparoscope
- Atraumatic bowel graspers
- Curved and right-angle dissectors
- Scissors
- Needle drivers
- Liver retractor
- Satinsky forceps
- Bipolar diathermy forceps
- Monopolar diathermy hook
- Articulated vascular linear stapler (30-mm and 45-mm vascular cartridges)
- Suction irrigator
- Umbilical tape
- Titanium clip appliers, small and medium
- EndoBag
- 5-0 monofilament sutures cut to 20–30 cm

38.4 Surgical Technique

38.4.1 Patient Position

The patient is positioned in the supine decubitus with the legs apart. The surgeon stands between the legs with one assistant on either side of the patient. Two monitors are positioned, one on either side of the patient's head. The bed is positioned in reverse Trendelenburg with a slight right inclination, which allows better exposition of the liver.

38.4.2 Port Placement

Pneumoperitoneum is made with an open umbilical access. Generally, we use four trocars, the first for the 30° camera in the umbilical port and the others positioned under direct visualization, two of 10-12 mm in the right and left flank and one trocar of 4 mm in the epigastric region: The fifth trocar is optional; it may be required for the Pringle maneuver and is positioned on the right anterior axillary line, 2 cm above the umbilical line (Fig. 38.1)

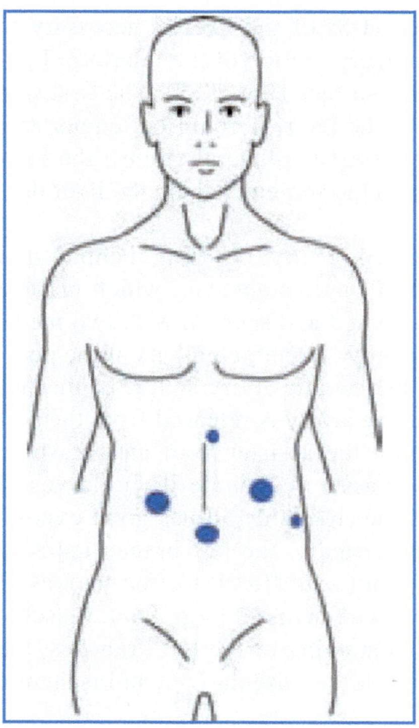

Fig. 38.1 Trocar positioning

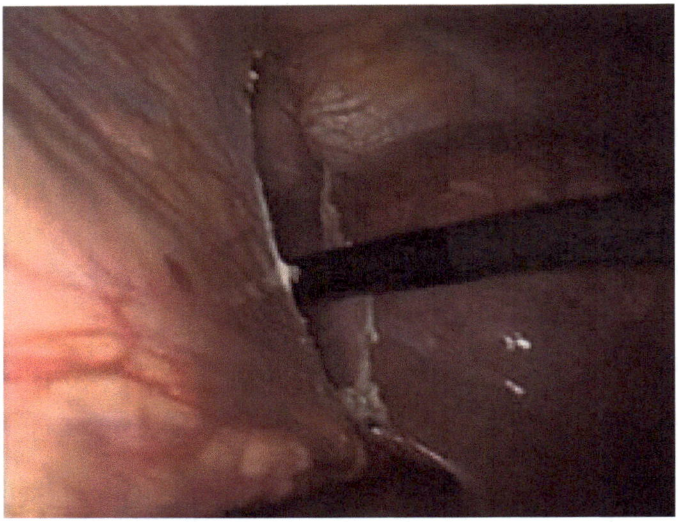

Fig. 38.2

38.4.3 Exploration and Pedicle Preparation

The procedure starts with complete exploration of the abdominal cavity and liver surface. An intraoperative ultrasound (US) of the liver is necessary to direct surgical strategy; it requires a preliminary section of the falciform ligament to allow exploration of the entire liver surface (Fig. 38.2). The first step is resection of the round ligament close to the liver. It could be temporarily saved, especially in cirrhotic patients, to preserve collateral circles; the ligament and the gallbladder will be used to help the surgeon handle the liver during parenchymal resection (Fig. 38.3).

The second step is to encircle the pedicle by inserting behind it a Roticulator Endo Grasp to prepare for the Pringle maneuver, which may be needed at any time. A vessel loop is then passed and secured with two medium clips (Fig. 38.3) We then proceed to identify the left elements of the pedicle (Fig. 38.4). First is the left branch of the hepatic artery and, if is present, left hepatic artery in the lesser omentum. The artery is released from the soft tissue of the hepatoduodenal ligament with the assistance of an ultrasonic dissector. Once the artery is released, we pass and encircle it with a vessel loop. A little traction on the loop toward the right side allows good exposition of the left branch of the portal vein. Usually, the left branch tends to have a longer extrahepatic course, so it is not so difficult to free the vessel from the connective soft tissue and wind it with a vessel loop. Each vessel is confirmed with US. Following temporary clamping of the left branches, the liver shows a clear demarcation line between the vascularized and ischemic

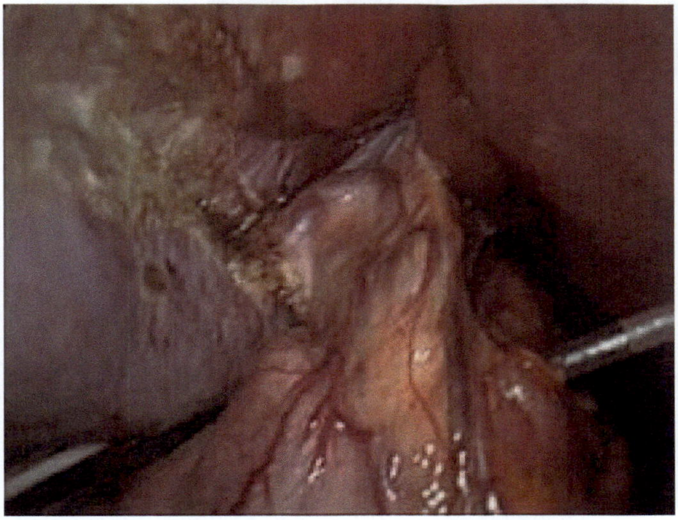

Fig. 38.3

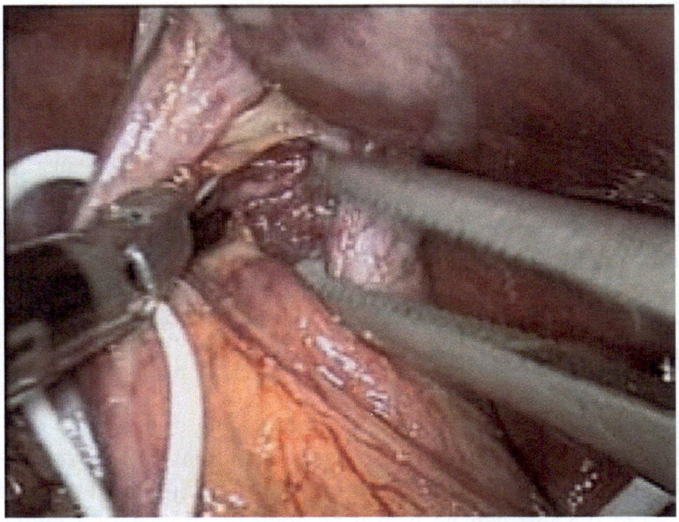

Fig. 38.4

segments. This prudent approach allows exclusion of any anatomical variant. The artery is then divided between two clips; the left branch of the portal vein is also divided between clips, or if large, by a linear vascular stapler.

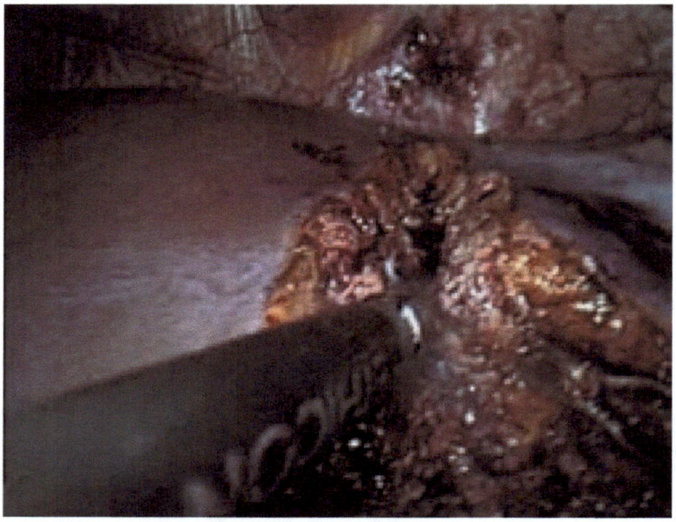

Fig. 38.5

38.5 Left Liver Mobilization

Left liver mobilization begins by dissecting the falciform ligament all the way up with the HARMONIC stapler to visualize merging of the MHV and LHV into the vena cava. Dissection continues along the left coronal and triangular ligaments. We do not suggest trying to isolate the common trunk because of the potential risk of massive and uncontrollable bleeding. Thanks to laparoscopic visualization, this dissection may be carried out easily. Liberation ends with the dissection of the lesser omentum.

38.6 Parenchymal Transection

After vascular demarcation, the section line is outlined using monopolar coagulation. We usually perform the transection under intermittent pedicle clamping, as in open surgery (15'+5'). Soft traction on the round ligament applied by the assistant toward the left can be useful to open the transection angle. Parenchymal tissue is crushed, and the biliary and vascular structures are skeletonized and coagulated by using the ultrasonic dissector if <3 mm. Larger vessels, such as venous branches tributaries of S4 are closed using a vascular linear stapler or clips and then cut by scissors (Fig. 38.5).

During declamping, we check the transection surface and coagulate all bleeding sources using a bipolar forceps continuously irrigated to improve hemostatic power. The left bile duct is then reached intraparenchymally and divided using two clips or, alternatively, a linear stapler. Finally, MHV and LHV are cut through using a linear 45/60-mm vascular stapler. Residual liver tissue is then sectioned using the ultrasonic dissector.

38.7 Specimen Extraction

The resected portion of the liver is placed in a high-capacity EndoBag and placed in the right flank. A Pfannenstiel incision is the better choice for retracting the specimen, particularly in the presence of previous open surgery. Alternatively, the umbilical incision, properly enlarged, may be used.

After reintroducing pneumoperitoneum, accurate hemostatic control and biliostasis is performed. Fibrin glue may be placed on resection surfaces to help preventing them from bleeding and biliary fistula in the postoperative course. A suction drain is positioned under the liver through the Winslow foramen. The umbilical port site is closed with separate stitches. Skin is then closed with nonabsorbable sutures.

References

1. Azagra JS, Goergen M, Gilbart E, Jacobs D (1996) Laparoscopic anatomical (hepatic) left lateral segmentectomy-technical aspects. Surg Endosc 10(7):758–761
2. Nguyen KT, Gamblin TC, Geller DA (2009) World review of laparoscopic liver resection-2,804 patients. Ann Surg 250(5):831–841
3. Dagher I, O'Rourke N, Geller DA et al (2009) Laparoscopic major hepatectomy: an evolution in standard of care. Ann Surg 250(5):856–860
4. Nguyen KT, Geller DA (2010) Laparoscopic liver resection-current update. Surg Clin North Am 90(4):749–760
5. Buell JF, Cherqui D, Geller DA et al (2009) The international position on laparoscopic liver surgery: The Louisville Statement, 2008. Ann Surg 250(5):825–830
6. Nguyen KT, Marsh JW, Tsung A, Steel JJ, Gamblin TC, Geller DA (2011) Comparative benefits of laparoscopic vs open hepatic resection: a critical appraisal. Arch Surg 146(3):348–556

Left Hepatectomy: Robot-Assisted Approach

39

Ugo Boggi, Stefano Signori, Fabio Caniglia, Mario Belluomini and Carla Cappelli

39.1 Introduction

Laparoscopy is now the standard approach for left lateral segmentectomy and is the preferred method for resection of liver tumors measuring <5 cm located in anterior segments [1, 2]. In properly selected patients managed at high-volume centers, laparoscopic liver surgery (LLS) is associated with reduced blood loss, diminished need for blood transfusions, low rate of microscopically positive margins, and improved outcome in cirrhotic patients [3]. Despite the fact that gifted hepatic surgeons facile with complex laparoscopic maneuvers have successfully performed posterior segmentectomies and major hepatectomies [1–3], the inherent limitations of the laparoscopic technique have significantly limited its acceptance for challenging hepatic resections [1, 2].

The da Vinci Surgical System (dVSS) (Intuitive Surgical Inc., Sunnyvale, CA, USA) is an electromechanical actuator transmitting the movements of the surgeon's hands to the tip of miniaturized instruments that have seven degrees of freedom, unlike the human wrist, and can articulate up to 90°. EndoWrist technology eliminates the fulcrum effect and provides the surgeon with the same level of dexterity as during open surgery. Further, the dVSS eliminates the surgeon's natural hand tremor and provides a steady, high-definition, stereoscopic, view, improves hand–eye coordination, allows movement scaling into micromotions, and provides optimal working ergonomics [4]. These tremendous technological improvements are particularly rewarding when fine dissection or suturing within deep and narrow spaces is required. In this respect, using the dVSS for major hepatectomy and/or resection of posterior

U. Boggi (✉)
Department of Oncology, Transplants, and New Technologies
University of Pisa, Pisa, Italy
e-mail: u.boggi@med.unipi.it

F. Calise and G. Casciola (eds.), *Minimally Invasive Surgery of the Liver,*
Updates in Surgery
DOI: 10.1007/978-88-470-2664-3_39, © Springer-Verlag Italia 2013

liver segments could be particularly useful. Robotic assistance is also expected to improve the surgeon's ability to deal with resection of the caudate lobe and/or safe dissection of large vessels in the presence of tumor adherence or anatomic distortion [4–9]. Finally, when patients requiring hepatic-vessel resection and reconstruction are still considered suitable for a minimally invasive hepatectomy, robot assistance may provide an opportunity for the surgeon to face this challenging scenario laparoscopically [10].

39.2 Patient Selection

Patients suitable for LLS, having either benign or malignant lesions, can be considered for robot-assisted hepatectomy. Robot assistance may be particularly useful when resection of caudate lobe is necessary or when challenging vascular dissection is anticipated. Although the dVSS allows biliary [5] and vascular [10] reconstructions, the high-grade biology of Klatskin tumors suggests that extrahepatic cholangiocarcinoma should not be considered for laparoscopic robot-assisted resection.

State-of-the-art preoperative imaging is mandatory. Generation of patient-specific 3D models (Fig. 39.1) is a useful adjunct and may be a step toward effective intraoperative navigation that could be further facilitated by the TilePro utility of the dVSS.

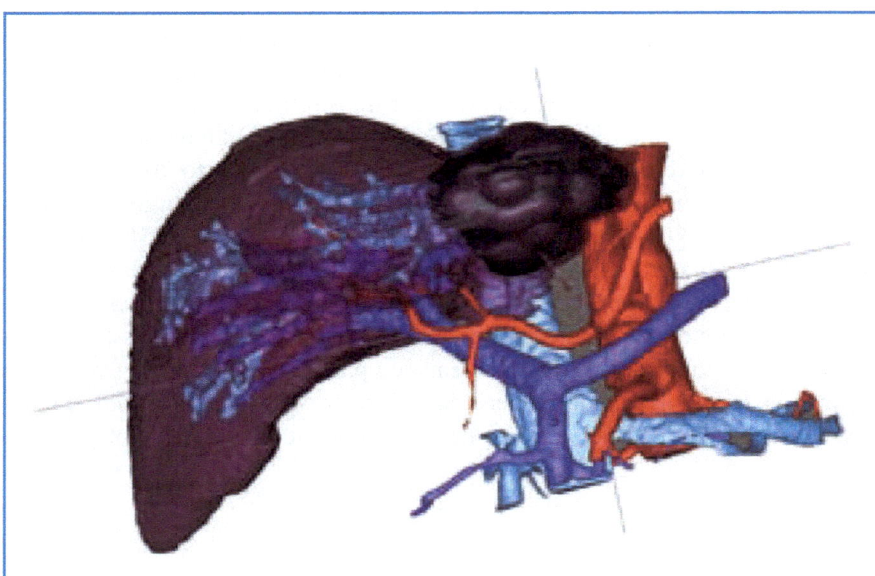

Fig. 39.1 Patient-specific 3D model providing segmenting radiological images

39.3 Surgical Technique

Patients are placed in the lithotomy position and secured to the operating table by wide bandings. The patient's arms are either tucked along the body or suspended above the head to avoid collision with robotic arms that could possibly result in nerve injury. The table is placed at 20° in the reverse – Trendelenburg position. The dVSS tower is docked directly over the patient's head, with two operating arms on the patient's left side. The robotic surgeon operates from the dVSS console, whereas the laparoscopic surgeon stands between the patient's legs to manage the suction irrigator, exchange instruments, and pass needles. A total of five ports are placed, as shown in Fig. 39.2. Pneumoperitoneum is maintained at 12 mmHg. Since the robotic tower's bulk limits accessibility to the patient by the anaesthesia team, airway control and vascular lines must be perfectly secured before starting surgery.

Surgery begins with visual exploration of the abdomen, followed by contact ultrasonography (US) of the liver. As soon as resectability is confirmed, the left lobe is freed from its surrounding peritoneal and diaphragmatic attachments. Next, the hepatoduodenal ligament is exposed by retracting upward and cephalad the gallbladder grasped with a Cadiere forceps driven by robotic arm 3. Left hepatic duct, artery, and portal vein are dissected and divided between ligatures. After vascular isolation of the portal triad supplying the left hepatic

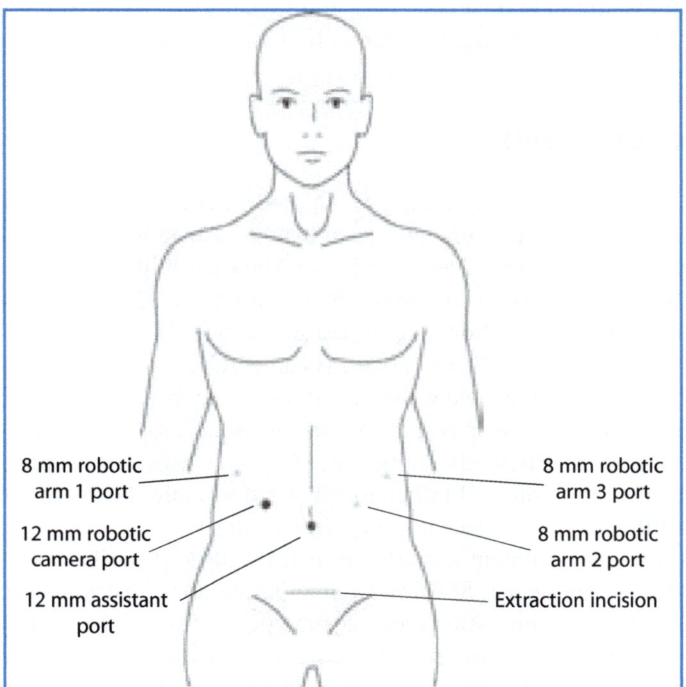

Fig. 39.2 Ports are placed along a semicircular line that has its lowest point just below the umbilicus

lobe, a vascular demarcation along the parenchymal transection plane becomes obvious. The gallbladder is resected en bloc with the specimen.

Before starting with parenchymal transection, the caudate lobe is dissected from the inferior vena cava. Small hepatic veins are individually secured by fine polypropylene sutures. Parenchymal transection is carried out using Maryland bipolar forceps (robotic arm 2) and monopolar curved scissors (robotic arm 1). During this phase, a Cadiere forceps (robotic arm 3) may be used to retract the specimen, to allow optimal alignment of the transection line with robotic scissors, or to tampon hemorrhage sites using a peanut sponge.

Hepatic veins are sealed and divided using an endoscopic stapler armed with a vascular cartridge. The stapler is usually inserted through the 12-mm assistant's port, as the alignment between the device and the target hepatic vein is usually optimal. In some patients, because of anatomic variations and/or distortion of conventional anatomical planes, the position of camera and assistant ports may be switched to insert the vascular stapler through the right-sided port originally used for the camera. Due to the risk of stapler malfunction [4], albeit exceedingly rare, the surgeon may prefer to use a TA endoscopic stapler or the newer no-knife GIA stapler, as these devices allow the surgeon to verify the correct positioning of the staple rows before dividing the vessels. The specimen is removed in an EndoBag via a transverse suprapubic incision measuring some 7 cm.

Before the operation is completed, the raw surface of the liver is carefully inspected. Bile leaks and bleeding sites are individually sealed by suture ligature. No topical hemostatic glue or sealants are employed. Two closed-suction Redon's drains (12 F) are placed near the transection surface of the liver.

39.4 Robotic Instruments

Robotic arm 3, used mostly for tissue retraction, is armed with a Cadiere forceps. Robotic arm 1, corresponding to the surgeon's right hand, is armed with monopolar curved scissors, and robotic arm 2 is armed with Maryland bipolar forceps. Most of the operation is carried our using these three instruments. The hook, for instance, is not necessary, as dissection can be performed equally well and precisely using scissors. The use of needle drivers is instead required to place sutures and ligate ties. The use of sophisticated energy devices, such as the robotic Harmonic shears, is not necessary in most patients, as the precision of robotic dissection allows the surgeon to efficiently control bleeding sites using less-sophisticated electromechanical technology. Further, as robotic Harmonic shears cannot articulate, the use of this device reduces the surgeon's dexterity.

The practical value of instrument articulation is difficult to describe and is best perceived by direct experience. Shifting from an articulating instrument to a straight one during the same operation enhances the perception of this difference. Finally, the surgeon's ability to securely achieve hemostasis is improved by the use of robotic Hem-o-lok clips. These clips are the same as those used in conventional laparoscopy, but the applier articulates, making delicate clip positioning easy even in the setting of suboptimal port-to-target alignment.

39.5 Surgical Dissection Planes

We prefer to carry out the resection strictly along anatomical planes. To do so, we prefer to ligate and divide the left portal pedicle before starting parenchymal transection. Careful preoperative imaging is of paramount importance to be prepared to face anatomic variations. If tumor type does not require lymphatic clearance of the hepatoduodenal ligament, dissection of the left hepatic artery and left branch of the portal vein begins just below the hilar plate. The bile duct may be divided intraparenchymally but, in most instances, can be safely dissected off, ligated, and divided before proceeding with parenchymal transection. The hepatic pedicle is not usually encircled in preparation for a Pringle maneuver, but this option may be considered if a difficult liver transection is anticipated.

After trial cross-clamping, the left hepatic artery and left branch of the portal vein are divided between ligatures. The use of staplers or clips is rarely required at this stage. The left hepatic vein, and occasionally the middle hepatic vein, may be encircled outside the liver. More often, however, to avoid the nonessential risks of tearing these delicate vessels, hepatic veins are taken intraparenchymally. The plane of liver transection follows the ischemic demarcation line. Intraoperative US may be further used as a guide to achieve negative margin resection when the tumor is close to the Cantlie line.

39.6 Conclusion

Robot-assisted surgery is here to stay, but it is not essential in patients having liver lesions that are easy to resect. High equipment cost is a further argument to restrict the use of robotics to the more challenging operations. Our experience with left, right, and extended right hepatectomies suggests that robot assistance greatly facilitates control of bleeding originating from large hepatic veins. As previously described, we were able to control a large caval injury breach caused by stapler misfiring on the right hepatic vein without consequences to the patient [4].

Although the advantages of robot-assisted laparoscopy compared with either open or conventional laparoscopic surgery remain to be determined, it seems reasonable to anticipate that technical improvements in the dVSS might facilitate safe dissection of paracaval lesions and/or bulky tumors involving the caudate lobe. More generally, however, winning back a level of surgical dexterity comparable with that experienced in open surgery could either improve the results of the standard laparoscopic left hepatectomy or enable surgeons to safely operate laparoscopically more patients who would otherwise require an open approach. As bleeding is the leading cause of conversion to open surgery in liver resections, using the dVSS could reduce the conversion rate without compromising patient safety.

References

1. Buell JF, Cherqui D, Geller DA et al (2009) The international position on laparoscopic liver surgery. The Louisville statement. Ann Surg 250:825–830
2. Nguyen KT, Gamblin TC, Geller DA (2009) World review of laparoscopic liver resection – 2,804 patients. Ann Surg 250:831–841
3. Rao A, Rao G, Ahmed I (2012) Laparoscopic or open liver resection? Let systematic review decide it. Am J Surg [Epub ahead of print, January 13, PMID: 22245507]
4. Boggi U, Moretto C, Vistoli F et al (2009) Robotic suture of a large caval injury caused by endo-GIA stapler malfunction during laparoscopic wedge resection of liver segments VII and VIII en-bloc with the right hepatic vein. Minim Invasive Ther Allied Technol 18:306–310
5. Giulianotti PC, Sbrana F, Bianco FM et al (2010) Robot-assisted laparoscopic extended right hepatectomy with biliary reconstruction. J Laparoendosc Adv Surg Tech A 20:159–163
6. Giulianotti PC, Coratti A, Sbrana F et al (2011) Robotic liver surgery: results for 70 resections. Surgery 149:29–39
7. Ji WB, Wang HG, Zhao ZM et al (2011) Robotic-assisted laparoscopic anatomic hepatectomy in China: initial experience. Ann Surg 253:342–348
8. Giulianotti PC, Tzvetanov I, Jeon H et al (2012) Robot-assisted right lobe donor hepatectomy. Transpl Int 25:e5–9
9. Lai ECH, Tang CN, Li MKW (2012) Robot-assisted laparoscopic hemi-hepatectomy: technique and surgical outcomes. Int J Surg 10:11–15
10. Giulianotti PC, Addeo P, Buchs NC et al (2011) Robotic extended pancreatectomy with vascular resection for locally advanced pancreatic tumors. Pancreas 40:1264–1270

Left Sectionectomy for Living Donor: Laparoscopic Approach

40

Marco Spada, Ugo Boggi, Calogero Ricotta,
Duilio Pagano and Salvatore Gruttadauria

40.1 Split-Liver and Living-Liver Donation

Before the introduction of split-liver and living donor liver transplantation, the mortality rate among children with end-stage liver disease exceeded 25% [1], and this precipitated the need for greater understanding of segmental liver anatomy and for improvements in surgical techniques. These improvements first led to the possibility of obtaining reduced-size allografts with autonomous arterial, portal, biliary, and venous drainage. Initially, split liver was applied for one pediatric and one adult recipient; its applications were progressively extended, with results similar to those of transplantation of the whole organ [2, 3]. The increasing age of cadaveric donors is hampering the use of this technique, with a significant lengthening of the waiting list for pediatric patients.

A different approach to the problem of organ shortage is represented by living donation. In this case, also, initial procedures were made in pediatric recipients: the first attempts were undertaken by Raia et al., in Brazil [4], and Strong et al. [5]. Since then, the technique has been widely applied, with satisfactory results in pediatric and, more recently, adult recipients [6, 7]. Between 1999 and 2002, five right-lobe living donor deaths (two in the USA and three in Europe) led to a more cautious approach to the use of this procedure [8]. As a result, there was a decline in the number of adult living donor liver transplants.

Donor morbidity has been correlated with the extent of hepatectomy: com-

M. Spada (✉)
Department of Abdominal and Transplantation Surgery, Istituto Mediterraneo per i Trapianti e Terapie ad Alta Specializzazione (IsMeTT) - University of Pittsburgh Medical Center (UPMC), Palermo, Italy
e-mail: mspada@ismett.edu

F. Calise and G. Casciola (eds.), *Minimally Invasive Surgery of the Liver*,
Updates in Surgery
DOI: 10.1007/978-88-470-2664-3_40, © Springer-Verlag Italia 2013

plications are significantly higher after right hepatectomy (20–40%) compared with left lobectomy or left lateral segmentectomy (10%). The most common donor complications are bile leak, bacterial infections, pleural effusions requiring intervention, wound infection, intra-abdominal abscess, and incisional hernias. The reported living donor morbidity rates range from 8% to 12 % [9].

For small pediatric recipients, the most commonly used graft is the left lateral segment, whereas for larger adolescents or small adults, a left-lobe allograft can be used for transplantation, as reported in some series [10].

40.2 Laparoscopic Liver Surgery

The rapid evolution of laparoscopic surgery has led to the feasibility of safe, minimally invasive hepatic resection. Gagner et al. reported the first laparoscopic liver resection in 1992 [11]. Azagra et al. were the first to perform a left lateral segmentectomy, in 1996 [12]. Since these initial reports, several surgical groups are now performing major hepatic resections laparoscopically [13], with reported morbidity and mortality rates equivalent to those of open surgery. The laparoscopic approach has resulted in decreased blood loss, shorter postoperative stay, reduced need for postoperative analgesic, and reduced risk of developing an incisional hernia [13].

From a technical point of view, minimally invasive hepatic resection includes pure laparoscopic, hand-assisted laparoscopic, and a laparoscopic-assisted open "hybrid" approach in which the operation is started laparoscopically and proceeds with a small laparotomy. It has been suggested that laparoscopic living donor hepatectomy may provide a new procedure with diminished donor morbidity and increased appeal for living liver donation.

40.3 Laparoscopic Living Donor Left Lateral Segmentectomy

A total of 34 living donor left lateral segmentectomies performed laparoscopically have been reported in peer-reviewed articles. Other case series are reported in brief abstracts. Cherqui et al. reported, in 2002, the first two cases of laparoscopic left lateral segmentectomy for living donation [14]. The same group, in 2006, reported an additional 14 cases, which were compared with 14 donors who underwent an open liver resection, demonstrating that this procedure was safe and reproducible, resulting in grafts similar to those obtained by open surgery [15]. In 2009, Troisi et al. reported one case in Belgium [16], and that same year, Coelho et al. reported a case in Brazil [17]. Five procedures of this type were done in the USA (Samstein, personal communication) and, at the 16th International Liver Transplant Society Congress, a French collaborative study described 53 procedures [18]. In 2011, Kim et al. reported 11 cases of laparoscopic left lateral segmentectomies, comparing outcomes with the open live-donor procedure [19]. In 2010, our group started the laparoscopic living donor hepatectomy program for pediatric transplantation. Since then, five procedures have been successfully performed. All procedures were done with a purely laparoscopic approach. The reported conversion rate was between 6% and 8%, whereas morbidity ranged from 10–30 % in the largest series. No donor death was reported. Table 40.1 depicts the main donor, operative, and recipient data of the reported cases, including our series. From an economic standpoint, the laparoscopic approach has been reported to be comparable with the open procedure in terms of out-of-pocket donor costs [19].

At our institute, donor workup for living donation includes triphasic liver computed tomography (CT) with volume measurement, magnetic resonance (MR) cholangiography, and percutaneous liver biopsy. Donors receive an exhaustive explanation of the advantages and disadvantages of open and laparoscopic donor hepatectomy, and since the beginning of our laparoscopic living donor program, none of them has refused the minimally invasive approach.

40.4 Surgical Technique

The donor is placed supine in the 30° reverse-Trendelenburg position, with the surgeon standing between the donor's legs. A Veress needle is placed above the umbilicus along the midline. Pneumoperitoneum at 10–12 mmHg is obtained using carbon dioxide. One 12-mm trocar is placed in the same site, and the 35° optical device is inserted to explore the abdominal cavity. Two to four additional 12- and 5-mm trocars are inserted in the upper abdominal quadrant. The round and falciform ligaments are divided with the HARMONIC scissors, and the left triangular ligament is dissected. The lesser omentum is opened through the *pars flaccida*. The left aspect of the hepatic hilum is then dissected with an electric hook to expose the left hepatic artery, which is dissected free and encircled with a vessel loop. If the arterial branch for segment 4 (S4) originates from the left hepatic artery, this is clipped and cut. The small parenchymal bridge connecting S3 and 4 is divided, exposing the round ligament. Its right side is dissected, isolating, clipping, and cutting the small portal branches to S4, until the left branch of the portal vein is exposed and taped. During this part of the procedure, bipolar electrocautery and nonabsorbable self-locking clips are used.

Small portal branches to the caudate lobe are clipped and cut to maximize the length of the left portal vein. The left lateral segment is rotated on the right side and the *ligamentum venosum* is dissected up to the left hepatic vein, which, if possible, is isolated and encircled with a cotton tape using an extraparenchymal approach. Parenchymal dissection is initiated along the falciform ligament with Harmonic or bipolar electrocautery. When the dissection reaches the hilar plate, the left biliary tree is cut sharply with scissors and its right stump clipped. The section of the parenchyma can be completed up to the ori-

Table 40.1 Laparoscopic living donor left lateral segmentectomies reported in the literature

Author	No.	Donor characteristics				Procedure data			Morbidity		Mortality	
		Age (years)	Sex (M/F)	Weight (kg)	Graft weight (g)	Operation time (min)	Blood loss (ml)	Hospital stay (days)	Donor	Recipient	Donor	Recipient
Cherqui et al. 2002 [14]	2	27–31	1/1	60–80	250–280	390 ± 30	300 ± 150	189 ± 44	–	–	–	–
Soubrane et al. 2006 [15]	16	29 ± 5	10/6	70 ± 10	253 ± 45	320 ± 67		7.5 ± 2.3	Bile leak with reoperation (1) Wound hematoma (2)	HAT (2) PVT with ReTX (1) Biliary complications (3)	Sepsis (1)	–
Troisi et al. 2009 [16]	1	31	0/1	NA	NA	NA	NA		–	–	–	–
Coelho et al. 2009 [17]	1	25	1/0	77	653	NA	NA		Abdominal collection (1)	NA	–	NA
Spada et al. (ahead of print)	5	27 ± 7	2/3	68 ± 15	280 ± 113	449 ± 48	NA	5.2 ± 1.1	–	PVT (1)	–	–
Stenard et al. 2010 [18]	53	NA	NA	NA	NA	285 ± 66	69 ± 98		Biliary leak (3) Abdominal collection (1)	Bleeding (1) HAT (3) Biliary leak (10) ReTX (4)	–	3
Kim et al. 2011 [19]	11	27 ± 6	1/10	NA	230 ± 43	330 ± 68	396 ± 72	6.9 ± 0.3	–	Portal vein stenosis (1) Biliary stenosis(1)	–	–

HAT, hepatic artery thrombosis; *NA*, not available; *PVT*, portal vein thrombosis; *ReTX*, retransplantation gin of the left hepatic vein. If the hepatic vein cannot be previously isolated extraparenchymally, its exposure is accomplished, completing the parenchymal transection. Care is taken not to jeopardize the middle hepatic vein, especially if a common trunk is present, and to identify and clip small hepatic vein branches crossing the transection plane.

A 7- to 8-cm suprapubic incision is obtained, a 15-mm EndoBag placed through this incision, and the left lateral segment putted into the EndoBag. Small bulldog clips are placed distally on the left hepatic artery and on the left branch of the portal vein, and these two vessels, along with the left hepatic vein, are sutured and cut with a unilateral linear stapler. The graft, already in the EndoBag, is rapidly retrieved, flushed on the back table with a 4 °C University of Wisconsin (UW) solution through the left portal vein, and prepared for subsequent transplantation. The suprapubic wound is then closed and pneumoperitoneum re-established in order to check for hemostasis and biliostasis on the resected edge of the liver.

40.5 Discussion

Laparoscopic intervention allows for minimally invasive surgery in otherwise healthy individuals and is a logical application in living liver transplantation. Donor selection and exclusion criteria are similar to those used in the open technique. Therefore, the laparoscopic approach does not limit the possibility of performing a living donation. In all reported experiences, in case of emergency transplantation, the laparoscopic technique is not proposed for logistical reasons and the open approach is preferred. Hepatic resection is performed using the same method as for open liver surgery. The "hanging-over" maneuvre, used by some authors [19], may facilitate parenchymal transection and a better view of the left bile duct. All authors paid attention to preserving left bile duct vascularization, cutting the bile duct sharply. During left lateral sectionectomy, no clamping is used; likewise, cholecystectomy and intraoperative cholangiography are not routinely performed. Warm ischemia time is always very short, especially with the adoption of stratagems such as those used by us. The lack of reports of primary nonfunction shows that the reported warm ischemia time had no negative impact on the outcome. Operative times are slightly longer than with the standard technique, especially in early experience. However, this had no negative impact on postoperative recovery and hospitalization times.

Concerns have been raised regarding the negative influence of pneumoperitoneum on graft functional recovery after transplantation [20]. In the reported series of living donor laparoscopic left lateral segmentectomy, there are no reported surgical, vascular, immunological, or infectious complications specifically resulting from the minimally invasive technique.

Overall, analysis of early results of the reported series shows comparable donor complications and grafts and recipient outcomes between laparoscopic left lateral segmentectomy and open donor procedures. The use laparoscopic surgery instead of an open procedure potentially provides living donors with a number of advantages, including: (1) reduced hemorrhaging, consequently reduced need for blood transfusion, (2) smaller incision, resulting in less post-operative scarring, (3) shorter recovery time, (4) less postoperative pain and disability, leading to reduced narcotic requirement, and 5) less risk of incisional hernia.

Living donor left lateral segmentectomy is a safe, feasible, and reproducible procedure and can be beneficial for the donor. Moreover, the procedure provides suitable grafts with early graft function and perioperative complication rates comparable with those from standard open procedures. We can therefore recommend minimally invasive donor hepatectomy as a safe new procedure in living donor liver transplants. Further analysis of outcomes, from larger case series and from multi-institutional registers, however, are essential in determining whether this new procedure will increase the appeal to and numbers of living liver donors.

Acknowledgements A contribution to this chapter is gratefully acknowledged to Dr. Bruno Gridelli.

References

1. Broelsch CE, Emond JC, Whitington PF, Thistlethwaite JR, Baker AL, Lichtor JL (1990) Application of reduced-size liver transplants as split grafts, auxiliary orthotopic grafts, and living related segmental transplants. Ann Surg 212:368–375
2. Spada M, Gridelli B, Colledan M et al (2000) Extensive use of split liver for pediatric liver transplantation: a single-center experience. Liver Transpl 6:415–428
3. Hong JC, Yersiz H, Busuttil RW (2011) Where are we today in split liver transplantation? Curr Opin Organ Transplant 16:269–273
4. Raia S, Nery JR, Mies S (1989) Liver transplantation from live donors. Lancet 2:1042–1043
5. Strong RW, Lynch SV, Ong TH et al (1990) Successful liver transplantation from a living donor to her son. N Engl J Med 322:1505–1507
6. Marcos A (2000) Right lobe living donor liver transplantation: a review. Liver Transplant 6:3–20
7. Gruttadauria S, Pagano D, Echeverri GJ, Cintorino D, Spada M, Gridelli B (2010) How to face organ shortage in liver transplantation in an area with low rate of deceased donation. Updates Surg 62:149–152
8. Trotter JF, Adam R, Lo CM, Kenison J (2006) Documented deaths of hepatic lobe donors for living donor liver transplantation. Liver Transpl 12:1485–1488
9. Azoulay D, Bhangui P, Andreani P et al (2011) Short- and long-term donor morbidity in right lobe living donor liver transplantation: 91 consecutive cases in a European center. Am J Transplantation 11:101–110
10. Neto JS, Pugliese R, Fonseca EA et al (2012) Four hundred thirty consecutive pediatric live donor liver transplants: variables associated with post-transplant patient and graft survival. Liver Transpl 18:577–584
11. Gagner M, Rheault M, Dubuc JE (1992) Laparoscopic partial hepatectomy for liver tumor. Surg Endosc 6:99
12. Azagra JS, Georgen M, Gilbart E, Jacobs D (1996) Laparoscopic anatomical (hepatic) left lateral segmentectomy. Technical aspects. Surg Endosc 10:758–761

13. Reddy SK, Tsung A, Geller DA (20119 Laparoscopic liver resection. World J Surg 35:1478–1486
14. Cherqui D, Soubrane O, Husson E et al (2002) Laparoscopic living donor hepatectomy for liver transplantation in children. Lancet 359:392–396
15. Soubrane O, Cherqui D, Scatton O et al (2006) Laparoscopic left lateral sectionectomy in living donors: safety and reproducibility of the technique in a single center. Ann Surg 244:815–820
16. Troisi R, Debruyne R, Rogiers X (2009) Laparoscopic living donor hepatectomy for pediatric liver transplantation. Acta Chir Belg 109:559–562
17. Coelho JC, Freitas AC, Mathias JE (2009) Laparoscopic resection of the left lateral segment of the liver in living donor liver transplantation. Rev Col Bras Cir 36:537–538
18. Stenard F, Boillot O, Scatton O et al (2010) Laparoscopic left lateral sectionectomy in living donor is the gold standard approach in France. Liver Transplant 16 Suppl 1:93 CE4
19. Kim KH, Jung DH, Park KM et al (2011) Comparison of open and laparoscopic live donor left lateral sectionectomy. British J Surg 98:1302–1308
20. Troppmann C, Osmond D, Perez RV (2003) Laparoscopic (vs open) live donor nephrectomy: a UNOS database analysis of early graft function and survival. Am J Transplant 3:1295–1301

Part V

Surgery: Outcome

Complications

41

Antonio Giuliani, Francesco Sicoli, Walter Santaniello,
Giangiacomo Nunzio Monti and Marcella Marracino

41.1 Introduction

Laparoscopic liver resection (LLR) is now considered as a safe and reproducible procedure but is reserved for surgeons trained in both liver and laparoscopic surgery [1–5]. When adequate training is provided, early surgical results are better and complications after LLR are reduced compared with open surgery [1–5]. Complication rates reported in the literature range from 0% to 50% [2]. This is probably to be explained by the differences in series in terms of surgical experience (initial or trained), number of patients, type of disease, and type of resections. In this chapter, we discuss both general and specific complications of LLR and the reasons for conversions to open procedures.

41.2 General Complications

Nguyen et al., in a review of 2,804 patients, reported nine deaths (0.3%) and 295 complications (10%). Interestingly, of the nine deaths, three were in cirrhotic patients (33.3%) and five (55.5%) were related to liver surgery [2]. The rate of general complications was 75.6 % (223/295), whereas the liver-related rate was 24.4% (72/295). If all 2,804 patients are considered, the liver-related complication rate was 2.5% while general and surgery-related complications were 7.9% [2]. The most frequently reported surgery-related complications are trocar-site bleeding (0.5%), wound infections (0.46%), intra-abdominal bleeding (0.35%), incisional hernias (0.35%), and intra-

A. Giuliani (✉)
Unit of Hepatobiliary Surgery and Liver Transplant Center, A. Cardarelli Hospital,
Naples, Italy
e-mail: giuldoc@hotmail.com

F. Calise and G. Casciola (eds.), *Minimally Invasive Surgery of the Liver,*
Updates in Surgery
DOI: 10.1007/978-88-470-2664-3_41, © Springer-Verlag Italia 2013

abdominal fluid collections (0.35%) [2]. The most frequent general complications involve the respiratory system (1.1%) (pleural effusion, pneumonia, respiratory failure, pneumothorax, atelectasis, pulmonary embolism) [2]. Patients with hepatocellular carcinoma (HCC) seem to be at a higher risk for postoperative complications than those with colorectal liver metastases (CRLM) (50% vs 11%, $p = 0.02$) [2].

In a cost–benefit comparison, complications after LLR seem to be more favorable than after open liver resection (OLR). Vanounou et al. reported that postoperative morbidity rate, according to Clavien classifications, is better for LLR ($p = 0.001$) [6]. In fact, 86% of patients with LLR showed no or minimal complications (grade I), with only 9% showing grade II complications (31% for open surgery). For grade III or IV complications, they reported 10% in patients in the open surgery compared with only 4% in the LLR group [6]. A more favorable outcome is confirmed for patients operated for malignancies through the laparoscopic approach ($p = 0.003$) [6]. The final median cost for LLR vs. OLR was US$15,260 vs. US$17,629 ($p = 0.03$), respectively [6]. Other authors report that days of required narcotic pain medication is decreased after LLR compared with OLS (1 day vs. 5 days; $p = 0.001$), as is the total amount of drugs required [7, 8].

The learning curve does not seem to have a role in determining morbidity after LLR. Cannon et al. compared 300 LLR patients divided in two groups according to the beginning and end of the surgeons' training experience, with no differences in terms of morbidity (11% vs. 14%, $p = 0.300$) and mortality (1% vs. 3%, $p = 0.625$) [9]. A different observation was made by Dagher et al. In their study on cirrhotic patients, general morbidity was higher in the early phase of experience (75 LLR) compared with the more recent phase (88 LLR) (16% vs. 5.7%; $p = 0.040$), whereas specific liver morbidity was lower in the early experience but was not significant (8% vs. 14.7%; $p = 0.224$) [4]. This was probably due to a more favorable patient selection with smaller lesions in the second phase (4.0 cm vs. 3.4 cm; $p = 0.044$) [4].

41.3 Liver-Related Complications

Buell et al. [10] reported a high rate of mortality (9.7%) in cirrhotic patients operated with LLR; a rate indeed too high when compared with the large series of patients operated with OLR [10–13]. An explanation of this high mortality is done by Donadon et al; they argue that this was probably due both to surgeons' initial experience and to an improper selection of patients, with Model for End-Stage Liver Disease (MELD) score >15 in some cases [14]. More recent series report a reduced mortality rate after LLR in cirrhotic patients, down to 1.6% [15]. The laparoscopic approach proves also to be better in patients with portal hypertension, as it avoids ligature of portocaval shunts, as must be performed in open surgery (see Chapter 18). Lai et al., reporting a 10-year experience of LLR for HCC, showed neither postoperative liver failure nor encephalopathy [16]. The world review by Nguyen et al. reported a rate of 2.5% of liver-related complications after LLR [2]. The most frequent were bile leaks (1.5%) and transient liver failure/ascites (1.0%). In the European experience of LLR for HCC, liver-specific complications occurred in 19 patients [4] (11.6%), including ascites in 14, bleeding in 4 requiring reoperation for hemostasis in 3, and one patients with biliary collection drained percutaneously [4]. Gustafson et al. compared 49 open and 27 LLR and found no significant differences between groups concerning major resections and number of lesions. They reported that bile leaks were reduced in the LLR groups (4 vs. 0) during hospital stay [17]. The risk of gas embolism due to pneumoperitoneum is very low. However, there are some sporadic case reports about delayed carbon dioxide cerebral embolism with ischemic cerebral lesions [18]. One case report described the spontaneous rupture of the splenic capsule subsequent to the Pringle maneuver during LLR [19]. This is an anecdotal observation, but it is important to point out that various types of complications can occur in LLR.

41.4 Conversions

The Louisville Statement indicates: "Conversion should be performed for difficult resections requiring extended operating times for patient safety, and should be considered prudent surgical practice rather than failure" [5].

In the review by Nguyen et al., the overall conversions rate was 4.1%, in most cases due to uncontrollable bleeding (1.4%) [2]. Other reasons were adhesions (0.4%) and anatomical limitation/inaccessible lesion location (0.4%); reasons for 0.8% of conversion were not documented [2]. Interestingly, if we consider only patients with CRLM, the conversion rate is higher (12%), as reported by Abu Hilal et al., which is probably related to previous open surgery [20]. This data is absolutely comparable with the conversion rate in other laparoscopic procedures in patients with previous abdominal surgery. In fact, Naguib et al. reported a 10.6% conversion rate in patients with previous abdominal surgery for laparoscopic colorectal resection [21]. In cirrhotic patients, a 6.8% conversion rate was reported for trained surgeons [4].

41.5 Conclusions

LLR seems to be related to a lower postoperative complication rate compared with OLR. This is probably due both to a more appropriate selection of patients suitable for a laparoscopic approach and to the dramatic decrease in general complications that occur following laparotomy. The conversion rate in LLR is similar to that observed in other major laparoscopic procedures and should not be considered as an unsuccessful event but good practice to avoid major complications.

References

1. Cherqui D (2003) Laparoscopic liver resection. Br J Surg 90:644–646
2. Nguyen KT, Gamblin TC, Geller DA (2009) World review of laparoscopic liver resection-2,804 patients. Ann Surg 250:831–841
3. Pilgrim C, TO H, Ustatoff V et al (2009) Laparoscopic hepatectomy is a safe procedure for cancer patients. HBP Surg 11:247–251
4. Dagher I, Belli G, Fantini C et al (2010) Laparoscopic hepatectomy for hepatocellular carcinoma: a European experience. J Am Coll Surg 211:16–23
5. Buell JF, Cherqui D, Geller DA et al (2009) World Consensus Conference on Laparoscopic Surgery. The international position on laparoscopic liver surgery: the Louisville Statement. Ann Surg 250:825–830
6. Vanounou T, Steel JL, Nguyen KT (2010) Comparing the clinical and economic impact of laparoscopic versus open liver resection. Ann Surg Oncol 17:998–1009
7. Mala T, Edwin B, Gladhaug I et al (2002) A comparative study of the short-term outcome following open and laparoscopic liver resection of colorectal metastases. Surg Endosc 16:1059–1063
8. Nguyen KT, Marsh JW, Tsung A et al (2011) Comparative benefits of laparoscopic vs open hepatic resection: a critical appraisal. Arch Surg 146:348–356
9. Cannon RM, Brock GN, Marvin MR et al (2011) Laparoscopic liver resection: an examination of our first 300 patients. J Am Coll Surg 213:501–507
10. Buell JF, Thomas MT, Rudich S et al (2008) Experience with more than 500 minimally invasive hepatic procedures. Ann Surg 248:475–486
11. Torzilli G, Makuuchi M, Inoue K et al (1999) No-mortality liver resection for hepatocellular carcinoma in cirrhotic and noncirrhotic patients: is there a way? A prospective analysis of our approach. Arch Surg 134:984–992
12. Fan ST, Lo CM, Liu CL et al (1999) Hepatectomy for hepatocellular carcinoma: toward zero hospital deaths. Ann Surg 229:322–330
13. Donadon M, Botea F, Belliappa V et al (2009) Experience with more than 500 minimally invasive hepatic procedures: a serious note of caution. Ann Surg 249:1064–1065
14. Donadon M, Botea F, Belliappa V et al (2009) Experience with more than 500 minimally invasive hepatic procedures: a serious note of caution. Ann Surg 249:1064–1065; author reply: 1065–1066
15. Lai EC, Tang CN, Ha JP et al (2009) Laparoscopic liver resection for hepatocellular carcinoma ten-year experience in a single center. Arch Surg 144:143–147
16. Gustafson JD, Fox JP, Ouellette JR et al (2012) Open versus laparoscopic liver resection: looking beyond the immediate postoperative period. Surg Endosc 26:468–472
17. Van Gorp V, Diltoer M, Delvaux G et al (2011) Delayed recovery following laparoscopic partial hepatectomy. Eur J Anaesthesiol 23:12–15
18. Low JK, Ayiomamitis GD, Hamoudi A et al (2011) Spontaneous rupture of the splenic capsule with massive bleeding subsequent to the pringle maneuver during laparoscopic liver resection. Surg Laparosc Endosc Percutan Tech 21:84–86
19. Abu Hilal M, Underwood T, Zuccaro M et al (2010) Short- and medium-term results of totally laparoscopic resection for colorectal liver metastases. Br J Surg 97:927–933
20. Naguib N, Saklani A, Shah P et al (2012) Short-term outcomes of laparoscopic colorectal resection in patients with previous abdominal operations. J Laparoendosc Adv Surg Tech 5:8

The Italian Experience in Minimally Invasive Surgery of the Liver: A National Survey

42

Luca Aldrighetti, Federica Cipriani, Francesca Ratti, Luciano Casciola and Fulvio Calise, on behalf of the Italian National Survey Study Group[1]

42.1 Introduction

In 2009, the most comprehensive review of published series of minimally invasive liver resection (MILR) reported nearly 3,000 cases performed worldwide, emphasizing an exponential growth in the application of this technique by surgeons experienced with both hepatic and laparoscopic surgery [1]. The goal of our survey was to provide an overview of the spread of the minimally invasive approach to liver resections in Italy.

[1] (Contributors are listed by number of cases): Belli Giulio, Ospedale S. M. Loreto Nuovo, Napoli; Aldrighetti Luca, Ospedale San Raffaele, Milano; Boni Luigi, A.O.U. Fondazione Macchi, Varese; Cillo Umberto, A.O.U. – Università degli Studi, Padova; Ettorre Giuseppe, A.O. San Camillo – Forlanini, Roma; De Carlis Luciano, A.O. Ospedale Niguarda Ca' Granda, Milano; Giuliani Antonio, A.O. Cardarelli, Napoli; Reggiani Paolo, Ospedale Maggiore Policlinico, Milano; Santambrogio Roberto, A.O. San Paolo, Milano; Spampinato Marcello, Policlinico di Abano Terme, Abano Terme; Morino Mario, Ospedale Molinette, Torino; Filauro Marco, Ospedale Galliera, Genova; Navarra Giuseppe, A.O.U. Policlinico Martino, Messina; Pinna Antonio, Policlinico S. Orsola - Malpighi, Bologna; Casciola Luciano, Ospedale San Matteo degli Infermi, Spoleto; Capussotti Lorenzo, Ospedale Mauriziano Umberto I, Torino; Casaccia Marco, A.O.U. San Martino, Genova; Nuzzo Gennaro, Policlinico Gemelli, Roma; Guerrieri Mario, A.O.U. Ospedali Riuniti, Ancona; Bassi Nicolò, Ospedale S. Maria di Ca' Foncello, Treviso; Brolese Alberto, Ospedale S. Chiara, Trento; Sgroi Giovanni, A.O. Treviglio – Caravaggio, Treviglio; Buonanno Maurizio, A.O. Rummo, Benevento; Jovine Elio, Ospedale Maggiore, Bologna; Spada Marco, IsMeTT, Palermo; Corcione Francesco, A.O. Monaldi, Napoli; Dalla Valle Raffaele, Ospedale Maggiore, Parma; Colledan Michele, A.O. Ospedali Riuniti, Bergamo; Gerunda Giorgio, A.O.U. Policlinico, Modena; Mezzatesta Pietro, Casa di Cura La Maddalena, Palermo; Di Somma Carmine Gianfranco, A.O.U. San Martino, Genova; Guglielmi Alfredo, Policlinico G.B. Rossi, Verona; Di Carlo Isidoro, A.O. Cannizzaro, Catania; Gruttadauria Salvatore, A.O.U. Policlinico Vittorio Emanuele, Catania; Antonucci Adelmo, Policlinico di Monza, Monza; Caldarera Goffredo, A.O. Garibaldi, Catania; Scuderi Vincenzo, Ospedale San Lazzaro, Alba; De Werra Carlo, Università degli Studi Federico II, Napoli; Maida Pietro, Ospedale Evangelico Villa Betania, Napoli.

L. Aldrighetti (✉)
Department of Surgery, Hepatobiliary Surgery Unit, San Raffaele Hospital,
Vita-Salute San Raffaele University, Milan. Italy
Email: aldrighetti.luca@hsr.it

F. Calise and G. Casciola (eds.), *Minimally Invasive Surgery of the Liver*, Updates in Surgery
DOI: 10.1007/978-88-470-2664-3_42,

42.2 Materials and Methods

All Italian surgical centers were contacted through the mailing list of the Italian Chapter of the International Hepato-Pancreato-Biliary Association (IHPBA) and the Società Italiana di Chirurgia [Italian Society of Surgery (SIC)]. The Survey announcement was included in the SIC on-line newsletter. Centers potentially interested in participating in the survey were directly contacted through personal e-mail addresses or phone numbers whenever not already included in other mailing lists or if it was felt they had not been previously reached (i.e., no answer received within 1 month of the announcement). The e-mail invitation to participate in the survey was sent with a letter of explanation and a specific Word format questionnaire regarding a series of items previously chosen by the promoting group (Luca Aldrighetti, Fulvio Calise, Luciano Casciola) to describe the clinical activity of any center performing MILR (see questionnaire at the end of the chapter). Specific questions were also added to provide a larger picture of the Italian experience in both robot-assisted liver resections and simultaneous resections of liver metastases and colorectal cancer. All returned questionnaires were screened to eliminate any double counting of patients, even if no apparent duplicate data were found in the total number of patients. A single surgeon was then identified for each center for any correspondence regarding the data or any need to complete missing data. No minimal cutoff in the number of cases was established for project inclusion so that even small series were taken into account.

Only liver resections performed through a minimally invasive technique were taken into consideration; these comprised totally laparoscopic, hand-assisted laparoscopic, single-port laparoscopic, and robot-assisted techniques. Tumour characteristics (benign, malignant), technical approaches (totally laparoscopic, hand-assisted, single-port, robotic), and extent of resection (minor, major, wedge) were assessed and analyzed in detail. Intraoperative data about technical devices, vascular control, and number of resections for each patient were recorded. Reasons for conversion to open surgery, perioperative deaths, and complications were also recorded. Finally, the beginning of MILR activity and the number of cases per year were taken into consideration for every center. All resections performed until 28 February 2012 were included. Data were recorded in an SPSS database and analyzed using statistical analysis.

42.3 Results

Questionnaires were obtained from 39 centers in 11 Italian regions. The centers reported a total of 1,677 MILR between 1 January 1995 and 28 February 2012. Participating centers and their regional distributions are shown in Fig. 42.1. The median number of patients per center was 27 (range 1–145). The world review of laparoscopic liver resections by Nguyen and colleagues in 2009 enrolled 2,804 liver resections performed over a 16-year period (1992–2008) [1].

Fig. 42.1 Regional distribution of the 39 centers participating in the survey

In this survey, MILR accounted for 10.3% of all liver resections performed during the same period in the same centers (n = 16,244) (Fig. 42.2). Among all centers, the rate of MILR over the full number of resections ranged between 0.9% and 58.0%. One center reported performing liver resections using the laparoscopic approach only. These data are consistent with the literature, which reports a range between 19% and 24.3% [2, 3].

The total number of patients requiring conversion was 180 of 1,677 approached by a minimally invasive technique, which means an Italian conversion rate of 10.7%, even if a wide difference in conversion rates was recorded among centers (range 0–30.9%). This rate appears to be in accordance with data from the literature, as a conversion rate between 2% and 15.0% has been reported. Viganò and Cherqui suggest that 60 cases are needed to complete the learning curve in MILR [4]. When only Italian series with >60 patients were considered (11 series), similar results were registered (10.5%, range 3.8–30.9%). The most common causes of conversion were: intraoperative hemorrhage in 62 cases (34.4%), concerns for oncological radicality in 47 (26.1%), and technical difficulties in 43 (23.8%). Other reasons were severe adhesions from previous surgery in 14 cases (7.7%), anaesthesiological problems in five (2.7%), and injury to adjacent organs in one (0.5%). The 180 liver resections converted to open surgery were excluded from statis-

Fig. 42.2 Minimally invasive liver resection (MILR) accounted for 10.3% of all liver resections performed during the same period in the same centers (n = 16,244)

tical analysis. Therefore, this survey comprises 1,497 liver resections completed using minimally invasive approaches. Analysis of the volume of cases throughout the years showed that Italy has experienced exponential growth in MILR, with a progressive increase and an almost linear trend (Fig. 42.3), which is significantly similar to the ever-growing volume of MILR performed every year in other Eastern and Western countries [1]. Moreover, no center abandoned the laparoscopic liver surgery (LLS) program once adopted. These findings likely reflect both an increased interest by surgeons and a wider acceptance by patients of the minimally invasive approach, even in the field of the liver resections.

Most centers began to perform MILR 9 years after starting programs of advanced laparoscopic surgery and after performing a significant amount of open liver resections. Moreover, in most centers, the number of MILR seems to have reached a fixed proportion of total annual resections after a mean of 4 years of activity. Only four centers simultaneously began advanced laparoscopic surgery and liver resections. It has been suggested that surgeons should have extensive previous experience in open liver surgery (OLS) and technical skill in advanced laparoscopic surgery before using MILR [4]. Data from our survey seem to demonstrate that Italian surgeons approach MILR with caution and follow general rules and guidelines.

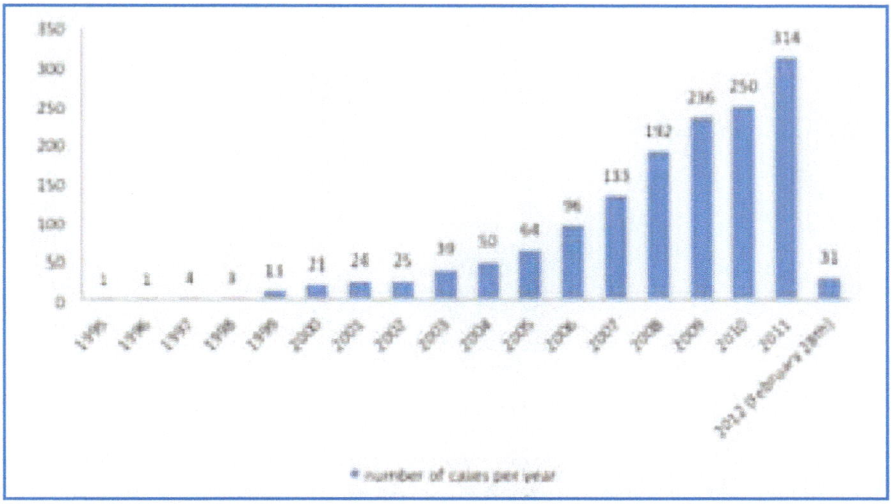

Fig. 42.3 In Italy, there has been exponential growth in minimally invasive liver resection (MILR) between 1995 and 2012, with a progressive increase and an almost linear trend

42.3.1 Indications

In Italy, MILR are usually performed following well-accepted indications for liver resections. Indeed, even though benign diseases were treated by laparoscopic approaches (412 cases, 27.5%), most LLR were performed for malignant diseases: 1,085 (72.5%) (Fig. 42.4). Interestingly, the world review by Nguyen et al. [1] showed a different relative distribution between benign and malignant diagnosis, with a significantly higher amount of benign cases (45% and 50.0%, respectively).

Benign neoplasms reported were focal nodular hyperplasia in 91 cases (22.0%), liver adenomas in 88 (21.3%), and hemangioma in 86 (20.8%). Benign cystic disease was present in 43 cases (10.4%) and included simple or complicated cysts, hydatid cysts, and cystoadenomas. In a minority of cases, liver resection was performed for intrahepatic lithiasis (16 cases, 3.8%) and regenerative or inflammatory nodules (14 cases, 3.3%).

Malignant lesions included hepatocellular carcinoma (608 cases, 56.0%), which was the most frequent indication and accounted for 40.6% of all MILR. Liver metastases were present in 432 cases (39.8%). There were 302 (69.9% of metastases) colorectal metastases; the remaining 130 were metastases from other sites: lung, breast, kidney, melanoma, gastrointestinal stromal tumor (GIST), squamous cell anal cancer, neuroendocrine tumor, and mesothelioma. The remaining 45 cases of malignant lesions (4.1%) were intrahepatic cholangiocellular carcinoma. Breakdown of the diagnoses is shown in Fig. 42.5 and compared with similar data from the world review [1]. In 115 cases (38.1% of colorectal metastases), secondary liver and primary colorectal cancer were

Fig. 42.4 Benign and malignant neoplasms resected through a laparoscopic approach

simultaneously resected through a minimally invasive approach (Fig. 42.6). Finally, laparoscopic left lateral sectionectomy was performed in four cases for living-donor liver procurement.

42.3.2 Extent of Liver Resections

According to the Brisbane 2000 System of Nomenclature [5], minor resections were performed in 1,391 cases (92.9% of total resections), with left lateral sectionectomy being the most widely performed procedure (357 cases, 25.7% of minor resections and 23.8% of total resections) (Fig. 42.7). Left lateral sectionectomy was also the most frequent procedure in all single series, indicating that the left lateral sectionectomy is becoming more frequently approached by laparoscopy. Wedge resections or single segmentectomies were performed in 870 cases (62.5% of all minor resections). Minor MILR involved mainly the so called laparoscopic segments: 2, 3, 4b, 5, 6 (784 cases, 90.1%), which are considered easier to approach by laparoscopic devices than the posterosuperior segments (1, 4a, 7, 8). Indeed, wedge resections or single segmentectomies involving posterosuperior segments were performed in only 80 cases (9.1%), with right sectionectomies being reported in an even smaller number of cases: right posterior sectionectomy in 33 cases (2.4%); right anterior sectionectomy in nine cases (0.6%) of minor resections. Resections of segment 1 (S1) were recorded in six cases. These results are consistent with indications for MILR provided by the Louisville Consensus Conference (2008), which included minor liver resections in patients with solitary lesions located in S2–6 [6].

Major liver resections were 106 (7.1% of total laparoscopic resections)

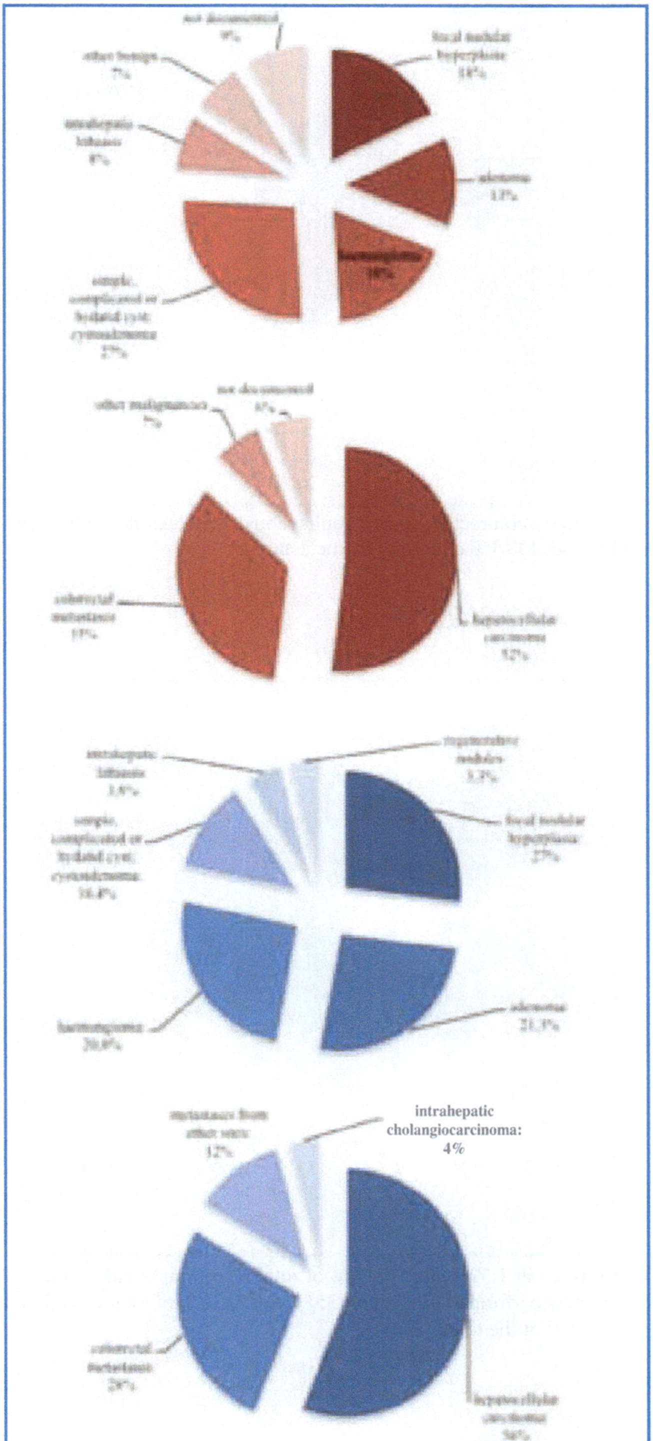

Fig. 42.5 Breakdown of diagnoses (benign, malignant) reported in the world review [1] (*red*) compared with our national survey (*blue*)

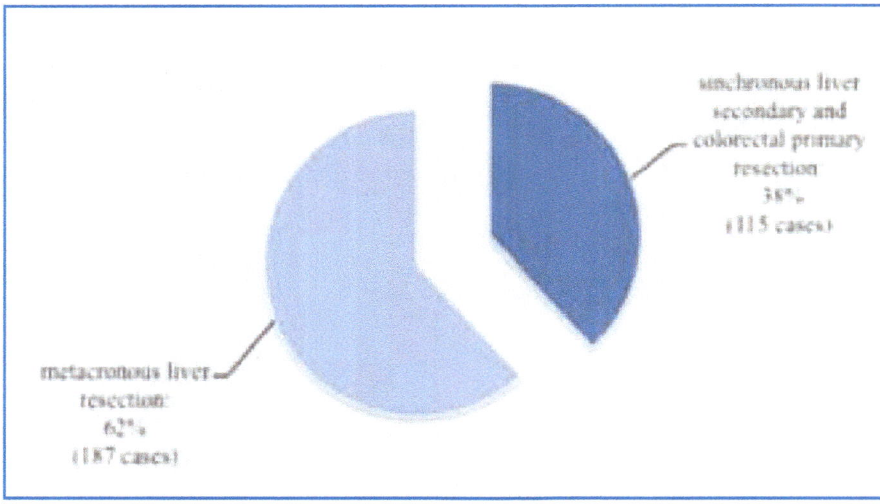

Fig. 42.6 Secondary liver and primary colorectal cancer simultaneously resected through a minimally invasive approach in 115 cases (38.1% of colorectal metastases)

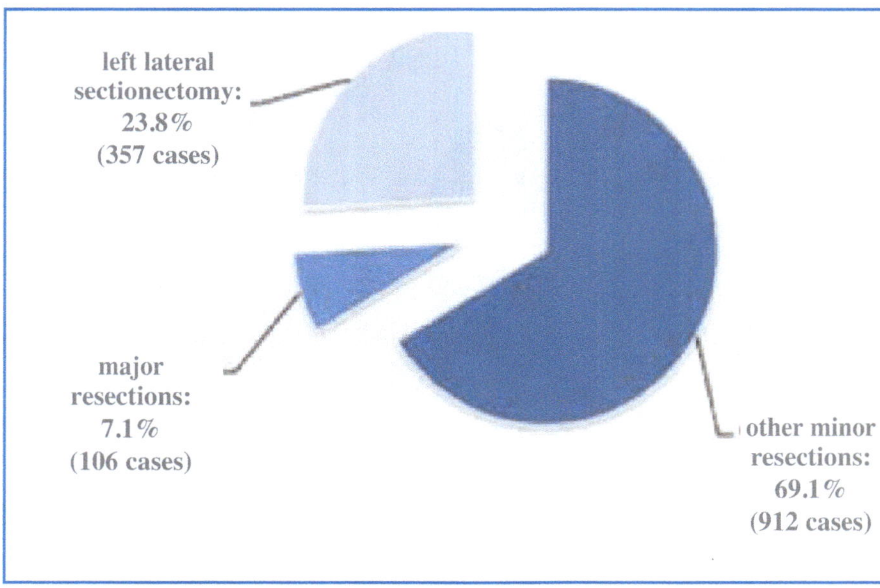

Fig. 42.7 Minor resections performed in 1,391 cases (92.9% of total resections), with left lateral sectionectomy being the most widely performed procedure (357 cases; 23.8% of total resections). Major resections accounted for 7.1% of the total

(Fig. 42.7). There were 63 left and 43 right hepatectomies, which accounted for 59.4% and 40.6% of all major resections, respectively. There were no central or extended right/left hepatectomies. Despite the fact that major hepatectomies are feasible using the laparoscopic approach [7], the small number of major liver resections found in our survey is substantially consistent with previous data from the literature (15.8% in the world review) [1] and probably reflects the low rate of candidates for major resections who can be approached laparoscopically. Indeed, major resections for gallbladder or Klatskin tumors are not as yet considered suitable for the minimally invasive approach, nor are major resections for large primary liver tumors or neoplasms involving major vascular pedicles (i.e., hepatic or portal vein thrombosis).

Multiple simultaneous resections were performed in 254 cases (16.9% of all resections). This finding is in accordance with the recent trend toward performing multiple simultaneous parenchyma-sparing resections instead of major hepatectomies whenever technically feasible in patients with colorectal liver metastases [8, 9]. Biliary or vascular reconstructions, which are generally considered contraindications to the minimally invasive approach even if technically feasible, were never performed.

42.3.3 Minimally Invasive Approaches

The majority of centers used a multiple-port totally laparoscopic approach, whereas hand-assisted and single-port resections were performed less frequently. Totally LLR accounted for 92.6% of the entire series (n = 1,386) (Fig. 42.8).

The hand-assisted approach, which is sometimes reported in the literature as being indicated for technically challenging cases, such as hemihepatectomy, resection of posterosuperior segments, or as an alternative to the conversion to open surgery, was reported in only 19 cases (1.3%).

Twenty-nine liver resections were performed with the single-port access (1.9%), which was used mainly in cases of left lateral sectionectomy that seems the best suitable MILR for such an approach, as the transection plane may be appropriately accomplished even without full triangulation among instruments [10].

Three centers recently performed robot-assisted hepatectomies: 63 robot-assisted liver resections, accounting for 4.2% of all liver resections, including 16 segmentectomies/sectionectomies involving the right posterior segments. No major resections were performed using the robot-assisted technique. It is reasonable to believe that the robot-assisted technique for liver resections will spread more extensively throughout Italian centers in the near future, covering a higher rate of the national pool of MILR and allowing a wider application of the minimally invasive approach to what are presently considered the "nonlaparoscopic segments" (right posterosuperior segments, cranial and posterior portion of S4, cranial portion of S1), as well as to major hemihepatectomies [11, 12].

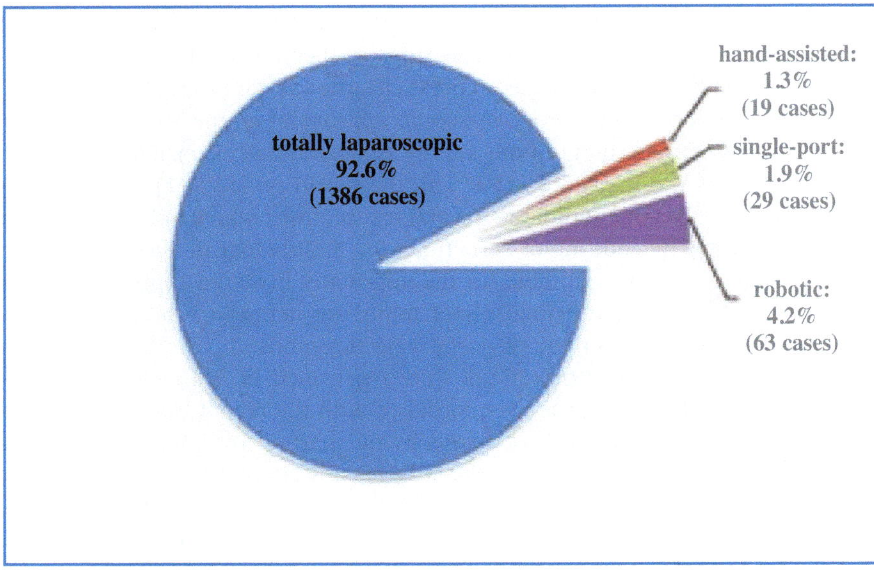

Fig. 42.8 The majority of centers used a totally laparoscopic multiple-port approach, whereas hand-assisted, single-port, and robot-assisted techniques were performed less frequently

42.3.4 Intraoperative Outcomes

Intraoperative details are described in Table 42.1. On the basis of the design of this survey, no specific analysis or comment on intraoperative outcomes can be reliably performed, as these data represent the cumulative outcomes from a heterogeneous group of MILR carried out by surgeons with wide-ranging experience in liver surgery and laparoscopic procedures. Further studies may be advisable to depict a nationwide picture of intraoperative performances in Italian centers using MILR.

Table 42.1 Intraoperative details

Parameter	Results
Intraoperative mortality, *n* (%)	0 (0 %)
Intraoperative blood loss mL (range)	125 (50–625)
Operating time[a] min (range)	145 (45–315)
Pringle maneuver • Performing centers, *n* (%) (continuous/intermittent). • Cases performed, *n* (%) • Total length[b]	 22 (56.4 %) (3/19) 199 (13.7 %) 22 (7–52)
Surgical drain placement, *n* (%)	1,392 (93.0 %)

[a]Weighted mean time; [b]weighted mean time of vascular occlusion.

42.3.5 Technical Features

Different technologies were used to transect the liver parenchyma in MILR, with ultrasound (US) or radiofrequency-energy-based devices being the most widely used (Ultracision, Sonosurg, Ligasure, Lotus), as 31 centers reported routinely performing MILR primarily through these devices. This is in accordance with trends shown in the literature [6], even though the choice among the various devices definitely depends on the personal experience and preference of each surgical team. Alternatives consisted of radiofrequency-based MILR (Habib 4X). One single center routinely used vascular staplers to perform the entire transection, whereas most centers use staplers only to transect the main portal branches and hepatic veins.

The more extensive application of technological devices in MILR than in conventional open liver resections – where Kelly-clamp crushing technique or ultrasonic dissection plus bipolar forceps or ligations are still the most common modalities of parenchymal transection – is understandable enough, as precise and complete progressive hemostasis is more critical during parenchymal transection in MILR than in open resections. Indeed, any – even minimal – bleeding from portal and hepatic vessels may be more challenging to control laparoscopically, as they impair optimal vision, which is an obstacle to safe progression of resection and ultimately requires conversion to open surgery.

In this survey, the Pringle maneuver was performed by 22 centers (56.4%), with a mean time of vascular control ranging from 9 to 57 min among centers. The Pringle maneuver was used continuously and intermittently in three and 19 centers, respectively. The weighted mean time of vascular occlusion was 22 min. Nevertheless, the number of LLR performed using the Pringle maneuver was only 199 (13.7%). These results confirm that MILR may be performed even without routine vascular control, as suggested by previous data from the literature [13]. On the other hand, routine encircling of the hepatic pedicle for Pringle maneuver has yet to be recommended as a precautionary measure whenever approaching an MILR.

Finally, the majority of centers still prefer to place an abdominal drain at the end of the liver resection (1,392 cases, 93.0%). This finding is somehow in contrast with the wide-spreading tendency toward fast-track and Enhanced Recovery After Surgery (ERAS) programs, which substantially ban the routine use of surgical drains.

42.3.6 Postoperative Outcomes

Postoperative details are listed in Table 42.2. There were no intraoperative deaths and three postoperative deaths (mortality rate 0.2%). This is in accordance with data reported in literature, where a postoperative mortality rate <1.0% is reported for MILR, thus confirming the safety of this approach even in the field of liver resections [1].

Table 42.2 Postoperative details

Parameter	Results
Postoperative mortality, *n* (%)	3 (0.2 %)
Postoperative morbidity: type of complication, *n* (total resections)	
• Ascites	81 (5.4 %)
• Pleural effusion	74 (4.9 %)
• Hemorrhage	61 (4.0 %)
• Bile leakage	35 (2.3 %)
• Fever	29 (1.9 %)
• Urinary tract infection	11 (0.7 %)
• Intra-abdominal infection	11 (0.7 %)
• Jaundice	5 (0.3 %)
• Pneumonia	5 (0.3 %)
• Bowel perforation	4 (0.2 %)
• Arrhythmia	3 (0.2 %)
• Bowel obstruction	1 (0.06)
• Pulmonary embolism	1 (0.06)
• Hypertensive peak	1 (0.06)
• Pneumothorax	1 (0.06)
• Acute respiratory failure	1 (0.06)
• Acute cardiac failure	1 (0.06)
• Peripheral paresthesia	1 (0.06)
Grade of severity (Clavien-Dindo classification), *n* (all complications)	
• Grade I	125 (36.5 %)
• Grade II	174 (50.8 %)
• Grade III	42 (12.2 %)
• Grade IV	2 (0.5 %)
• Grade V	3 (0.2 %)
Postoperative course	5.6 postoperative days (2–12)

In this survey, 342 postoperative complications were reported (22.8%); ascites occurred in 81 cases (23.6% of all complications, 5.4% of all MILR); pleural effusion was reported in 74 cases (21.6% of all complications, 4.9% of all MILR); postoperative hemorrhage occurred in 61 patients (17.8% of all complications, 4.0% of all MILR); postoperative bile leakage was reported in 35 cases (10.2% of all complications, 2.3% of all MILR). The complete list of complications is reported in Table 42.2, together with their stratification by severity according to the Clavien-Dindo classification [14]. Most complications were grade II (174; 50.8%); 125 were grade I (36.5%); 42 were grade III (12.2%) requiring surgical, endoscopic, or radiological interventions; two were grade IV. In the literature 10–15% morbidity is reported, with a mean postoperative bile leakage incidence of 1.5% for MILR [1].

During the postoperative course, patients returned to unrestricted diet a weighted mean of 2 days after surgery. Postoperative hospital stay ranged between 2 and 12 days, with a wide variability likely due both to the great heterogeneity of MILR and to the variety of postoperative patient management among centers. However, the weighted mean hospital stay was 6 (range 3–15) days, whereas weighted mean postoperative course lasted 5 (range 3–12) days. No significant discrepancy was noted between these data and previous results from multiple series, systematic reviews, and meta-analyses reported in the literature.

42.4 Conclusion

In recent years, the minimally invasive approach to liver resections has increased significantly in Italy. About 1,500 minimally invasive liver resections have been carried out across the country, and 39 centers from 11 regions contributed to our data collection survey. Several centers definitively completed the learning curve, as attested by clinical results consistent with major series from Western and Eastern countries. Our survey demonstrates the overriding importance that our country plays in the spread of this technique. National collaborative projects may result in Italian groups playing a significant role in the international scenario for future advances in minimally invasive approaches to liver resections.

Laparoscopic Resective Liver Surgery: National Survey

General information
Institution
Address
Contact person

Temporal references

	Year
When did laparoscopic surgery start to be performed?	
When did laparoscopic liver resections start to be performed?	

Surgical activities

	Cases
Total of liver resections (open + LPS)	
Total of laparoscopic liver resections (conversions included)	
Laparoscopic liver resections completed	
Non-resective laparoscopic liver procedures:	
• Unroofing	
• Pericystectomy	
• Radiofrequency ablation	
• …	

Synchronous colonic-hepatic resection

	Cases
Total of synchronous colonic-hepatic resections	
Total of synchronous colonic-hepatic laparoscopic resections *	
Synchronous colonic-hepatic laparoscopic resections completed *	
Hepatic resection:	
• Right hepatectomy	
• Left hepatectomy	
• Segmentectomy (specify segment)	
• Wedge resection	
(Enter only the cases with laparoscopic liver resection).*	

Distribution of resection per year

Year	Total of liver resections	Open liver resections	Laparoscopic liver resections (intention-to-treat)	Laparoscopic liver resections completed	Major laparo-scopic liver resections (*)

(*) Right/left/extended hepatectomy.

Robotic liver resections
NO
YES:

Year
Total
Major liver resections

Diagnosis (only liver resections)

	Cases
Malignant neoplasms	
Hepatocellular carcinoma	
Cholangiocellular carcinoma	
Other malignant primaries	
Colorectal metastases	
Metastases from neuroendocrine neoplasm	
Metastases from other primaries	
Benign neoplasms	
Haemangioma	
Adenoma	
Focal nodular hyperplasia	
Intra-hepatic lithiasis	
Other benign neoplasms	

Liver resection

Type of resection (indicate the main procedure in case of multiple resections)	Cases
Right hepatectomy	
Left hepatectomy	
Extended right hepatectomy	
Extended left hepatectomy	
Central hepatectomy	
Left lateral sectionectomy	
Right posterior sectionectomy (6-7)	
Right anterior sectionectomy (5-8)	
Segmentectomy	
• 1	
• 2	
• 3	
• 4	
• 5	
• 6	
• 7	
• 8	

Type of resection	Cases
Single	
Multiple	

Technique of resection	Cases
Totally laparoscopic	
Hand assisted	
Single port	

Transection devices (indicate which device(s) were adopted and the number of cases)	Cases
Pringle Maneuver	Cases
YES	
NO	
Drainage	
YES	
NO	

Conversions
Total of laparoscopic liver resection that were converted
Causes:

Intraoperative data
Length of operation (min)
- Median (range)
- Mean (±SD)

Blood losses (ml)
- Median (range)
- Mean (±SD)

Pringle maneuver
- Mean length (min)
- Type of vascular clamping: intermittent or continuous?

Postoperative data
Length of hospitalization
- Median (range)
- Mean (±SD)

Postoperative stay
- Median (range)
- Mean (±SD)

Refeeding
- Median (range)

Complications (Dindo-Clavien classification)

Type of complication	Severity grade	Cases

Classification of surgical complications according to Dindo-Clavien

Grades	Definition
Grade I:	Any deviation from the normal postoperative course without the need for pharmacological treatment or surgical, endoscopic and radiological interventions. Acceptable therapeutic regimens are: drugs as antiemetics, antipyretics, analgetics, diuretics and electrolytes and physiotherapy. This grade also includes wound infections opened at the bedside.

Grade II:	Requiring pharmacological treatment with drugs other than such allowed for grade I complications.
	Blood transfusions and total parenteral nutrition are also included.
Grade III:	Requiring surgical, endoscopic or radiological intervention
Grade III-a:	Intervention not under general anesthesia
Grade III-b:	Intervention under general anesthesia
Grade IV:	Life-threatening complication (including CNS complications)[1] requiring IC/ICU management
Grade IV-a:	Single-organ dysfunction (including dialysis)
Grade IV-b:	Multiorgan dysfunction
Grade V:	Death of a patient
Suffix 'd':	If the patient suffers from a complication at the time of discharge (see examples in Appendix B, http://Links.Lww-.com/SLA/A3), the suffix "d" (for 'disability') is added to the respective grade of complication. This label indicates the need for a follow-up to fully evaluate the complication.

[1]Brain hemorrhage, ischemic stroke, subarrachnoidal bleeding, but excluding transient ischemic attacks (TIA); *IC,* intermediate care; *ICU,* intensive care unit.
www.surgicalcomplication.info
SD, standard deviation; *min,* minutes.

Comments

Suggestions

References

1. Nguyen KT, Gambin TC, Geller DA (2009) World review of laparoscopic liver resection – 2,804 patients. Ann Surg 250:831–841
2. Morino M, Morra I, Rosso E, Miglietta C, Garrone C (2003) Laparoscopic vs open hepatic resection. Surg Endosc 17:1914–1918
3. Nguyen KT, Marsh JW, Tsung A, Steel JJ, Gamblin TC, Geller DA (2011) Comparative benefits of laparoscopic vs open hepatic resection: a critical appraisal. Arch Surg 146:348–356
4. Viganò L, Laurent A, Tayar C, Tomatis M, Ponti A, Cherqui D (2009) The learning curve in laparoscopic liver resection. Improved feasibility and reproducibility. Ann Surg 250:772–782
5. Terminology Committee of the International Hepato-Pancreato-Biliary Association (2000) The IHPBA Brisbane 2000 terminology of liver anatomy and resections. HPB Surg 2:333–339
6. Buell JF, Cherqui D, Geller DA et al; World Consensus Conference on Laparoscopic Surgery (2009) The international position on laparoscopic liver surgery: The Louisville statement, 2008. Ann Surg 250(5):825–830
7. Dagher I, O'Rourke N, Geller DA et al (2009) Laparoscopic major hepatectomy: an evolution in standard of care. Ann Surg 250:856–860
8. Torzilli G, Montorsi M, Donadon M (2005) "Radical but conservative" is the main goal of ultrasonography-guided liver resection: prospective validation of this approach. J Am Coll Surg 201:517–528

9. Ferrero A, Viganò L, Lo Tesoriere N, Russolillo N, Sgotto E, Capussotti L (2009) Bisegmentectomies as alternative to right hepatectomy in the treatment of colorectal liver metastases. Hepatogastroenterology 56:1429–1435
10. Aldrighetti L, Ratti F, Catena M, Pulitanò C, Cipriani F, Ferla G (2012) Laparoendoscopic single site (LESS) surgery for left-lateral hepatic sectionectomy as an alternative to traditional laparoscopy: case-matched analysis from a single center. Surg Endosc [Epub ahead of print, January 26, PMID: 2278101]
11. Casciola L, Patriti A, Ceccarelli G, Bartoli A, Ceribelli C, Spaziani A (2011) Robot-assisted parenchymal-sparing liver surgery including lesions located in the posterosuperior segments. Surg Endosc 25:3815–3824
12. Giulianotti PC, Sbrana F, Bianco FM, Addeo P (2010) Robot-assisted laparoscopic extended right hepatectomy with biliary reconstruction. J Laparoendosc Adv Surg Tech A 20:159–163
13. Pulitanò C, Catena M, Arru M, Guzzetti E, Comotti L, Ferla G, Aldrighetti L (2008) Laparoscopic liver resection without portal clamping: a prospective evaluation. Surg Endosc 22:2196–2200
14. Clavien PA, Barkun J, de Oliveira ML et al (2009) The Clavien-Dindo classification of surgical complications: five-year experience. Ann Surg 250:187–196

Perspectives: Where Shall We Be 20 Years from Now?

43

Antonio Giuliani and Alberto Patriti

43.1 A Premise

In the 1990 at the dawn of the laparoscopic surgery era, Miller published a review on Gastroenterology titled "Laparoscopic cholecystectomy: passing fancy or legitimate treatment option?" [1]. A few years after the first laparoscopic cholecystectomy by Mouret in the 1987, some doubts arose about the future of this new surgical approach. However, against every criticism, the revolution began; only 2 years after Soper et al. published an article titled "Laparoscopic cholecystectomy. The new 'gold standard'?" [2]. Quickly, the historical cliché "the bigger the incision, the better the surgeon" yielded to "the smaller the incision, the better the surgeon".

The world of all surgeons changed daily: their hands were no longer "in" the patient's abdomen but "over" it; the surgeon could no longer "touch" the organs but had to learn how to "feel" them. The new era was named "the laparoscopic revolution," but probably the correct definition is "laparoscopic evolution," as laparoscopy dated back to many years earlier. This evolution continues dramatically, with the surgeon's hands no longer over the patient but a considerable distance away, possibly even in another room with a remote control: this is the robot era. Now, the surgeon must learn how to get used to a 3D vision with the camera as a "medium messenger." Twenty five years after the laparoscopic revolution, it must be admitted that laparoscopic cholecystectomy is not a passing fancy. If laparoscopic surgery is considered in some cases to still be an option, in other cases the surgeon must consider it a duty.

A. Giuliani (✉)
Department of Surgery, Unit of Hepatobiliary Surgery and Transplant Center,
A. Cardarelli Hospital,
Naples, Italy
Email: giuldoc@hotmail.com

F. Calise and G. Casciola (eds.), *Minimally Invasive Surgery of the Liver*,
Updates in Surgery
DOI: 10.1007/978-88-470-2664-3_43, © Springer-Verlag Italia 2013

The development of new instruments, new approaches such as as single-incision laparoscopic surgery (SILS), and new technologies continuously transform this technique, raising fundamental questions about the learning curve and economic sustainability. However, if the road to the laparoscopic revolution was smooth and limitless, in many surgical procedures, the road to laparoscopic liver resections (LLR) was bumpy, as it has been in liver surgery since its beginnings [3].

The axiom *noli tangere* regarding the liver was changed by Luigi Berta, an Italian surgeon who in 1716 performed the first liver resection on a protruding part of the liver from a stab wound. Only 170 years later, an experimental phase was initiated by Lins, who performed a liver resection and began experiments on rabbits. Contemporarily, Hjortsjo in 1950 and Couinaud in 1957 independently described the human liver anatomy and recognized its functional division in segments, which began the modern era of liver surgery. Fifty years later, LLR entered an area "of resistance to the dramatic diffusion of laparoscopic surgery" [3]. Cherqui concluded his article with many doubts about LLR reproducibility, patient selection, and oncological indications while considering LLS as an appealing option even for a problematic reconstruction after resection [3]. Less than 10 years later, many studies answer Cherqui's questions, with more than 4,200 LLR performed worldwide [4]. Similarly, concerns about oncological results following LLR seem to be dispelled by considerable positive evidence.

The feasibility, efficacy, and reproducibility of LLR are clear, with many advantages for patients in terms of surgical results in the short-, medium-, and long-term follow-up compared with open liver surgery. Moreover, LLR shows some clear advantages for cirrhotic patients, who were always considered borderline for the laparoscopic approach. LLR has better results with less postoperative liver failure compared with open surgery due to its lesser invasiveness, reduced blood loss, and caval port shunt sparing [5].

43.2 Where Are We Going?

This is where we are now: curiously, in 2012, the ancient principle of *noli tangere*, with minimally invasive surgery, and the procedure for rapid liver regeneration after split-liver surgery in situ [6] – reminding us of the Prometheus myth – seems to be the future of liver surgery.

However, some limitations seem to remain, such as the possibility of performing multiple wedge resections to spare liver parenchyma, the approach to posterior segments, and the possibility of performing vascular and biliary anastomoses.

43.3 The Learning Curve

The future scenario depends on how the next generation of surgeons will face the basic and advanced tasks of their clinical practice. In most general surgery departments in the Western world, >50% of interventions are laparoscopic. About half of the surgical activity in liver surgery units is devoted to liver metastases, mainly from colorectal cancer. In the Italian survey of LLR by Aldrighetti and colleagues (Chap. 42), more than 1,500 LLR were performed in more than 30 centers in our country, with only a minority of them fully dedicated to hepatobiliary surgery. LLR, at least for minor interventions (up to two segments), seems to be moving forward.

It is already a established that many surgeons learn to perform minor LLR without wide experience in open liver resections (OLR) but with consistent experience in laparoscopic abdominal surgery only. This trend will continue and expand, and treatment of synchronous metastases will increasingly become the domain of the general surgeon.

Current laws – namely, the Italian legislation – will need to cope with the increasing need for the experimental learning curve (Chap. 3): pigs and human cadavers will be required more frequently to allow young surgeons to practice. The learning curve is strictly related to the robot's future and destiny, as well.

43.4 The Robot

Technological progress is unpredictable, particularly in the case of the robot, the future of which is strictly related to the development of many sectors of research: aeronautics, mechanical industry, computer technology, molecular biology. The global economic crisis could place limits to rapid improvements but it is easy to foresee that, at the deadline of original patents, economies of the Third World will quickly be able to provide much less expensive and even more advanced instruments, ready to invade the worldwide market. If one takes into account that the robot is a "transversal" instrument to be correctly used in symbiosis among several specialties, i.e. general surgery, thoracic surgery, gynecology, urology etc., the economic burden often flaunted in opposition to the use of robot could fade away.

It is noteworthy that today, Intuitive, Inc., which holds patents for the robot hardware, does not provide instruments specifically designed for robot-assisted liver surgery. They will need to change and begin to invest, as matter of fact, in the two main expected developments in robotics applied to liver surgery: image integration, and new tools for parenchymal transection. Both open and minimally invasive liver surgery largely depends on preoperative imaging (computer tomography and magnetic resonance imaging) and intraoperative ultrasonography (IOUS) to evaluate vascular and biliary anatomy, identify known and occult intrahepatic lesions, and aid operative planning. A software (TilePro; Intuitive Surgical, Sunnyvale, CA, USA), working with the da Vinci

Si platform, may just show data sources in the robotic operative view. TilePro delivers information from up to two data sources simultaneously. Data are superimposed on the surgical view and could be delivered via S-video or digital visual interface (DVI) input into the console. Six data sources may be toggled to present any two at one time to the console surgeon. Additionally, data can be broadcast to all monitors in the room so the assistants, nurses, and anesthesia teams can view the simulcast. The TilePro "inbox" in the console view may be scaled depending on surgeon preference, with a 50% display dimension having a resolution of 512 × 384 pixels. Data sources are limitless: static images, graphic data, dynamic video, intraoperative ultrasound (IUS), can all be input. The console surgeon with one click of the camera foot pedal may perform changes or remove information [6, 7]. The possibilities of TilePro are actually weakened by the lack of a robotic US. IUS is generally performed by the on-table surgeon and, if not trained in US, by the console surgeon, who moves from the console to the table with consequent time wasting and loss of information. By collaboration between Johns Hopkins University and Intuitive, a prototype of a new robotic device for laparoscopic US has been developed, which promises to overcome the limitations of IUS for robot-assisted liver surgery. The instrument is composed of a linear transducer installed on an articulated probe controlled from the master tool manipulators of the surgical console [8]. In preclinical trials, the probe was used with dedicated open-source software, enabling the US images to be displayed in different ways: the split-screen display mode, in which the surgeon has a side-by-side view of the endoscopic and US images; the "picture-in-a-picture" display mode, that insets the US image into the endoscopic view; and the "flashlight" display mode, in which the US image is overlaid onto a 3D representation of the imaging plane in the stereo view of the console. The effect of this mode is to display the US image in the plane in which it is physically acquired by the transducer. Image magnification and camera filtering are other evolutions of the robotic optic system with interesting future clinical applications. The digital image at the console can be manipulated, and the camera can be adapted to be sensitive to different light frequencies. Intuitive Inc. has recently cleared for the European market new system components, including a surgical endoscope capable of visible light and near-infrared imaging. The 3D high-definition (HD) stereoscopic camera provides visible light and near infrared illumination through the surgical endoscope via a flexible light guide. The system allows surgeons to view high-resolution, near-infrared images of blood flow in vessels and microvessels, tissue and organ perfusion, and biliary excretion in real-time during minimally invasive surgical procedures. The da Vinci robotic system with this new camera has been just adopted in clinical studies to replace intraoperative cholangiography using indocyanine green to delineate biliary anatomy and drive resection of kidney and prostate tumors [9, 10].

The other limitation of robot-assisted liver surgery is the lack of dedicated instruments. Parenchymal transection is generally carried out with the bipolar Precise forceps, the Harmonic scalpel, or the Plasma Trisector Gyrus [11, 12].

None of the aforementioned instruments were designed for liver surgery. Improvements in transection devices are eagerly awaited, from the simple humidified bipolar forceps to the more complex articulated Harmonic scalpel. Articulating ultrasonically activated devices have been described and tested in preclinical studies, but there are as yet no experimental or clinical applications in robotic surgery [13].

43.5 Conclusions

Where shall we be 20 years from now? We are witness to the rising popularity of minimally invasive surgery, which seems to join perfectly with new trends in medical oncology. If surgery may still have a place for treating liver malignancies, it will be due to preponderant use of minimally invasive surgery. Continuous updates and advancements in laparoscopic and robotic technologies, possible integration of these technologies with new medical anticancer drugs, and finally, patient choice, will be driving forces behind the minimally invasive approach to resecting liver malignancies.

The more fascinating aspect of the possible evolution of minimally invasive surgery is miniaturization and complete integration of the robotic devices in the operating room, along with imaging modalities and medical therapies. Quantum dots are tiny crystals that glow when stimulated by ultraviolet (UV) light. The wavelength, or color, of the light depends on the size of the crystal. Latex beads filled with these crystals can be designed to bind to specific DNA sequences. By combining different sized quantum dots within a single bead, scientists can create probes that release distinct colors and intensities of light. When the crystals are stimulated by UV light, each bead emits light that serves as a sort of spectral bar code, identifying a particular region of DNA. This implies that even early tumors will be detectable, overcoming the limitations of X-ray-and US-based modalities. The use of optic digital systems in the operating room to detect small tumors by integrating nanotechnologies and articulated, miniaturized surgical devices to remove the tumors with conservative operations could be one possible future of surgical oncology.

Some might argue that surgery of the future will be considerably distant from the attitude of the surgeon and that an extreme technocracy will be arrested by the need of the surgeon to be the first artificer of patient healing. However, it is our personal opinion that when the surgeon understands that some technologies may expand his/her senses and improve his/her abilities, he or she will be the main promoter of this evolution.

Probably, we shall observe more LLR and other laparoscopic liver procedures (portal vein ligations, treatment of cysts) in general surgery units, especially in those with the availability of robotic technology. The surgeon may even be in another room, but there will always be an assistant (human? robotic?) near the patient. Young surgeons will learn liver surgery, probably at a distance from the patient, while seated in front of a console with the tutor by his or her side.

The surgeon's hands will be guided by technology, visualizing in the monitor (with an indicator) how to dissect, ligate, and coagulate in order to perform the optimal procedure for each individual patient. We should be able to treat other pathologies also, such as Klatskin tumors, and vascular and biliary resections with ductal anastomosis will be routinely performed in most advanced centers.

Another fascinating technology will change the scenario: nanotechnology, that is, objects <100 nm in size. Experimental microrobots are already available, and it seems likely that in one or two decades, microrobots will be commonly used in surgery before their eventual replacement by nanorobots. We shall witness a gradual transition from surgical repair to disease prevention with nanotechnologies. Nanoparticle-assisted surgery already illuminates cancers so that surgeons can completely remove them

Another aspect of the interposition of a digital media between the patient and the surgeon are the advantages for education. Young surgeons will benefit first of the new technology. Learning liver surgery will be extremely different due to the possibility of interaction in virtual reality with the anatomy of the liver. The role of the proctor will therefore change dramatically in this scenario. Probably, the expert surgeon will be able to provide real-time feedback, supervision, and graduated responsibility, even if away from the patient or the hospital. Using telementoring and telemedicine models, a single surgeon will proctor more residents or fellows, avoiding continuous and expensive travels Finally let us all make an effort to provide for our young surgeons the opportunity for adequate training prior to entering the operating room: this will help them advance more quickly toward innovation and benefit to their patients.

Of course, all this will have an impact on our training as surgeons: expect to be surpassed by surgical advances in your lifetime. Therefore, be vigilant about upcoming trends, and prepare to adapt to them. Do not underestimate any new technique; instead, learn about it and study the trend.

References

1. Miller TA (1990) Laparoscopic cholecystectomy: passing fancy or legitimate treatment option? Gastroenterology 99(5):1527–1529
2. Soper NJ, Stockmann PT, Dunnegan DL, Ashley SW (1992) Laparoscopic cholecystectomy. The new 'gold standard'? Arch Surg 127(8):917–921; discussion 921–923
3. Cherqui D (2003) Laparoscopic liver resection. Br J Surg 90(6):644–646. doi:10.1002/bjs.4197
4. Nguyen KT, Gamblin TC, Geller DA (2009) World review of laparoscopic liver resection-2,804 patients. Ann Surg 250(5):831–841. doi:10.1097/SLA.0b013e3181b0c4df
5. Dagher I, Belli G, Fantini C et al (2010) Laparoscopic hepatectomy for hepatocellular carcinoma: a European experience. J Am Coll Surg 211(1):16–23. doi:10.1016/j.jamcollsurg.2010.03.012
6. Volonté F, Pugin F, Buchs NC et al (2012) Console-integrated stereoscopic OsiriX 3D volume-rendered images for da Vinci colorectal robotic surgery. Surg Innov. doi:10.1177/1553350612446353
7. Volonté F, Pugin F, Bucher P, Sugimoto M, Ratib O, Morel P (2011) Augmented reality and image overlay navigation with OsiriX in laparoscopic and robotic surgery: not only a matter of fashion. J Hepatobiliary Pancreat Sci 18(4):506–509. doi:10.1007/s00534-011-0385-6

8. Schneider CM, Peng PD, Taylor RH et al (2012) Robot-assisted laparoscopic ultrasonography for hepatic surgery. Surgery 151(5):756–762. doi:10.1016/j.surg.2011.07.040
9. Brouwer OR, Buckle T, Bunschoten A et al (2012) Image navigation as a means to expand the boundaries of fluorescence-guided surgery. Phys Med Biol 57(10):3123–3136. doi:10.1088/0031-9155/57/10/3123
10. Tobis S, Knopf JK, Silvers C et al (2012) Robot-assisted and laparoscopic partial nephrectomy with near infrared fluorescence imaging. J Endourol. doi:10.1089/end.2011.0604
11. Casciola L, Patriti A, Ceccarelli G, Bartoli A, Ceribelli C, Spaziani A (2011) Robot-assisted parenchymal-sparing liver surgery including lesions located in the posterosuperior segments. Surg Endosc 25(12):3815–3824. doi:10.1007/s00464-011-1796-9
12. Giulianotti PC, Coratti A, Sbrana F et al (2011) Robotic liver surgery: results for 70 resections. Surgery 149(1):29–39. doi:10.1016/j.surg.2010.04.002
13. Ogura G, Nakamura R, Muragaki Y, Hashizume M, Iseki H (2009) Development of an articulating ultrasonically activated device for laparoscopic surgery. Surg Endosc 23(9):2138–2142. doi:10.1007/s00464-008-0248-7

Index

Printed in September 2012

MIX
Papier aus verantwortungsvollen Quellen
Paper from responsible sources
FSC® C105338

If you have any concerns about our products,
you can contact us on
ProductSafety@springernature.com

In case Publisher is established outside the EU,
the EU authorized representative is:
Springer Nature Customer Service Center GmbH
Europaplatz 3, 69115 Heidelberg, Germany

Printed by Libri Plureos GmbH
in Hamburg, Germany